Reinhart Job
**Electrochemical Energy Storage**

# Also of interest

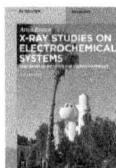

Reinhart Job

# Electrochemical Energy Storage

Physics and Chemistry of Batteries

2nd Edition

**DE GRUYTER**

**Author**
Prof. Dr. Reinhart Job
Department of Electrical Engineering and Computer Science
FH Münster
University of Applied Sciences
Stegerwaldstr. 39
48565 Steinfurt
Germany
reinhart.job@fh-muenster.de

ISBN 978-3-11-914971-6
e-ISBN (PDF) 978-3-11-161853-1
e-ISBN (EPUB) 978-3-11-161864-7

**Library of Congress Control Number: 2025941311**

**Bibliographic information published by the Deutsche Nationalbibliothek**
The Deutsche Nationalbibliothek lists this publication in the Deutsche Nationalbibliografie;
detailed bibliographic data are available on the internet at http://dnb.dnb.de.

www.degruyterbrill.com
Questions about General Product Safety Regulation:
productsafety@degruyterbrill.com

——

Health is the greatest gift,
contentment the greatest wealth,
faithfulness the best relationship

(Buddha)

To my wife
Barbara

# Preface

The upcoming climate change catastrophe stands for immense challenges for the global community and societies. In particular, to slowdown and reduce the steadily growing concentrations of climate gases like carbon dioxide, methane and others in our atmosphere and to stop the fatal increase of the global mean temperatures, an urgent transformation of our classical carbon-based energy supply and economy to a sustainable energy economy with renewable energies is imperative. In this regard, the modern well-developed technology based and driven societies are in great demand for the near-term realization of an effective and sustainable energy supply transformation, since historically, they are simply responsible for the global problems we are actually facing.

The transformation to a sustainable global energy economy requires huge technological efforts. Therefore, the future engineers and scientists, who have to realize a sustainable global energy supply and economy, have to be very confident with energy related technologies – especially the prospects and limits have to be assessed carefully. Therefore, energy related questions have to in the focus of a modern scientific engineering education.

Due to a nonuniform and sometimes unsteady energy yield, a large-scale renewable energy supply requires an increase and broadening of centralized and decentralized energy storage systems. According to these requirements, there are high expectations on batteries with regard to various energy storage systems in the near future. However, in engineering – especially in electrical engineering, the batteries are mostly regarded as a black box, though, this point of view is not sufficient and reasonable for the energy storage demands of the near future. Therefore, it is necessary that the knowledge about batteries – with regard to the basic physical and chemical mechanisms as well as the advantages and limitations of batteries – is strongly enhanced in frame of the qualification of modern engineers. By this book, the author targets this goal and intents to contribute to bridge the gap in engineering education – especially if applied engineering education is focused.

The book is written for graduate students in the disciplines physics and chemistry and especially in the fields of engineering and industrial engineering studies with a focus on electrical and mechanical engineering, power engineering and energy management. It is not intended to provide a specialized book for distinguished researchers or experts in electrochemistry and highly sophisticated battery technology.

The book is divided into five chapters, where Chapter 1 comprises introductory remarks. The main content of the book on electrochemical energy storage is presented in following four Chapters 2 to 5.

In Chapter 2, some basics of thermodynamics are summarized starting with the explanation of relevant terms and definitions like systems, state variables, thermodynamic potentials, state functions and thermodynamic processes. Moreover, the thermodynamic laws are discussed including the Carnot process, although the Carnot pro-

https://doi.org/10.1515/9783111618531-202

cess is not directly to electrochemical energy storage, but – from my point of view – should not be missed in a brief summary of thermodynamics. Finally, in Chapter 2 the chemical potential is explained, too.

In Chapter 3, the basics of electrochemistry are elucidated. First, the terms and definitions in the context of electrolytes are discussed: Beside the definition and description of strong and weak electrolytes, general characteristics of solutions and solvation are described. Moreover, the physical properties like ionic mobility and conductivity, molar conductivity and ion diffusion are discussed, too. Then electrochemical electrodes are discussed in detail. In this context, the interface region between the electrode and the electrolyte is described, i.e. in particular the so-called Helmholtz layer and the Helmholtz double layer. Then, the standard hydrogen electrode is presented that is very important for the establishment of the galvanic series and the standard electrode potential. In the next step of the discussions in Chapter 3, the description of electrochemical cells leads to the definition of galvanic cells. The thermodynamic basics of electrochemical cells (including galvanic cells), the Nernst equation, the galvanic series and the standard electrode potential are discussed in detail. Finally, the principles of electrochemical reactions and energy storage are elucidated; and the principle mechanisms of galvanic cells are discussed with the help of a copper-silver galvanic cell as a first example.

Chapter 4 is devoted to batteries. We start with comprehensive definitions of battery parameters, i.e. the description of the nominal current, the discharging and charging currents, the nominal discharging time, the capacity and nominal capacity, $C$-rate, nominal voltage, nominal specific or volumetric energy and power densities, efficiencies, lifetime, self-discharge, etc. The Ragone diagram for the comparison of various battery types is also presented. After these detailed introductory remarks, the mechanisms and characteristics of batteries are discussed in detail starting with historical primary and secondary batteries like the voltaic pile, the Daniell cell, the Leclanché cell and the Planté cell. After this historical excursion, technically important primary batteries (zinc-carbon and zinc-chloride cells, alkaline batteries, mercury and silver-oxide batteries, zinc-air and aluminum-air batteries, lithium batteries, lithium iron phosphate and sodium-batteries) and secondary batteries are discussed (lead-acid batteries, nickel-cadmium- batteries, nickel-metal-hydride batteries, lithium-ion batteries, sodium-ion batteries). Moreover, overcharging and overdischarging processes are discussed for several important batteries as well as thermal runaway events in lithium-ion and sodium batteries, too.

In Chapter 5, a few aspects will be discussed that are not directly related to physics and chemistry of batteries. However, those aspects have a strong relevance concerning the steady growth of battery fabrication. In particular, in Chapter 5 we want to emphasize for the example of lithium-ion batteries the exploitation and production of the important electrode materials; and we give a brief overview of the materials reserves and resources for the future demand of lithium-ion batteries. The exploitation and production of lithium is described for the two extraction process routes, i.e.

the extraction from minerals and from brine, where lithium production from brine is the significant one. The peculiarities of the exploitation and production of the important materials for the positive electrode from various ores (i.e. cobalt, nickel, manganese), and the exploitation and production of graphite for the negative electrode are discussed, too.

At this point, it remains for me to acknowledge the awesome support of my former student assistant: Many thanks to Anna Michel (née Fuchs) for the layout and preparation of the most of the graphs and figures. I am also very thankful to my peers at the De Gruyter Brill publishing house. In particular, I want to thank Karin Sora (Vice President STEM – Strategy and Publishing Program Development) and Nadja Schedensack (Project Editor) for their encouragement, professional support and advice.

Finally, and most important, I want to say thank you very much to my wife Barbara, who – as always – strongly supported me by her encouragement and patience.

Reinhart Job, May 2025

# Contents

# 1 Introduction

Energy supply and consumption is one of the biggest challenges of modern societies. Focusing on efficient economic growth with a liberal economic order and advanced technologies, the energy demand of the modern rich societies is steadily increasing. Moreover, the emerging countries and economies require an even higher increase of energy demands, if they want to reach the same economic and social level as the rich countries. Considering also the poor countries with their very legitimate aim and right for a much better life for their people, the expected worldwide energy demand will dramatically increase in the future. Although the number of people without access to electricity recently falls below 1 billion, still hundreds of million people have no access to electric power (in 2020 more than 800 million people). According to the "World Energy Outlook (2019)" of the International Energy Agency, it is expected that the global electricity demand will increase by about 25% per decade for the next two decades. For the most part, the energy supply and, in particular, the electric power supply are still based on fossil fuels like coal, oil and natural gas. In addition, nuclear power also plays a significant role for the electricity supply. On the other hand, it is to annotate that wood is a very important source of energy in poor countries and in poor or remote regions of the world, where burned wood is used as fuel for cooking and heating. Thus, in the poor countries – and not only there – there is a big backlog concerning the countries' general infrastructure. Hence, infrastructure backlog also drives a strong necessity for an increasing energy supply.

The classical energy resources (coal, oil, natural gas and uranium) face and cause serious problems – already present nowadays and may even intensify in the future. First, the classical energy resources are not sustainable. In the medium-term, shortages will occur. The peak oil phenomenon and the consequences arising from this problem are already evident. Sooner or later, similar shortages – but to a somewhat lesser extend – will be faced with the other classical energy resources. There is still a lot of coal and natural gas on earth; however, thinking about climate change, it would be fatal, if we consume all the carbon-based fossil resources that are still lying in the ground.

Therefore, concerning energy supply with the classical fossil energy resources, one has to speak about climate change and environmental degradation. Climate change and environmental degradation already cause tremendous problems and costs. Continuing the tendency of a steadily increasing energy supply on the base of classical fossil energy resources, the global environmental problems will eventually not be manageable anymore – especially for the poorer countries and probably sooner than generally expected. The upcoming problems, that can be fatal for our modern technological societies as well as for the global climate and environment, flora and fauna, only allow for one solution: the actual classical energy supply has to be transformed – in fact as fast as possible – into a sustainable energy economy based on renewable energies.

https://doi.org/10.1515/9783111618531-001

The term – or better the concept – sustainability can be traced back to the Club of Rome, that is, to the study *The Limits of Growth* published in 1972, where the necessity to reach a sustainable ecological and economic stability was emphasized. The study simulated the consequences of interactions between the environment and natural resources of the planet Earth and human systems upon a steadily increasing population and economy. In particular, the consumption of nonrenewable natural resources was taken as a variable into account – besides four other variables, that is, population growth, food production, industrialization and environmental pollution. The study first gave insights into the limited resilience of our planet Earth upon steadily growing population numbers and continuing economic growth; that is, the constraints were disclosed that a limited planet put on population numbers and human activities.

Nowadays, the term sustainability comprises three categories that are integrated in the model of the so-called triple bottom line (abbreviated TBL or 3BL), that is, an accounting framework with an ecological, a social and an economic or financial part, respectively (Figure 1.1). Sometimes, one also speaks about the three pillars of sustainability. The concept is based on the term sustainable development that was defined in 1987 by the Brundtland Commission of the United Nations. The ecological pillar defines the requirements and goals with regard to nature, ecology, environment and climate. The social pillar defines the development of future societies that satisfy all people in all regions of the world, promoting a peaceful and worth living global community. The economic pillar establishes a durable – that is, sustainable – base for the production and acquisition of products avoiding the exhaustion of natural resources. Sustainability can be achieved if the ecological, the economical and the social pillars are respected.

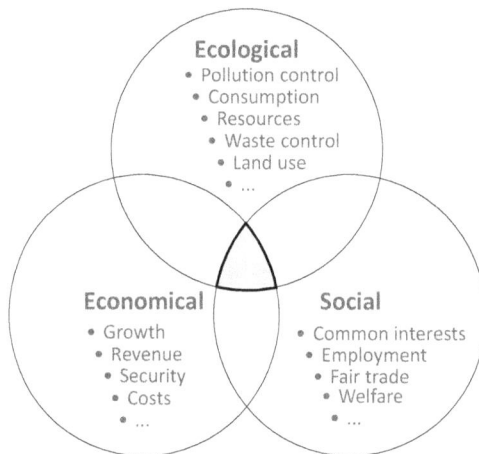

**Figure 1.1:** Model of the triple bottom line or the three pillars of sustainability: sustainability comprises three categories, that is, an ecological part, an economical part and a social part. Sustainability can be achieved if the ecological, the economical and the social pillars are respected, that is, in the overlapping areas in the center of the graph.

Speaking about a sustainable energy sector, several significant thematic key aspects have to be considered, that is, usable energy and reasonable energy supply and resources, further development and optimization of technological concepts, systemic challenges and social transformations, consumption and efficiency and transportation and storage. Usable energy resources are manifold and comprise the classical fossil energy resources (coal, gas, oil, uranium, etc.) and prospective energy resources like hydrogen or methane hydrates as well as the renewable energies. If we do not want to concentrate on nuclear energy, with regard to climate change, renewable energies are the only solution for a sustainable energy supply.

Besides solar and wind energy, the renewable energies also comprise hydropower, tidal power and ocean wave energy, biomass energy and geothermal energy. Hydropower is well developed; however, in densely populated countries, a further expansion of hydropower is difficult and often impossible. Tidal power requires appropriate geographic locations, and ocean wave energy does not play a big role up to now (and probably also not in the future). The application of geothermal power is promising, but is not yet a well-developed technology. Moreover, the usage of biomass is not convincing for several reasons (erosion of land, ethically questionable, etc.).

Solar and wind energy exhibit a great potential for the medium- and long-term energy supply in the future. The solar and wind power technologies are well developed; however, energy storage is required due to variable energy yield. Hence, according to an increasing fraction of wind and solar energy supply, a reliable energy distribution also requires increasing storage capacities. Large battery systems, especially rechargeable battery systems, are an option to counteract power grid instabilities. Moreover, a growing trend toward decentralized energy supply and grid structure – that is, so-called microgrids – requires energy storage, too. Microgrids are a localized group of electrical power sources and loads that are usually connected to a conventional wide area synchronous grid and that are synchronous with this macrogrid. However, microgrids can be also operated in an island mode, that is, if it is disconnected from the macrogrid. Microgrids can integrate various distributed energy sources in an efficient way, especially renewable energy sources like wind or solar power. For microgrids, energy storage is essential since the power quality has to be ensured – including frequency and voltage regulation, smoothing of the electrical output of the renewable energy sources and, last but not least, to provide a backup power for the system. Battery systems can highly support such storage requirements.

Considering energy storage with regard to electrical energy supply at large, a number of particular duties and challenges have to be considered depending on the actual situation. In general, electric energy supply has to be secure and guaranteed. Electrical energy for peak demand as well as for the compensation of the load fluctuations in the grids has to be kept at hand. In addition, energy storage can be useful for an efficient energy conversion and consumption. Moreover, electrical energy storage can even support climate protection and sustainability.

Thus, electrical energy storage distinctly varies depending on the particular purpose and application. Important parameters are, for instance, the amount of energy that has to be stored, the necessary storage dynamics and power gradients that are possible during charging and discharging. Other important factors are the lifetime of storage devices or the number of storage cycles (charging and discharging cycles) that are possible. Moreover, location dependencies, stationary or mobile operation, have to be considered too. Finally, it is also of consequence and relevant, if the electrical energy storage is considered for short-term, medium-term or long-term operation or even for seasonal storage applications.

In the last decades, electrochemical energy storage and conversion, that is, batteries, became more important for electrical energy storage. Mobile communication and computing have driven the development of powerful battery types with high capacities (lithium–ion batteries). Moreover, due to the advent of electromobility, that is, the transformation of gasoline- or diesel-driven cars to hybrid and electric cars, it can be highly expected that the demand for batteries will dramatically increase. Finally, large batteries or battery systems for electrical energy storage for a reliable renewable energy supply (solar and wind power) will also strongly increase in the medium term. In parallel, the expectations of the customers with regard to battery performances, durability, energy content, weight and so on are also steadily increasing.

Batteries can be divided into two groups, that is, into primary batteries that are not rechargeable and into rechargeable secondary batteries. Fuel cells can be designated as tertiary batteries, but they are not a subject of matter in this book. Batteries are galvanic cells that spontaneously transform chemical energy into electroelectrical energy. In particular, they operate as a voltage source. The principle mechanisms of batteries are based on redox reactions (reduction–oxidation reactions), where the oxidation and reduction reactions occur in two spatially separated half-cells. The half-cells are internally connected by an ion conductor and externally by electrical wiring a current can flow that can be used for some application as desired. The detail will be discussed in Chapter 3. In general, a galvanic cell can be constructed by any combination of two different electrodes (i.e., an anode and a cathode) and an appropriate electrolyte in between; and it works as a battery. Upon discharge in the anodic half-cell, an oxidation reaction occurs, and at the cathode, a reduction reaction. Galvanic cells – batteries – deliver an electrical current until the electrochemical equilibrium is reached, and then the electrochemical reactions come to a stop. In case of primary cells, the battery is dead if the electrochemical equilibrium is reached. Secondary cells or batteries can be recharged again. Recharging transfers the secondary battery into a nonequilibrium state again and prepares the cell for spontaneous discharge again.

Batteries for the consumer market are low cost mass products with a compact, mechanically robust and low-weight setup. They are well protected against leakage and/or gassing. A large variety of primary and secondary batteries are available. For example, the geometries and nominal voltages of some typical alkaline battery types are shown in Table 1.1.

**Table 1.1:** Typical geometries and voltages of common alkaline batteries.

| ANSI | IEC-AM | Size (mm) | Voltage (V) |
|------|--------|-----------|-------------|
| A | LR20 | 61 × 34 (⌀) | 1.5 |
| C | LR14 | 50 × 26 (⌀) | 1.5 |
| AA | LR6 | 50 × 14 (⌀) | 1.5 |
| AAA | LR03 | 44 × 10 (⌀) | 1.5 |
| AAAA | LR8D425 | 42.5 × 8.3 (⌀) | 1.5 |
| N | LR1 | 30 × 12 (⌀) | 1.5 |
| 9V | 6LR61 | 48 × 26 × 17 | 9.0 |

Note: The labels of the American National Standards Institute (ANSI) and of the International Electrotechnical Commission for alkaline manganese batteries (IEC-AM) are shown. The first six battery types have a cylindrical shape (length multiplied by the diameter); the last one has a rectangular shape. The listed voltages are nominal voltages.

Speaking about energy storage, at this point, we have to define the term energy carriers. An energy carrier is a system or substance or a fuel that contains energy. Of course, energy carriers do not produce energy; they simply contain energy that can be transformed to other energy forms with more or less good efficiencies. Thus, stored energy within the energy carrier can be transformed later – when it is needed – into useful energy forms like heat, mechanical energy and kinetic energy (e.g., driving a car). Examples for energy carriers are, for instance, petroleum, coal, wood, biomass, natural gas hydrogen, capacitors and batteries. The definition of energy carriers can be more precisely specified, that is, we can distinguish primary and secondary energy carriers. Primary energy carriers are those substances or materials that can be extracted directly from nature: (a) fossil energy carriers like raw petroleum or crude oil, coal and natural gas; (b) nuclear fuels like uranium or thorium; (c) biomass and (d) food (fat, carbohydrates and proteins). Secondary energy carriers are, for instance, fuels (gasoline, diesel fuel) that are produced in an oil refinery. Other examples for secondary energy carriers are ethanol or hydrogen. Ethanol can be obtained by fermentation from biomass; hydrogen can be obtained by electrolysis or chemical extraction processes from crude oil. Sometimes, electrical energy is commonly assigned as an energy carrier.

Table 1.2 shows typical mean values of the energy densities of various energy carriers. Variations from the values shown in Table 1.2 can arise from materials or substance qualities or specific system settings in the case of batteries. It can be seen that the energy densities that can be stored in batteries are much lower than those of classical energy storage substance or fuels like gasoline, diesel, coal, methane or, in particular, uranium. Moreover, the energy densities of rechargeable secondary batteries are smaller than those of primary batteries. The reason is that primary batteries can be better optimized with regard to a maximal energy density since the electrochemical

**Table 1.2:** Energy densities of various substances and fuels are compared with the energy densities of primary and secondary batteries.

| Substance/fuel | | | Batteries | |
|---|---|---|---|---|
| Trinitrotoluol | 4.6 MJ/kg | Primary batteries | Zinc–carbon | 0.23 MJ/kg |
| Brown coal | 11.3 MJ/kg | | Alkaline | 0.45 MJ/kg |
| Hard coal | 34.0 MJ/kg | | Zinc–air | 1.2 MJ/kg |
| Methanol | 19.7 MJ/kg | | Lithium–air | 1.2 MJ/kg |
| Ethanol | 26.7 MJ/kg | | Aluminum–air | 4.7 MJ/kg |
| Gasoline | 40–42 MJ/kg | Secondary batteries | Lead-acid | 0.11 MJ/kg |
| Diesel | 43 MJ/kg | | Nickel–cadmium | 0.14 MJ/kg |
| Methane | 50 MJ/kg | | Nickel–metal hydride | 0.28 MJ/kg |
| Atomic hydrogen | 216 MJ/kg | | Lithium–polymer | 0.54 MJ/kg |
| Natural uranium | 648,000 MJ/kg | | Lithium–ion | 0.65 MJ/kg |

Note: The energy density values of the batteries are typical values that can slightly differ depending on the specific battery types and producers.

reaction upon discharging occurs only in one direction, that is, toward thermodynamic equilibrium. Self-discharge can also be better controlled in primary batteries. Later on, these properties are emphasized in Chapter 4.

Some specific rechargeable batteries exhibit somewhat higher energy densities, for example, rechargeable lithium–sulfur batteries (1.3 MJ/kg), lithium–air batteries (1.6 MJ/kg) or other metal–air batteries. However, most of those battery types are still under investigation and only used well in the laboratories or they can be used only for special applications and are not suitable for a low-cost mass production and/or large-scale mass applications for the consumer market.

Anyway, although those special rechargeable batteries exhibit significantly higher energy densities than the standard secondary batteries, their energy densities are still rather low when compared to classical energy carriers and fuels. Moreover, it cannot be expected that in the future, significant improvements with an increase higher than one order of magnitude are possible. The properties of batteries are directly related to the characteristics of the two different electrodes that define the electrochemical operation mechanism of a battery (together with the electrolyte that operates as an ion conductor). In particular, the battery voltage is related to the physical properties of the two electrodes, that is, the difference of the voltage potentials of the two different electrodes define the voltage that a battery can deliver. Details will be discussed in Chapter 3.

It is to underline that battery research is well experienced; especially after the Second World War, a large amount of research has been performed and tremendous efforts for battery improvements were undertaken, first driven by the cold war (batteries for military applications) and later driven by extended demands in mobile computation and communication. Recently, the advent of large-scale electromobility actually drives a third wave of battery research and development. However, with regard to feasible electrode materials in principle all appropriate elements of the periodic table were already checked in detail; however, the number of elements in periodic table is limited. Hence, sudden extraordinary cell improvements with energy densities exhibiting the same order than those for the classical gasoline or diesel fuels definitely cannot be expected.

With this regard, this book is intended to provide an understanding of the physical, chemical and electrochemical basics of batteries, that is, an understanding that allows for an assessment of the advantages and limits of batteries. Against this background, we start in Chapter 2 with a discussion of physical basics, that is, with thermodynamic considerations. In Chapter 3, the basics of electrochemistry are discussed with regard to battery applications. In Chapter 4, the most important types of batteries are discussed in detail starting with the most important historical cells; then – without making no claim to be complete – technically important nonrechargeable primary and rechargeable secondary batteries are presented and discussed, where we finally end with the lithium–ion battery and its successors, the lithium iron phosphate and the sodium–ion batteries. In Chapter 5, exemplary for lithium–ion batteries, we also briefly comment on the problem of raw material supply for an increasing battery market in the future.

# 2 Thermodynamics

Thermodynamics is a fundamental part of classical physics. It is a closed theory describing macroscopic physical properties – such as the internal energy, entropy, pressure and volume – of matter (and radiation). Thermodynamics shows that the physical properties of matter are subject to general constraints that are common to all substances and materials; anyhow, the appearances of particular properties of specific substances and materials are different. The fundamental constraints are expressed in the thermodynamic laws. These laws completely rule our macroscopic world and they are described further briefly in this chapter.

Thermodynamics is the base of a large variety of topics in science and engineering. It is fundamental for mechanical and chemical engineering. In particular, it describes physical and chemical equilibrium states, and therefore, it is very important for chemistry and chemical reactions. With the help of experimentally measured thermal data, energetic changes of matter and systems can be calculated, and hence, a change of equilibrium states or phase transitions can be predicted and simulated. The same holds for the chemical reactions and processes. Altogether, thermodynamics has a very strong practical relevance in chemistry and engineering.

Before we discuss the four fundamental laws of thermodynamics, in the following subsection, we introduce a variety of physical terms and definitions, which are necessary for the later explanations.

## 2.1 Terms and definitions

Dealing with thermodynamics requires a set of definitions and a good understanding of the physical terms and wordings, which has to be clarified before further explanations and discussions are presented.

### 2.1.1 Systems

A system is a macroscopically spatially well-defined part of the universe. It is the special part of the universe that is to be studied or analyzed, and it is separated from the surroundings (Figure 2.1). The surroundings are the part of the universe that lay outside the boundaries of the system, that is, the surroundings are the rest of the universe.

Systems are characterized by their content of energy, matter and radiation and can be described by the thermodynamic state variables, for instance, mass, temperature, pressure, volume, internal energy, and entropy. Depending on the boundary, a system can exchange mater and/or energy with the outer world. Usually, one distinguishes between open and closed systems depending on whether the system can inter-

https://doi.org/10.1515/9783111618531-002

act – in the sense that energy or mass is exchanged – with the surroundings. However, this terminology is not very precise. In thermodynamics, a better and more detailed discrimination can be done if we differentiate isolated systems, closed systems and open systems.

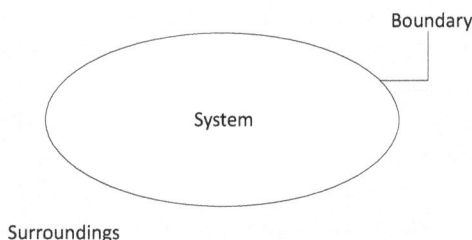

**Figure 2.1:** A system is encircled by a surrounding space (or universe). A well-defined boundary separates the system from the surroundings.

In this connection, an isolated system does not exchange any matter or energy with the surroundings. It is completely isolated from the outer world and does not interact with its surroundings at all. Nothing can be transferred into the isolated system, and nothing can be transferred out of it. Isolated systems cannot be reached by long-range external forces, for instance, gravity. However, internally, they can encounter such forces. They also obey the conservation law, that is, the total energy of the isolated system is constant. Ideal isolated systems does not occur in nature. However, they might be useful hypothetical concepts to study thermodynamics and can serve as idealized model for the approximation of real-world phenomena.

In case of a closed system, the exchange of energy – including heat and work – is possible with the surroundings, but no matter can be exchanged. For example, we can consider a simple closed system, which is composed of a single type of particles – either atoms or molecules. Then such a closed system has and keeps a constant number of particles, although it might exchange energy (heat or work) with the outer world. We now want to look at a somewhat more complex situation where within the closed system chemical reactions can occur. Hence, various kinds of molecules might be aggregated and/or disaggregated during the reaction process. The fact that the considered system where the reactions occur is now closed means that the total number of the involved elemental atoms remains constant. In general, in physics – especially in relativistic physics – energy and mass are equivalent according to the famous Einstein equation $E = m \cdot c^2$. However, in consideration of closed systems, one has to keep in mind that in thermodynamics energy and mass are usually (or at least very often) treated separately as conserved. Therefore, in closed thermodynamic systems, the mass keeps constant.

An open system can interact with its surroundings. It can exchange energy as well as matter with the outer world. More generally, one also can state that open systems can exchange information with the surroundings – in the sense that entropy can be regarded as a kind of information. If an open system is in equilibrium, all kinds of flows between the system and the surroundings come to a standstill. This holds for both energetic as well as mass flows.

If a system is in equilibrium, no changes occur on the macroscopic scale. However, on the microscopic scale, changes can happen, but the so-called principle of the detailed balance holds in this case and has to be obeyed. This means that each microscopic change or process is balanced by its opposite change or process. Most systems in nature – and especially in engineering – are not in thermodynamic equilibrium. However, in many cases such systems are rather close to the equilibrium state so that the approximation of a thermal equilibrium can provide a quite useful understanding of a system or a process.

### 2.1.2 State variables

State variables are measurable physical properties of a system. One has to consider two kinds of state variables, that is, intensive and extensive state variables.

The intensive state variables are not changing if the observed system is varying in its size. In this connection, one has to discriminate two types of such intensive state variables, that is, the system immanent intensive state variables and the material or substance immanent intensive state variables. The first ones include physical variables describing the system – they are system relevant. Examples are, for instance, the temperature $T$ and the pressure $p$. The second ones, that is, the material or substance-related intensive variables, include molar and specific variables. A molar intensive state variable is dependent on the particle number or the amount of the substance, for instance the molar volume $V_m$, the molar mass $M_m$ and so on. Specific intensive variables include, for example, the specific heat capacity $c$, the specific enthalpy $h$ and so on.

On the other hand, the extensive state variables are changing with the size of the system. Typical examples are the volume $V$, the mass $m$ or the entropy $S$. Other examples for extensive state variables are the thermodynamic potentials, that is, for instance, the internal energy $U$, the Helmholtz free energy $F$, the enthalpy $H$ and the Gibbs free energy $G$. Thermodynamic potentials are described in the next section.

### 2.1.3 Thermodynamic potentials

Thermodynamic potentials are scalar quantities that completely describe the equilibrium state of thermodynamic systems. They are extensive state variables. The four most important thermodynamic potentials are the internal energy $U$, the Helmholtz free energy $F$, the enthalpy $H$ and the Gibbs free energy $G$. They can be measured against an arbitrary origin, that is, an arbitrarily chosen zero point or initial state. Thermodynamic potentials differ from each other by their state variables, which are on their part also state functions. They can be transformed into each other by Legendre transformations.

To clarify this point, we start with the internal energy $U$ as one of the main thermodynamic potentials. In general, the internal energy can be written as a function that depends on a variety of state variables $Y_i$. In case of the internal energy, those state variables are all extensive:

$$U = U(Y_i) \tag{2.1}$$

If we consider nonmagnetic multicomponent systems, the specific extensive state variables which have to be considered are the entropy $S$, the volume $V$ and the amount of substance $N_i$ (where $N_i = N_1, \ldots, N_k$, for $k$ different components of the substance). Then we obtain

$$U = U(S, V, N_1, \ldots, N_k) \tag{2.2}$$

For a system that consists only of a single component, eq. (2.2) reduces to $U = U(S, V, N)$. Sometimes, eq. (2.1) is denominated as the fundamental thermodynamic relation of the internal energy $U$. It describes the amount of all equilibrium points or equilibrium states of a thermodynamic system as a function of $U$ that in turn depends on extensive state variables $Y_i$.

Very often in thermodynamics, the fundamental relation is expressed by the total differential $dU$, that is, as the infinitesimal change of the internal energy:

$$dU = \left(\frac{\partial U}{\partial S}\right)_{V,\{N_i\}} dS + \left(\frac{\partial U}{\partial V}\right)_{S,\{N_i\}} dV + \sum_{i=1}^{k} \left(\frac{\partial U}{\partial N_i}\right)_{S,V} dN_i \tag{2.3}$$

The subscripts indicate those parameters that are kept constant at the partial derivation. We first assume a constant amount of substance. Furthermore, we consider the definitions of temperature $T$ and pressure $p$, that is:

$$\left(\frac{\partial U}{\partial S}\right)_{V,\{N_i\}} = T, \quad \left(\frac{\partial U}{\partial V}\right)_{S,\{N_i\}} = -p \tag{2.4}$$

Then eq. (2.3) simplifies, and we get the quite prominent simple form of the fundamental relation:

$$dU = T \cdot dS - p \cdot dV \tag{2.5}$$

Now we consider a nonconstant amount of substance. In addition, we look at the complicated case of a multicomponent system using the definition for the chemical potential $\mu_i$ for the component $i$ as follows:

$$\mu_i = \left(\frac{\partial U}{\partial N_i}\right)_{S,V,N_{j\neq i}} \tag{2.6}$$

In Section 2.3, the chemical potential will be discussed more in detail. With the chemical potential, eq. (2.5) extends to eq. (2.7):

$$dU = T \cdot dS - p \cdot dV + \sum_{i=1}^{k} \mu_i dN_i \tag{2.7}$$

The entropy $S$, the volume $V$ and all particle numbers $\{N_i\}$ are the so-called natural variables of the internal energy. In case of the internal energy, those are all extensive variables. Other thermodynamic potentials can also have intensive natural variables.

With the natural variables, all properties of the thermodynamic system can be obtained. This requires that the thermodynamic potential is determined as a function of its natural variables. Then all of the thermodynamic properties of the system can be calculated by the partial derivatives of the thermodynamic potentials with respect to the natural variables.

The volume $V$ in eq. (2.7) is – as mentioned previously – an intensive state variable. In general, a system can have several or many of such external state variables. Then instead of eq. (2.7), we have to use the more generalized form of the fundamental thermodynamic relation:

$$dU = T \cdot dS - \sum_{j=1}^{n} X_j \, dx_j + \sum_{i=1}^{k} \mu_i \, dN_i \tag{2.8}$$

The $X_j$ in eq. (2.8) are called generalized forces. They correspond to the external intensive state variables $x_j$. In the simple case of eq. (2.7) with only one intensive state variable, the pressure $p$ belongs to the generalized force and the volume $V$ consequently is the external state variable. Equation (2.8) is the generalized standard form of the total differential of the internal energy. It holds for any reversible and nonreversible changes of a system.

As mentioned already, thermodynamic potentials can be transformed into each other by Legendre transformations. Legendre transformations can be used to change from an original function – for instance, for the internal energy as shown in eq. (2.7) that depends on the variable $S$ (entropy) – to a new function, that is the Helmholtz free energy $F$, depending on a new variable. The new variable is the conjugate of the original $S$, and it is the partial derivative of original function (internal energy) with respect to the original variable $S$, that is, the new variable is the temperature $T$. Here, the entropy $S$ and the temperature $T$ are the conjugate pair. The new thermodynamic potential, that is, the Helmholtz free energy $F$, can be defined by the difference between the original function, that is, the internal energy, and the product of the new and the original variable, that is, $T \cdot S$ (see eq. (2.11)).

For this simple procedure, we use the product rule for differentiable functions:

$$d(T \cdot S) = T \cdot dS + S \cdot dT \tag{2.9a}$$

$$\Leftrightarrow T \cdot dS = d(T \cdot S) - S \cdot dT \tag{2.9b}$$

Equation (2.9b) can be inserted into eq. (2.7), and we obtain

$$dU = d(T \cdot S) - S \cdot dT - p \cdot dV + \sum_{i=1}^{k} \mu_i dN_i \tag{2.10}$$

Then we define the "new" thermodynamic potential $F$ as follows:

$$dF = dU - d(T \cdot S) \tag{2.11}$$

Finally, we get the total differential of the Helmholtz free energy $F$ that is the Legendre transform of the internal energy $U(S, V, \{N_i\})$:

$$dF = -S \cdot dT - p \cdot dV + \sum_{i=1}^{k} \mu_i dN_i \tag{2.12}$$

The natural variables of the Helmholtz free energy $F$ are the temperature $T$, that is, an intensive natural variable, the volume $V$ and the particle numbers $\{N_i\}$. The latter two ones are extensive natural variables.

In the same manner, the application of the Legendre transformations yields the other thermodynamic potentials. For the enthalpy $H$, the following total differential holds:

$$dH = T \cdot dS + V \cdot dp + \sum_{i=1}^{k} \mu_i dN_i \tag{2.13}$$

The natural variables of enthalpy $H$ are the intensive pressure $p$ and the extensive natural variables $S$ and $\{N_i\}$.

Finally, for the total differential of the Gibbs free energy $G$, we obtain

$$dG = -S \cdot dT + V \cdot dp + \sum_{i=1}^{k} \mu_i dN_i \tag{2.14}$$

The natural variables of Gibbs free energy $G$ are the two intensive natural variables $p$ and $T$ and the extensive particle numbers $\{N_i\}$.

Equations (2.12)–(2.14) are the fundamental thermodynamic relations of the Helmholtz free energy $F$, the enthalpy $H$ and the Gibbs free energy $G$. Here, we only looked at the four most important thermodynamic potentials, since a thorough description would go far beyond the scope of this book. However, we can state that for all other thermodynamic potentials of a system analog equations can be derived; and always, a fundamental relation for any thermodynamic potential of the system can be expressed.

### 2.1.4 State functions

State functions are mathematical relations between state variables, which describe the state of a system under specific physical conditions. Therefore, state functions can be also named equations of state. State functions are characteristic for the actual equilibrium state of a system. The equilibrium state of a system is the state that is reached after a sufficient period of time, when the state variables are not changing any more. The state functions do not depend on the path inside the state space that is followed to reach the actual state of the system, as will be shown next. The state space is an abstract space where different positions represent states of a physical or thermodynamic system – not literal locations.

The thermodynamic state functions can be obtained from the first derivations of the thermodynamic potentials. The equations for the thermodynamic potentials, as shown in eqs. (2.7) or (2.8) and (2.12)–(2.14), can be generally written in the following form:

$$d\Phi = \sum_i y_i\, dx_i \tag{2.15}$$

In this case, $\Phi$ stands for any thermodynamic potential and $x_i$ and $y_i$ are the conjugate pairs and $x_i$ stands for the natural variable. The partial derivation of an arbitrary thermodynamic potential with respect to the natural variables yields:

$$y_j = \left(\frac{\partial \Phi}{\partial x_j}\right)_{x_{i \neq j}} \tag{2.16}$$

Equation (2.16) provides an expression for any thermodynamic parameter $y_j$ as the partial derivative of the thermodynamic potential with respect to its natural variable $x_j$. All other natural variables $x_{i \neq j}$ are kept constant. Functions, as they were shown in eq. (2.16), are called the state function of a thermodynamic system. Sometimes, they are also named equations of state, since they describe the physical parameters of the thermodynamic state of a system.

If we consider the four most important thermodynamic potentials – the internal energy $U$, the Helmholtz free energy $F$, the enthalpy $H$ and the Gibbs free energy $G$ – we get the following collection of thermodynamic state functions. We start with the partial derivations of the internal energy $U$:

$$T = \left(\frac{\partial U}{\partial S}\right)_{V,\{N_i\}}, \qquad -p = \left(\frac{\partial U}{\partial V}\right)_{S,\{N_i\}} \tag{2.17}$$

By the way, we have already used these expressions as definitions for the temperature and the pressure in eq. (2.4). In case of the internal energy $U$, the state functions for the system describe all their changes with constant temperature and pressure. In fact, this is the result of the combination of the first and second law of thermodynamics (Section 2.2). Now we look at the partial derivations of the Helmholtz free energy $F$:

$$-p = \left( \frac{\partial F}{\partial V} \right)_{T,\{N_i\}}, \qquad -S = \left( \frac{\partial F}{\partial T} \right)_{V,\{N_i\}} \qquad (2.18)$$

Equation (2.18) shows that in case of the Helmholtz free energy $F$, the state functions describe their changes with constant pressure and entropy. Looking at the partial derivations of the enthalpy $H$, we obtain

$$T = \left( \frac{\partial H}{\partial S} \right)_{p,\{N_i\}}, \qquad V = \left( \frac{\partial H}{\partial p} \right)_{S,\{N_i\}} \qquad (2.19)$$

Hence, for the enthalpy $H$ the state functions describe their changes with constant temperature and volume. Finally, the partial derivations of the Gibbs free energy $G$ yield:

$$V = \left( \frac{\partial G}{\partial p} \right)_{T,\{N_i\}}, \qquad -S = \left( \frac{\partial G}{\partial T} \right)_{p,\{N_i\}} \qquad (2.20)$$

We see that for the Gibbs free energy $G$ the state functions describe their changes with constant volume and entropy.

We can find an analog expression for the chemical potential, if we define $\Phi$ as any of the thermodynamic potentials, that is, the internal energy $U$, the Helmholtz free energy $F$, the enthalpy $H$ or the Gibbs free energy $G$, and if $X$, $Y$, $\{N_{i \neq j}\}$ are the natural variables of the thermodynamic potential ($N_i$ is excluded). Then we obtain

$$\mu_j = \left( \frac{\partial \Phi}{\partial N_j} \right)_{X,Y,\{N_{i \neq j}\}} \qquad (2.21)$$

In general, state functions $y(x_1, x_2, \ldots)$ can be uniquely described by its independent variables $x_1, x_2, \ldots$. From state functions, a total differential can be formed that is the sum of the partial derivatives of $y(x_1, x_2, \ldots)$:

$$dy = \left( \frac{\partial y}{\partial x_1} \right)_{x_j, j \neq 1} dx_1 + \left( \frac{\partial y}{\partial x_2} \right)_{x_j, j \neq 2} dx_2 + \cdots \qquad (2.22)$$

The partial derivatives can be differentiated again with respect to the independent variables. In doing so, according to the Schwarz's or Young's theorem, the order of taking the partial derivatives can be interchanged (symmetry of the second derivatives) and we get:

$$\frac{\partial}{\partial x_i} \left( \frac{\partial y}{\partial x_j} \right) = \frac{\partial^2 y}{\partial x_i \partial x_j} = \frac{\partial^2 y}{\partial x_j \partial x_i} = \frac{\partial}{\partial x_j} \left( \frac{\partial y}{\partial x_i} \right) \qquad (2.23)$$

From eq. (2.23) one can conclude that it does not matter along which path the state function $y$ is changed. The state function $y$ can be changed first along coordinate $x_i$

and then along coordinate $x_j$, or vice versa, first along $x_j$ and then along $x_i$. There is no difference in the result. In other words, the integral with respect to d$y$ is independent on the path of integration. The state function of an equilibrium state of a system is independent on how the system has reached this equilibrium state.

### 2.1.5 Thermodynamic processes

A thermodynamic process is the transition from one state of a thermodynamic system to another state, that is, it describes the change of the system from an initial state to a final state. Usually, the thermodynamic process is defined by the initial and the final state of the system whereas the course of the process is usually not of interest. Thermodynamic processes can occur under various constraints.

Reversible processes can be completely reversed in such a way that the initial state can be reached again without any changes of the system and the surroundings. Such processes do not result in an increase of the entropy, that is, the entropy keeps constant and $\Delta S = 0$. During the course of reversible processes, no loss or energy dissipation occurs. Energy dissipation is correlated to heat dissipation. Since a change of the entropy $\Delta S$ is proportional to the exchange of heat $Q$ at constant temperature, one directly can see that the entropy is constant ($\Delta S = 0$) in case of reversible processes without any heat dissipation ($Q = 0$):

$$\Delta S = \frac{Q}{T} \tag{2.24}$$

Hence, reversible processes can also be interpreted as isentropic processes, where the entropy remains constant during the course of the process. They are always in equilibrium. In nature, completely reversible processes are an idealization. Only under some circumstances, one might speak about a quasireversible process if the system can come again very close to its initial state, but strictly speaking, this is a simplification that is not correct.

Irreversible processes are not in equilibrium. During irreversible processes, energy dissipation or losses occur (e.g., friction, thermal losses or others) resulting in an increase of the entropy inside the system, that is, $\Delta S > 0$. Irreversible processes never can be completely reversed. The initial state of the system never can be reached again after an irreversible process has occurred. In thermodynamics, all processes that approach an equilibrium state are irreversible and cannot be reversed. Finally, one can state that all natural processes are irreversible.

Isothermal processes occur at constant temperature during the course of the process, that is, no temperature changes occur. This situation can appear if the system is in contact to an infinite thermal reservoir or heat bath. If the process is sufficiently slowly elapsing, then the system can permanently adjust to the temperature of the reservoir through heat exchange, and therefore, its temperature keeps constant. Then the system undergoes an isothermal process. Of course, the isothermal process is also an idealization that can only be approximated in real nature.

Isobaric processes occur at constant pressure during the course of the process. Isobaric processes implying temperature changes can never be executed in closed volumes. This can be seen, for instance, from the ideal gas laws.

On the other hand, isochoric processes occur at constant volume, that is, $\Delta V = 0$ during the course of the process. Also based on the ideal gas laws, it can be stated that isochoric processes can be never executed at constant pressure, if temperature changes occur and vice versa.

Adiabatic processes occur without any heat exchange between the thermodynamic system that is changing and its surroundings, that is, $Q = 0$. During adiabatic processes, energy is transferred only in the form of work. This includes the fact that during an adiabatic process, no matter is transferred from the system to its surroundings and vice versa. The assumption of an adiabatic process is of course again an idealistic assumption that can only be an approximation of natural processes.

Isentropic processes occur at constant entropy, that is, $\Delta S = 0$ during the course of the process. The constant entropy can be interpreted in two ways. First, an isentropic process can be understood as ideal reversible adiabatic process, where work is transferred completely frictionless without any losses. The process is quasistatic and reversible. Hence, the entropy keeps constant. This is an idealized understanding of the isentropic process that cannot occur in nature. Nevertheless, it can be a useful model for the simplified understanding of real processes. On the other hand, isentropic processes can be regarded as system changes, where irreversible work is carried out at or by the system, and in parallel, heat is exchanged with the surroundings of the system in a way that the heat exchange just compensates for the irreversibility of the work so that entropy finally keeps constant during the process. This understanding does not imply the idealized adiabatic process as a basis and is closer to natural processes, since heat is transferred.

Finally, we also want to mention that the isenthalpic processes occur at constant enthalpy, that is, $\Delta H = 0$ during the course of the process. In case of an ideal gas, isenthalpic processes define the isotherms, that is, the lines of constant temperature in a phase diagram. Reversible isothermal processes are always isenthalpic. However, the converse statement does not hold.

## 2.2 Thermodynamic laws

The four laws of thermodynamics define fundamental macroscopic physical quantities that characterize thermodynamic systems. They completely rule and describe our macroscopic world. The first and second laws of thermodynamics are the most prominent ones. They exclude the existence of perpetual motion machines of the first and second kind. The second thermodynamic law also provides a direction for all natural processes. Hence, it is essential even for the evolution of the universe. Besides the first and second law of thermodynamics, we also know a zeroth and a third law that complete the whole set to the four basic laws of thermodynamics. These laws are briefly discussed in the following sections.

### 2.2.1 Zeroth law of thermodynamics

The zeroth law of thermodynamics introduces the temperature as a physical quantity. It says that, if two systems are in thermodynamic equilibrium with a third system, then they are in thermodynamic equilibrium with each other (Figure 2.2). The state variable that matches all three systems is the temperature. Hence, we also can say: If the systems A and B are in thermodynamic equilibrium and have the same temperature, and the systems A and C are in thermodynamic equilibrium having also the same temperature, then also the systems B and C are in thermodynamic equilibrium and have the same temperature, that is, all systems have the same temperature. The temperature is a scalar, and it is an intensive state variable that is not changing if the considered system is varying its size.

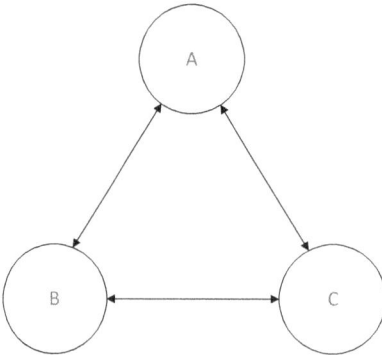

**Figure 2.2:** If two systems A and B that are in thermodynamic equilibrium with a third system C, the systems A and B are also in thermal equilibrium with each other.

A thermodynamic equilibrium between two systems is a transitive relation. Both systems have the same temperature; therefore, heat exchange between these systems does not occur – it is physically not possible at all. In mathematical words, we can generalize that a binary relation $R$ over a set $X$ is transitive, if whenever an element $a$ is related to an element $b$, and $b$ is in turn related to an element $c$, then $a$ is also related to $c$. The formal mathematical definition of transitivity is given as follows:

$$\forall a, b, c \in X: (aRb \wedge bRc) \Rightarrow aRc \tag{2.25}$$

Transitivity is a premise for equivalence and partial order relations.

## 2.2.2 First law of thermodynamics

The conservation of energy is one of the strongest principles in physics. The first law of thermodynamics deals with the conservation of energy and applies it to thermodynamic systems.

The first law of thermodynamics means that the total energy $E$ of an isolated system is constant. The total energy $E$ is given by the sum of the kinetic energy $E_{kin}$, the potential energy $E_{pot}$ and the internal energy $U$ of the system:

$$E = E_{kin} + E_{pot} + U \tag{2.26}$$

Here we can assume that the kinetic and the potential energies are constant. Kinetic energies and potential energies as a whole might occur due to the fields of external forces, but usually they do not play a role in thermodynamic considerations. Hence, from eq. (2.26), we can state that the internal energy of an isolated system is constant.

As mentioned earlier, in case of open systems, energy in form of heat and work – but no matter – can be exchanged with the surroundings. The internal energy $U$ can be changed by work $W$ done on or by the system. In the same manner, $U$ can be changed by heat $Q$ that is transferred to the system or out of the system. Under these conditions, the first law of thermodynamics can be written in the form:

$$\Delta U = Q + W \tag{2.27}$$

The heat $Q$ is positive, if heat is transferred to the system, and negative, if heat is transferred out of the system. The same holds for the work. If work is done on the system, $W$ is positive; and if work is done by the system, $W$ is negative.

In words, the first law of thermodynamics – that is, eq. (2.27) – can be expressed as follows: The total amounts of heat and work that are exchanged between a system and its surroundings are equal to the increase or decrease of the system's internal energy. In detail, the total amount of heat and work that are transferred to a system is equal to the increase of the system's internal energy. The reversal is, of course, also correct: The total amount of heat and work that is transferred from a system to its surroundings is equal to the decrease of the system's internal energy.

The first law of thermodynamics can be also formulated in such a way that the change of the internal energy of a closed system is equal to the amount of heat that is transferred to the system, minus the amount of work that is done by the system on its surroundings. This formulation implies that the perpetual motion machine of the first kind does not exist – it is physically impossible.

If heat is transferred to the system, the internal energy $U$ of the system will be increased. In this case, an increase of $U$ can be related to an increase of the systems temperature $T$ according to the formula:

$$\Delta U = c_V \cdot m \cdot \delta T \tag{2.28}$$

where $c_v$ is the specific heat capacity at constant volume, $m$ the mass of the system and $\delta T$ the temperature change due to the delivered heat. However, changes of the internal energy of a system might also result from changes of the system's chemical energy, electrical energy or others.

If the changes of the internal energy are very small or infinitesimally small, the first law of thermodynamics – eq. (2.27) – can be written as

$$dU = \delta Q + \delta W \qquad (2.29)$$

The internal energy is a state function, as we have mentioned before. Infinitesimal changes of a state function are expressed by the symbol d, that is, the total differential. However, heat and work are not state functions. Thus, small changes of heat or work cannot be expressed by total differentials – that is, the total differentials $dQ$ or $dW$ does not exist. Therefore, small changes of heat and work are expressed by the symbol $\delta$. In this context, the symbol $\Delta$ describes the difference between two states, that is, $\Delta U = U_2 - U_1$ is the difference of the internal energies of a system in states 1 and 2.

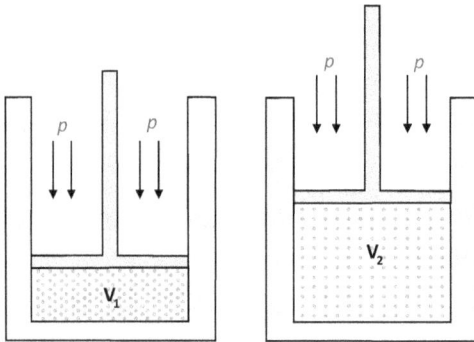

**Figure 2.3:** A piston compresses an ideal gas inside a cylinder in an engine.

We now want to consider only pressure–volume work $\delta W = p \cdot dV$, which occurs, if within a system the volume changes. The system could be for instance an ideal gas in a cylinder with variable volume like in an engine (Figure 2.3). If the volume expands against an external force, the gas supplies work $\delta W$ to the surroundings. The work is done by the system, and therefore $\delta W$ is negative. Concerning only such pressure–volume work, the first law of thermodynamics can be written in the form:

$$dU = \delta Q - pdV \qquad (2.30)$$

If we recall chemical reactions, we can consider two situations, that is, chemical reactions, which are taking place at a constant volume – for instance in a closed process chamber – or at a constant pressure that usually is the normal pressure. We start with the discussion of the first case, that is, the reaction in an autoclave, where the volume of the reaction chamber is constant, that is, $dV = 0$. The chemical reaction in an autoclave is an isochoric process. From eq. (2.30), we see that in this case, $dU = \delta Q$ – under ideal conditions. Therefore, the heat that is exchanged between the system and the surroundings during the chemical reaction in an autoclave correlates directly to the change of the internal energy. The internal energy $U$ is a function of the state variables $T$ and $V$, and therefore, the total differential of $U$ can be written as

$$dU = \left(\frac{\partial U}{\partial T}\right)_V dT + \left(\frac{\partial U}{\partial V}\right)_T dV \tag{2.31}$$

The partial derivative of the internal energy with respect to the temperature at constant volume is called the heat capacity $C_V$ at constant volume, that is:

$$C_V = \left(\frac{\partial U}{\partial T}\right)_V \tag{2.32}$$

Hence, in case of the ideal isochoric process during the chemical reaction, in an autoclave with $dV = 0$, we obtain

$$dU = C_V dT \tag{2.33}$$

By the way, this equation holds for any transformation of an ideal gas.

Now we want to discuss the case of a chemical reaction taking place at constant pressure, for example, at normal pressure in an open beaker glass. For this purpose, we have to take another thermodynamic potential or state function into account. That is the enthalpy $H$, which is related to the internal energy by the following formula:

$$H = U + p \cdot V \tag{2.34}$$

In case of a chemical reaction that occurs at constant pressure and considering eq. (2.30), the total differential of the enthalpy

$$dH = dU + pdV + Vdp \tag{2.35}$$

can be reduced to the formula:

$$dH = \delta Q \tag{2.36}$$

In case of an isobaric process, the heat that is exchanged during the chemical reaction at constant pressure between the system and the surroundings is just the change of the enthalpy. The enthalpy $H$ is a function of the state variables $T$ and $p$, and therefore, the total differential of $H$ can be written as

$$dH = \left(\frac{\partial H}{\partial T}\right)_p dT + \left(\frac{\partial H}{\partial p}\right)_T dp \tag{2.37}$$

The partial derivative of the enthalpy with respect to the temperature and at constant pressure is called the heat capacity $C_p$ at constant pressure, that is:

$$C_p = \left( \frac{\partial H}{\partial T} \right)_p \tag{2.38}$$

Therefore, in case of the ideal isobaric process during a chemical reaction which occurs for instance in an open beaker glass at normal pressure with $dp = 0$, we get:

$$dH = C_p dT \tag{2.39}$$

This equation is also true for any transformation of an ideal gas.

We now want to look a little more in detail at the simplification of the ideal gas. An ideal gas consists of particles that have no volume and that practically do not interact with each other, that is, no interaction forces between the particles exist. Those gas particles can simply be regarded as randomly moving mass points, which collide elastically. The elastic collisions are the only interaction of the ideal gas particles. For the ideal gas, the partial derivative of the internal energy $U$ with respect to the volume at constant pressure is zero, and the internal energy is independent from the volume. Then according to eq. (2.31), we can state that for an ideal gas in case of an isothermal process – with $dT = 0$ – the internal energy $U$ remains constant, that is, $dU = 0$. In a more real consideration, where interaction forces occur between the gas particles, the partial derivative of the internal energy $U$ with respect to the volume at constant pressure is not zero, and it is called the internal pressure of the system (or of the gas). This internal pressure is a measure of the cohesive forces within a system.

Similar considerations can be done with regard to the enthalpy. In case of an ideal gas, the partial derivate of the enthalpy with respect to the pressure at constant temperature is zero. Only if interaction forces exist between the gas particles, a change of the gas pressure can cause a change of the enthalpy.

### 2.2.3 Second law of thermodynamics

As briefly described in the last section, the first law of thermodynamics deals with the change of the internal energy upon a transition between two states of a system. In this context, theoretically, the two types of energy named heat and work can be completely converted into each other. According to the first law of thermodynamics, such a conversion is not dependent on the direction of the chemical or physical process, that is, the conversion of heat into work and vice versa could occur in any order – there are no restrictions on base of the first thermodynamic law. However, in nature, real chemical or physical processes occur only in a certain direction.

For example, if during a thunderstorm a roof tile is falling to the ground, that is, potential energy is converted to kinetic energy that in turn is finally converted to heat, after the tile came to a stop on the ground. According to the first law of thermodynamics, the reversal process could be also possible: A tile lying on the ground could absorb heat from the surroundings that subsequently could be transformed to potential energy, that is, the tile could spontaneously jump from the ground to the roof. Of course, such a process does not occur in nature. Perpetual motion machines of the second kind – that are machines which spontaneously convert thermal energy into mechanical work – are impossible. Hence, the thermodynamic laws have to be extended; and the second law of thermodynamics comes into play. This law describes the elotrophy or directionality that cannot be explained by the first thermodynamic law.

The second law of thermodynamics is strongly related to the entropy. In addition, the entropy can be deduced from the Carnot cycle, that is, a thermodynamic cycle that defines an upper limit of the efficiency of any heat engine that operates on base of the conversion of heat and work. The Carnot heat engine is an ideal theoretical thermodynamic engine. It receives and delivers heat energy from and to the surroundings, and it provides work to the surroundings. A schematic diagram of the Carnot engine is shown in Figure 2.4. For the Carnot engine, all thermodynamic processes proceed reversibly. Therefore, the ideal Carnot engine defines the theoretical maximum of work that can be supplied to the surrounding world by heat engines. Its ratio of received heat and provided work defines the ideal maximal efficiency of thermodynamic heat engines at a given temperature. In the following discussion, we want to study the Carnot cycle a little more in detail and draw some conclusions that are interesting for further studies.

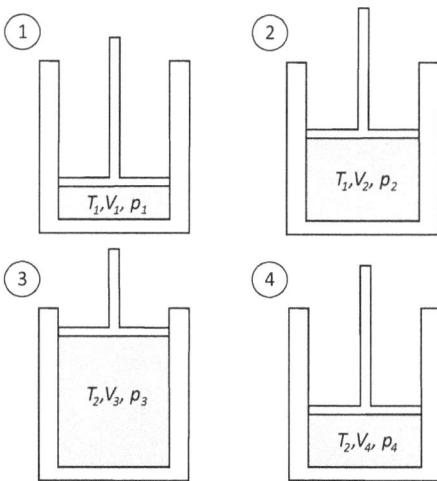

**Figure 2.4:** Carnot engine.

For the Carnot engine shown in Figure 2.4, the four changes of state that an ideal gas has to pass through during the complete thermodynamic cycle are shown in a pressure–volume diagram ($pV$ diagram, Figure 2.5). Two isothermal and two adiabatic or isentropic changes of state occur during one complete cycle.

We start the clockwise circulation of Carnot cycle at state 1, where the ideal gas in the Carnot engine has a volume $V_1$ and a pressure $p_1$. During the first process step

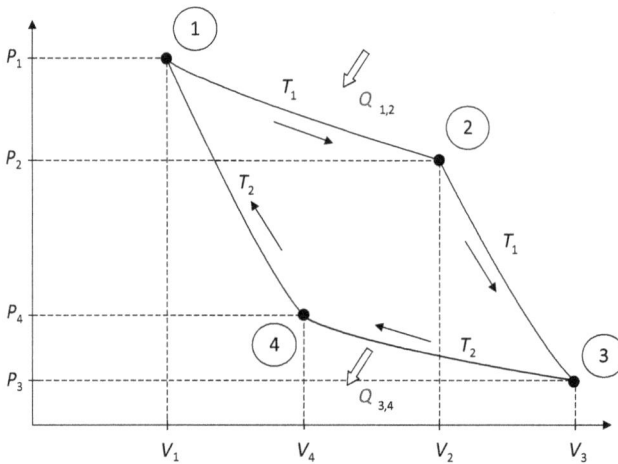

**Figure 2.5:** Carnot cycle (pV diagram).

from state 1 to state 2, a reversible isothermal expansion of the ideal gas occurs. At state 2, the gas has a volume $V_2 > V_1$ and a pressure $p_2 < p_1$. The isothermal expansion is executed at the higher temperature $T_1$ ($>T_2$). The ideal gas receives so much heat from the surroundings – or from a hot reservoir – that the temperature remains constant during this process step, that is, it expands at constant temperature $T_1$ and does work $W_{1,2}$ on the surroundings. The expansion of the ideal gas is caused by the absorption of heat energy $Q_{1,2}$ from the hot temperature reservoir or – that is equivalent according to eq. (2.24) – by the absorption of entropy $\Delta S_{1,2}$. The internal energy of the ideal gas is not altered during the change from state 1 to state 2:

$$\Delta U_{1,2} = Q_{1,2} + W_{1,2} \tag{2.40a}$$

$$\Leftrightarrow Q_{1,2} = -W_{1,2} \tag{2.40b}$$

Using $W_{1,2} = -p \cdot dV$ and $p = n \cdot R \cdot T_1/V$, where $n$ is the amount of the ideal gas in mol and $R$ is the universal gas constant ($R \approx 8.3145$ J/mol K), we can calculate the work that is done on the surroundings during the change from state 1 to state 2:

$$W_{1,2} = -\int_{V_1}^{V_2} \frac{n \cdot R \cdot T_1}{V} dV = -n \cdot R \cdot T_1 \cdot \ln\left(\frac{V_2}{V_1}\right) \tag{2.41}$$

$W_{1,2}$ is negative, and the ideal gas expands at constant temperature $T_1$; that is, during the first state change, work $W_{1,2}$ is done on the surroundings by the ideal gas.

The second state change from state 2 to state 3 is an adiabatic or isentropic process step. The ideal gas expands from the volume $V_2$ to a larger volume $V_3$ ($V_3 > V_2$), while its pressure is reduced from $p_2$ to a lower pressure $p_3$ ($p_3 < p_2$). Hence, the ideal gas continues to expand and does work on the surroundings losing an equivalent amount of internal energy. During the expansion of the ideal gas it cools down to the lower temperature $T_2$, while the entropy remains constant, since in accordance to eq. (2.24), no heat exchange occurs ($Q_{2,3} = 0$). Here, the change of the internal energy equals the work that is done on the surroundings.

Furthermore, we already have discussed in the last section that for the ideal gas the partial derivative of the internal energy $U$ with respect to the volume at constant pressure is zero. Hence, the internal energy of an ideal gas is independent from the volume. In this case, the general equation (2.31) is reduced, and we get:

$$dU = \left(\frac{\partial U}{\partial T}\right)_V dT = C_V dT = \delta W \qquad (2.42)$$

This expression is in accordance with eqs. (2.29) or (2.30) since $\delta Q = 0$ due to the considered adiabatic state change. Finally, with eq. (2.42), we obtain the work $W_{2,3}$ by integration:

$$W_{2,3} = \int_{T_1}^{T_2} C_V dT = C_V \cdot (T_2 - T_1) = -C_V \cdot (T_1 - T_2) \qquad (2.43)$$

$W_{2,3}$ is again negative, that is, during the second state change work is done on the surroundings.

The third state change from state 3 to state 4 is again an isothermal process step. The ideal gas is compressed from the volume $V_3$ to a smaller volume $V_4$ ($V_4 > V_1$), while its pressure is enhanced from $p_3$ to a higher pressure $p_4$ ($p_4 > p_3$). The isothermal compression is done at the lower temperature $T_2$ ($<T_1$). This is done by work done on the ideal gas, and hence, the corresponding amount of heat energy $Q_{3,4}$ is transferred to the surroundings – or to a cooler reservoir – so that the temperature remains constant during this process step. The work done on the ideal gas during this isothermal process step is given by the following equation:

$$W_{3,4} = -\int_{V_3}^{V_4} \frac{n \cdot R \cdot T_2}{V} dV = -n \cdot R \cdot T_2 \cdot \ln\left(\frac{V_4}{V_3}\right) = n \cdot R \cdot T_2 \cdot \ln\left(\frac{V_3}{V_4}\right) \qquad (2.44)$$

$W_{3,4}$ is positive, and the ideal gas is compressed at constant temperature $T_2$; that is, the work $W_{3,4}$ is done on the ideal gas by the surroundings.

The fourth state change from state 4 back to state 1 closes the Carnot cycle; it is again an adiabatic or isentropic process step. The ideal gas is now compressed from the volume $V_4$ to the smaller volume $V_1$ ($V_1 < V_4$), while its pressure is enhanced from $p_4$ to the higher pressure $p_1$ ($p_1 > p_4$). No heat is exchanged during this state change, that is, $Q_{4,1} = 0$. After the third process step, that is, during the fourth process step, the surroundings still do work on the ideal gas and its compression continues – but now under adiabatic conditions. During the compression the ideal gas heats up to the higher temperature $T_1$, while the entropy remains constant. In reversed analogy to the second adiabatic state change from state 2 to state 3, now the work that is done by the surroundings on the ideal gas equals the change of its internal energy, and we obtain positive work $W_{4,1}$:

$$W_{4,1} = \int_{T_2}^{T_1} C_V dT = C_V \cdot (T_1 - T_2) \tag{2.45}$$

Now we can calculate the balance of work for the Carnot engine. The total amount of work that could be rated is simply the sum of the single parts of work that appeared during the four process steps of the Carnot cycle:

$$W = W_{1,2} + W_{2,3} + W_{3,4} + W_{4,1} \tag{2.46a}$$

If we insert eqs. (2.41), (2.43), (2.44) and (2.45) for $W_{1,2} \ldots W_{4,1}$ into eq. (2.46a), we obtain

$$W = -n \cdot R \cdot T_1 \cdot \ln\left(\frac{V_2}{V_1}\right) - C_V \cdot (T_1 - T_2) - n \cdot R \cdot T_2 \cdot \ln\left(\frac{V_4}{V_3}\right) + C_V \cdot (T_1 - T_2) \tag{2.46b}$$

$W_{2,3}$ and $W_{4,1}$ eliminate each other. Hence, we obtain

$$W = -n \cdot R \cdot T_1 \cdot \ln\left(\frac{V_2}{V_1}\right) - n \cdot R \cdot T_2 \cdot \ln\left(\frac{V_4}{V_3}\right) \tag{2.46c}$$

A further simplification of (2.46c) can be achieved with the equation that holds for ideal gases that undergo reversible adiabatic processes:

$$T \cdot V^{\gamma-1} = \text{const.}, \gamma = C_p / C_V \tag{2.47}$$

Two adiabatic process steps are involved in the Carnot cycle, that is, the change of state 2 to state 3 and the change of state 4 to state 1. Hence, according to eq. (2.47), we get the two following relations:

$$T_1 \cdot V_2^{\gamma-1} = T_2 \cdot V_3^{\gamma-1} \tag{2.48a}$$

$$T_1 \cdot V_1^{\gamma-1} = T_2 \cdot V_4^{\gamma-1} \tag{2.48b}$$

The division of eqs. (2.48a) by (2.48b) yields:

$$\frac{V_2}{V_1} = \frac{V_3}{V_4} \tag{2.49}$$

Hence, eq. (2.46b) can be written as follows:

$$W = -n \cdot R \cdot (T_1 - T_2) \cdot \ln\left(\frac{V_2}{V_1}\right) \tag{2.50}$$

Equation (2.50) specifies the work that can be supplied by the Carnot engine. Its efficiency can be calculated from the ratio of this negative work and the heat $Q_{1,2}$ that is transferred from the surroundings to the Carnot engine. Using eqs. (2.40b) and (2.41), we get the following expression for the efficiency $\eta$ of the Carnot engine:

$$\eta = -\frac{W}{Q_{1,2}} = \frac{n \cdot R \cdot (T_1 - T_2) \cdot \ln(V_2/V_1)}{n \cdot R \cdot T_1 \cdot \ln(V_2/V_1)}. \tag{2.51a}$$

or rather:

$$\Leftrightarrow \eta = \frac{T_1 - T_2}{T_1} = 1 - \frac{T_2}{T_1} < 1 \tag{2.51b}$$

From eq. (2.51b), we can see that the efficiency is high, if the temperature difference between the cold reservoir ($T_2$) and the hot reservoir ($T_1$) is large. This holds not only for the ideal Carnot engine but also for real heat engines. However, the most important conclusion is that the efficiency of heat engines can never be 100% – not even for the ideal Carnot engine. Hence, heat can never be completely converted into work. This is one of the formulations of the second law of thermodynamics.

Passing through the Carnot cycle, the heat portions $Q_{1,2}$ and $Q_{3,4}$ are exchanged between the surroundings and the Carnot engine at the temperatures $T_1$ and $T_2$, respectively (Figure 2.5). Since the corresponding changes of states are reversible, we get:

$$\frac{Q_{1,2}}{T_1} + \frac{Q_{3,4}}{T_2} = 0 \tag{2.52}$$

In general, we can write for any reversible thermodynamic cycle:

$$\sum_i \frac{Q_i}{T_i} = 0 \tag{2.53a}$$

or rather, in the infinitesimal form:

$$\oint \frac{\delta Q}{T} = 0 \tag{2.53b}$$

Because the sum of $Q_i/T_i$ equals zero when passing through the whole cycle, we deduce, from eq. (2.53), that the heat divided by the temperature, where the heat exchange occurs, can be regarded as changes of a new state variable – the entropy $S$:

$$\Delta S = \frac{Q}{T} \tag{2.54a}$$

or rather, for the total differential d$S$:

$$dS = \frac{\delta Q}{T} \tag{2.54b}$$

For reversible thermodynamic cycles, the entropy is constant, hence:

$$\Delta S = 0, \quad dS = 0 \tag{2.55}$$

In case of irreversible state changes during thermodynamic cycles, the entropy is increasing, and we obtain:

$$\Delta S > 0, \quad dS > 0 \tag{2.56}$$

These inequalities hold for isolated systems, where no heat exchange occurs ($\delta Q = 0$).

Now we can rephrase the second thermodynamic law: In isolated systems, the state variable $S$ – entropy – can only remain constant or it increases. Irreversible processes cause an increase of the entropy. Chemical or physical processes can spontaneously occur if the entropy is incréasing.

This principle defines a direction of all processes that occur in isolated systems, since the entropy cannot decrease. According to the second law of thermodynamics, an isolated system has its highest entropy in case of the thermodynamic equilibrium when all processes came to an end and no temperature differences occur at all within the system. Therefore, one can state that under the assumption that our universe is an isolated system, this also defines the end of universe – sometimes called the big freeze – and especially the end of life. However, there is some hope, since it is not hundred percent clear yet that our universe is an isolated system.

We now want to have a closer look at irreversible processes. Those processes typically proceed within closed systems at – for instance – constant temperature or constant pressure. Irreversible processes start at an initial state that is out of equilibrium and end at a final state that is the equilibrium state. In case of open systems, where heat exchange with the surroundings can occur, irreversible state changes are described by

$$\Delta S > \frac{Q}{T}, dS > \frac{\delta Q}{T} \tag{2.57}$$

During the course of the irreversible process, the entropy is steadily increasing; and finally it achieves a maximal value when the process has reached the equilibrium state and comes to a stop.

According to eq. (2.30), that is, according the first law of thermodynamics, in eq. (2.57) we can replace the exchanged heat $\delta Q$ by $dU + p \cdot dV$ considering only volume work. Under these conditions, inequality (2.57) can be converted into the following expression:

$$dU + p \cdot dV - T \cdot dS < 0 \tag{2.58}$$

This inequality can now be analyzed under various process conditions; and from this, we can draw some general conclusions. We start with the assumptions that in the closed system the irreversible process proceeds at constant volume and constant entropy, that is, $dS = 0$ and $dV = 0$. In this case, inequality (2.58) is reduced to the expression:

$$(dU)_{S,V} < 0 \tag{2.59}$$

From inequality (2.59), we can conclude that in the closed system the internal energy is decreasing until the irreversible process comes to a stop. Then the internal energy has reached a minimum value.

If the irreversible process proceeds at constant entropy and constant pressure, that is, $dS = 0$ and $dp = 0$, the inequality (2.58) can be converted without any changes by adding $V \cdot dp = 0$ resulting in the following inequality:

$$dU + p \cdot dV + V \cdot dp < 0 \tag{2.60}$$

Hence, we get an inequality, where the total differential $d(U + p \cdot V)$ is smaller than zero:

$$d(U + p \cdot V) < 0 \tag{2.61}$$

Here $U + p \cdot V$ is the enthalpy $H$, as we already know from eq. (2.34). This is in accordance with eq. (2.13), since we actually are dealing with closed systems, where no exchange of matter occurs, and therefore the number of particles remains constant, that is, $dN_i = 0$. At this point, we can deduce that for the irreversible process under isentropic and isobaric conditions, the enthalpy $H$ tends to a minimum, that is:

$$(dH)_{S,p} < 0 \tag{2.62}$$

The enthalpy $H$ can be regarded as a measure of the total energy of a system. It is an energy that comprises two kinds of energy forms. On one side, it includes the internal energy $U$, i.e. the part of energy that is required to create the system. On the other side, it includes the amount of energy that is required to provide space for the system; this part is represented by the product $p \cdot V$. The space with a certain pressure and volume that is needed for the system is provided by displacement from the environmental space. Sometimes the enthalpy $H$ is also called the heat content of a system. It describes the conversion of energy during a chemical reaction. The change of the heat content during a chemical reaction is called latent heat.

If a state change of a system occurs, that is, a chemical reaction, the enthalpy $H$ is changing, too. If the change of the enthalpy $\Delta H$ is positive, heat – that is, energy – is transferred to the system or in other words, the system absorbs heat. A chemical reaction with $\Delta H > 0$ is called an endothermic reaction. If heat is transferred from the system to the surroundings, that is, $\Delta H < 0$, the reaction is called exothermic. Only exothermic reactions can spontaneously occur.

If the irreversible process proceeds at constant temperature and constant pressure, that is, $dT = 0$ and $dp = 0$, inequality (2.58) can be extended in the following way:

$$dU + p \cdot dV + V \cdot dp - T \cdot dS - S \cdot dT < 0 \tag{2.63}$$

Inequality (2.63) was obtained by adding $V \cdot dp$ and subtracting $S \cdot dT$, where both terms are equal to zero due to constant temperature and pressure. Keeping in mind that for a closed system $H = U + p \cdot V$, or rather $dH = d(U + p \cdot V)$, inequality (2.63) results in the total differential:

$$d(H - T \cdot S) < 0 \tag{2.64}$$

For the Gibbs free energy, we can write:

$$G = U + p \cdot V - T \cdot S \tag{2.65a}$$

or rather

$$G = H - T \cdot S \tag{2.65b}$$

Hence, from inequality (2.64), we can conclude that for an irreversible process occurring at constant temperature and pressure also the Gibbs free energy $G$ tends to a minimum in a closed system, that is:

$$(dG)_{T,p} < 0 \tag{2.66}$$

Usually chemical reactions occur at constant temperatures and constant pressure. Hence, with regard to chemical processes, inequality (2.66) is essential. The reason is that at constant temperatures and constant pressure, chemical reactions only occur spontaneously if the Gibbs free energy $G$ is reduced during the course of this process. The change of the Gibbs free energy during a chemical process is the driving force for a chemical reaction. As already mentioned, $H$ describes the conversion of energy during a chemical reaction, but the Gibbs free energy is the amount of energy that can be used for work, and it is smaller than the enthalpy $H$.

Finally, we consider an irreversible process in a closed system proceeding at constant temperature and constant volume, that is, $dT = 0$ and $dV = 0$. Now, inequality (2.58) can be extended without any changes by subtracting $S \cdot dT$. Hence, we obtain:

$$dU - T \cdot dS - S \cdot dT < 0 \tag{2.67a}$$

With the Helmholtz free energy $F = U - T \cdot S$ and its total differential $dF = d(U - T \cdot S)$ – compare to eq. (2.11) – the inequality can be converted into

$$(dF)_{T,V} < 0 \tag{2.67b}$$

Equation (2.67b) shows that the Helmholtz free energy $F$ becomes smaller during the course of an irreversible process within a closed system. The Helmholtz free energy is a measure of the amount of useful work that can be gained from closed systems. In thermodynamic equilibrium, the Helmholtz free energy has reached its minimum.

Overall, we can summarize that in closed systems irreversible processes cause a minimization of the internal energy $U$, the enthalpy $H$, the Gibbs free energy $G$ and the Helmholtz free energy $F$. If the system has reached its final state, that is, the equilibrium state, the processes come to a stop, and $U$, $H$, $G$, and $F$ take their minimal values.

## 2.2.4 Third law of thermodynamics

If the temperature of a system is decreasing, the entropy is also becoming smaller. At $T = 0$ K – that is, at the lower limit or absolute zero of the thermodynamic temperature scale – the system is in a state with the smallest possible energy. The entropy $S$ of a system gets independent on the thermodynamic parameters and converges to a constant value $S_0$, if the temperature converges to absolute zero, that is:

$$\lim_{T \to 0} S(T, p, V, \ldots) = S_0 \tag{2.68}$$

If a system only has one energetic minimum state, the degree of degeneracy of the system, or in other words, the number of the microstates in the ground state of the system is one, that is, $\Omega_0 = 1$. Under these conditions, the entropy converges to zero:

$$\lim_{T \to 0} S(T, p, V, \ldots) = 0 \tag{2.69}$$

This is the case for an ideal perfect crystal at absolute zero.

Equations (2.68) and (2.69) represent the third law of thermodynamics and can be proved by quantum statistical methods. The third law of thermodynamics can be also formulated in a way that it is impossible to reach absolute zero at $T = 0$ K.

## 2.3 The chemical potential

Because of its importance, at this point we want to discuss a little more in detail the chemical potential $\mu$, which is an intensive state variable and strongly connected to the Gibbs free energy $G$.

As shown in eq. (2.14), the Gibbs free energy is not only a function of its two natural variables, that is, the temperature $T$ and the pressure $p$, but also of the amount of substance of all components of a system (here we speak about open systems). Coming back to eq. (2.14) and using eqs. (2.20) and (2.21), we can summarize:

$$dG = \left(\frac{\partial G}{\partial T}\right)_{p, \{N_i\}} dT + \left(\frac{\partial G}{\partial p}\right)_{T, \{N_i\}} dp + \sum_i \left(\frac{\partial G}{\partial N_i}\right)_{T, p, \{N_{j \neq i}\}} dN_i \tag{2.70}$$

Comparing eqs. (2.14) and (2.70), we find

$$\mu_i = \left(\frac{\partial G(T, p, N_j)}{\partial N_i}\right)_{T, p, \{N_{j \neq i}\}} \tag{2.71}$$

From eq. (2.71), the chemical potential $\mu_i$ is the partial derivative of the Gibbs free energy with respect to the amount of the species or component $i$ of the system, where the concentrations of all other components remain constant during the process or reaction – just as the constancy of the temperature and the pressure.

The chemical potential is a kind of potential energy that characterizes chemical reactions, phase transitions and diffusion phenomena. A chemical reaction can only spontaneously occur, if the chemical potential of the final state (after the reaction) is smaller than the chemical potential of the initial state (before the reaction), that is,

$$\Delta\mu = \mu_{\text{initial}} - \mu_{\text{final}} > 0 \Leftrightarrow \mu_{\text{final}} < \mu_{\text{initial}} \tag{2.72}$$

Otherwise, that is, if $\mu_{\text{final}} > \mu_{\text{initial}}$, a reaction cannot spontaneously occur.

For a pure substance, the chemical potential is the change of the Gibbs free energy upon a change of the amount of the substance $n$, that is:

$$\mu = \frac{dG}{dn} = \frac{G}{n} \Leftrightarrow G = n \cdot \mu \tag{2.73}$$

The expression for $G$ for pure substances – as shown in eq. (2.73) – can be extended for the case of mixtures with several components $i$. Hence, in an analog way, the Gibbs free energy of mixtures is summed up by the amount $n_i$ and the chemical potentials $\mu_i$ of all components of the mixture:

$$G = \sum_i n_i \cdot \mu_i \tag{2.74}$$

The chemical potentials $\mu_i$ of the components of a mixture are changing upon a variation of the composition of the mixture.

We introduce the molar Gibbs free energy $G_m$, where $n$ is the amount of substance:

$$G_m = \frac{G}{n} \tag{2.75}$$

In practice, $G_m$ is used, if the number of particles in a system is constant ($dN = 0$). The unit of $G_m$ is J/mol, according to definition (2.75), where the Gibbs free energy is normalized with the amount of substance $n$.

For any chemical reaction, the change of the molar Gibbs free energy can be calculated from the chemical potentials of the components or substances that take part in the reaction:

$$\Delta G_m = \sum_i v_i \cdot \mu_i \tag{2.76}$$

Here $v_i$ is the stoichiometric coefficient of the substance $i$. It is negative for the starting materials of the chemical reaction, that is, for the so-called educts or reactants, and positive for the reaction products. $\Delta G_m$ is the difference between the molar Gibbs free energy after and before the chemical reaction ($\Delta G_m = G_{m\text{-final}} - G_{m\text{-initial}}$).

An arbitrary chemical reaction with starting materials A and B and reaction products C and D can be described by the following formula:

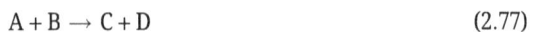

$$A + B \rightarrow C + D \tag{2.77}$$

A spontaneous chemical reaction, that is, an irreversible reaction that ends up in stable products, is only possible if the molar Gibbs free energy becomes smaller during the reaction. In this case, the reaction shown in formula (2.77) is spontaneously occurring from the left-hand side (educts) to the right-hand side (products) – but only if $\Delta G_m < 0$. Hence, in total the products have a smaller molar Gibbs free energy than the starting materials before the chemical reaction. In this case, we speak about an exergonic reaction or processes. For exergonic processes a positive flow of energy occurs from the systems to the surroundings (Figure 2.6) – for example, an exergonic chemical reaction occurs spontaneously and emits heat to the surroundings.

In the opposite case, where $\Delta G_m > 0$, the reaction cannot spontaneously occur. This case is called endergonic process or reaction; and now a positive flow of energy occurs from the surroundings to the system (Figure 2.6). The reaction occurs from the right-hand side to the left-hand side, that is, in opposite direction to the example shown in formula (2.77). This only can happen, if energy – for instance in form of heat – is delivered from the surroundings to the system.

Chemical reactions can occur on its own, as long as the condition $\Delta G_m < 0$ holds. A chemical reaction comes to a stop, if $\Delta G_m$ becomes zero; and the state of equilibrium is reached.

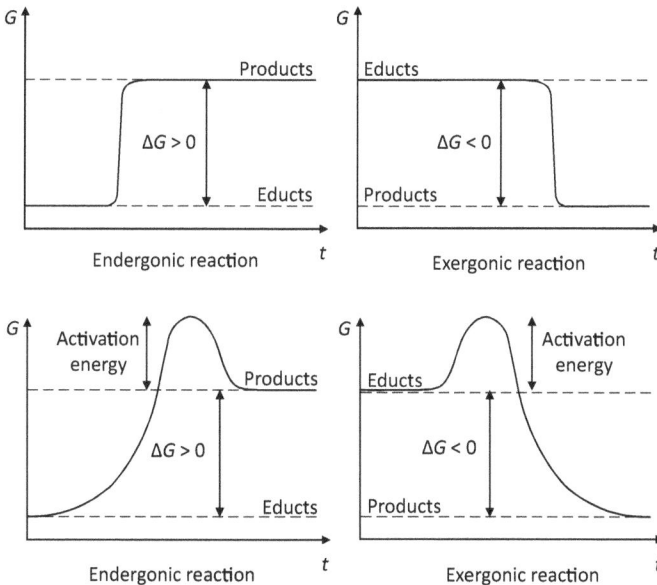

Figure 2.6: Endergonic and exergonic reactions. The upper graphs show the ideal simplified case without activation energy; the lower graphs show the case that an activation energy is necessary to start the reaction (this is the usual case).

As shown in eq. (2.21), the chemical potential $\mu_i$ can be obtained by the partial derivative of any thermodynamic potential, that is:

$$\mu_i = \left(\frac{\partial U}{\partial n_i}\right)_{S,V,\{n_{j\neq i}\}} = \left(\frac{\partial H}{\partial n_i}\right)_{S,p,\{n_{j\neq i}\}} = \left(\frac{\partial F}{\partial n_i}\right)_{T,V,\{n_{j\neq i}\}} = \left(\frac{\partial G}{\partial n_i}\right)_{T,p,\{n_{j\neq i}\}} \tag{2.78}$$

Equations (2.7), (2.12–2.14) – that were presented already in Section 2.1.3 – and expression (2.78) for the chemical potential $\mu_i$ are called the fundamental equations of Gibbs. For the definition of $\mu_i$ usually only the partial derivative of the Gibbs free energy is used, since here the temperature $T$ and the pressure $p$ are the natural variables that are kept constant – this is more practical than the other options.

$$\mu_i := \left(\frac{\partial G}{\partial n_i}\right)_{T,p,\,\{n_{j\neq i}\}} \tag{2.79}$$

We now consider chemical mixtures under the assumption that the components of the mixture do not react. Any extensive state variable $Z$ is a function of the amount $n_i$ of the components $i$ of the mixture. If we keep the temperature $T$ and the pressure $p$ constant, the total differential of the arbitrary state variable is given by

$$dZ = \sum_i \left(\frac{\partial Z}{\partial n_i}\right)_{T,p,\,\{n_{j\neq i}\}} \cdot dn_i \tag{2.80}$$

This equation can be compared, for instance, with eq. (2.14) for the total differential of the Gibbs free energy $G$ considering that now – in eq. (2.80) – we kept the temperature and pressure constant and that here we used the amount $n_i$ of the components instead of the extensive particle numbers. We define $Z_i$ as the partial derivative of the arbitrary state variable for the component $i$ of the mixture, that is:

$$Z_i = \left(\frac{\partial Z}{\partial n_i}\right)_{T,p,\,\{n_{j\neq i}\}} \tag{2.81}$$

This partial molar property describes the change of any state variable $Z$ in dependence on the change of the amount $n_i$ of the component $i$ of the mixture. Using eq. (2.81), eq. (2.80) transforms into

$$dZ = Z_i \cdot dn_i \tag{2.82}$$

As we have briefly mentioned in Section 2.1.2, extensive state variables are, for instance, the Gibbs free energy $G$ and others (the volume $V$, the mass $m$, the entropy $S$, the internal energy $U$, the Helmholtz free energy $F$, the enthalpy $H$, etc.). We consider the Gibbs free energy $G$; and in accordance with eq. (2.79), we now get the partial molar Gibbs free energy $G_i$, that is:

$$G_i = \left(\frac{\partial G}{\partial n_i}\right)_{T,p,\,\{n_{j\neq i}\}} = \mu_i \tag{2.83}$$

and in analogy to eq. (2.82), we obtain

$$dG = G_i \cdot dn_i = \mu_i \cdot dn_i \tag{2.84}$$

In case of a binary mixture with only two components, the total differential of the Gibbs free energy $G$ can be written as follows – in compliance with eq. (2.81):

$$dG = \mu_1 \cdot dn_1 + \mu_2 \cdot dn_2 \tag{2.85}$$

Integration yields

$$G = \mu_1 \cdot n_1 + \mu_2 \cdot n_2 \tag{2.86}$$

The corresponding total differential is

$$dG = \mu_1 \cdot dn_1 + n_1 \cdot d\mu_1 + \mu_2 \cdot dn_2 + n_2 \cdot d\mu_2 \tag{2.87}$$

Combining eqs. (2.85) and (2.87), we obtain

$$\mu_1 \cdot dn_1 + \mu_2 \cdot dn_2 = \mu_1 \cdot dn_1 + n_1 \cdot d\mu_1 + \mu_2 \cdot dn_2 + n_2 \cdot d\mu_2 \tag{2.88a}$$

$$\Leftrightarrow 0 = n_1 \cdot d\mu_1 + n_2 \cdot d\mu_2 \tag{2.88b}$$

This is the so-called Gibbs–Duhem relationship – here for the case of a chemical mixture with two components. The total amount $n$ of the mixture is, of course, given by

$$n = n_1 + n_2 \tag{2.89}$$

We define the mole fraction or molar fraction $x_i$ by

$$x_i = \frac{n_i}{n} \tag{2.90}$$

Then we get for the case of the binary mixture:

$$0 = x_1 \cdot d\mu_1 + x_2 \cdot d\mu_2 \tag{2.91}$$

This is the Gibbs–Duhem relationship for the two-component mixture in another form.

In general, for mixtures with any number of components, we get for the Gibbs–Duhem relationship for the Gibbs free energy:

$$0 = \sum_i n_i \cdot d\mu_i \tag{2.92}$$

or rather:

$$0 = \sum_i x_i \cdot d\mu_i \tag{2.93}$$

Equation (2.91) can be extended to the so-called Gibbs–Duhem–Margules equation that describes the relation of the two chemical potentials of the binary mixture, if the composition is changed:

$$0 = x_1 \cdot \left(\frac{\partial \mu_1}{\partial x_1}\right)_{T,p,n_2} + x_2 \cdot \left(\frac{\partial \mu_2}{\partial x_2}\right)_{T,p,n_1} \tag{2.94}$$

If the dependence of the chemical potential on the concentration of one component of the mixture is known, then the dependence of the chemical potential on the concentration of the other component can be calculated.

And finally, we can write the Gibbs–Duhem relationships – that is, eqs. (2.95) and (2.96) – for mixtures with any number $i$ of components in the following generalized forms:

$$0 = \sum_i n_i \cdot d\mu_i \tag{2.95}$$

or rather:

$$0 = \sum_i x_i \cdot d\mu_i \tag{2.96}$$

One has to annotate that the Gibbs–Duhem relationship can only be used for sufficiently large systems or mixtures. If the system is too small, surface effects and other microscopic phenomena come into the play that cannot be described by this relationship.

For the corresponding Gibbs–Duhem–Margules relationships – that is, eqs. (2.97) and (2.98) – we get for the generalized forms:

$$0 = \sum_i n_i \cdot \left(\frac{\partial \mu_i}{n_i}\right)_{T,p,\{n_{j\neq i}\}} \tag{2.97}$$

or rather:

$$0 = \sum_i x_i \cdot \left(\frac{\partial \mu_i}{x_i}\right)_{T,p,\{n_{j\neq i}\}} \tag{2.98}$$

We now extend our considerations to the case of a nonconstant pressure – whereas the volume $V$ and the temperature $T$ are kept constant. From eq. (2.66), we get the total differential of the Gibbs free energy:

$$dG = dH - T \cdot dS - S \cdot dT \tag{2.99}$$

The total differential $dH$ of the enthalpy was given in eq. (2.35). Furthermore, we use eq. (2.30) for the total differential $dU$ of the internal energy, that is, $dU = \delta Q - p \cdot dV$. If we insert eq. (2.54b) into (2.30), we get

$$dU = T \cdot dS - p \cdot dV \tag{2.100}$$

Equation (2.100) can be inserted into eq. (2.35), and we get for the total differential of the enthalpy:

$$dH = T \cdot dS - p \cdot dV + p \cdot dV + V \cdot dp = T \cdot dS + V \cdot dp \tag{2.101}$$

Therefore, $dG$ can be written as follows:

$$dG = T \cdot dS + p \cdot dV - T \cdot dS - S \cdot dT = V \cdot dp - S \cdot dT \tag{2.102}$$

Since the temperature is kept constant, we finally receive:

$$dG = V \cdot dp \tag{2.103}$$

Hence, in case of ideal conditions, that is, for the ideal gas, $p \cdot V = n \cdot R \cdot T$ or rather $V = n \cdot R \cdot T/p$, we obtain:

$$dG = \frac{n \cdot R \cdot T}{p} \cdot dp \tag{2.104}$$

$R$ is the universal gas constant. Using eq. (2.73) and keeping the amount of substance constant, we get:

$$dG = n \cdot d\mu \tag{2.105}$$

If we combine eqs. (2.104) and (2.105), the total differential of the chemical potential is given by:

$$d\mu = \frac{R \cdot T}{p} \cdot dp \tag{2.106}$$

This equation can be integrated:

$$\int_{\mu^0}^{\mu} d\mu = R \cdot T \cdot \int_{p^0}^{p} \frac{1}{p} \cdot dp \tag{2.107}$$

Here $\mu^0$ is the so-called chemical standard potential of the ideal gas at the standard pressure $p^0 = 1$ bar; $\mu^0$ itself is not dependent on the pressure. However, integration yields the pressure-dependent relation for the chemical potential $\mu$:

$$\mu = \mu^0 + R \cdot T \cdot \ln\frac{p}{p^0} \tag{2.108}$$

This holds – as mentioned earlier – for the ideal gas.

Now we consider the dependence of the chemical potential on the pressure for the case of a multicomponent system or rather an ideal gas with $i$ components. The equation for the multicomponent mixture looks quite similar to eq. (2.108). However, the chemical potential $\mu_i$ of the component $i$ – the partial molar Gibbs free energy –

now is dependent on the partial pressure $p_i$ of the corresponding component; hence, we get:

$$\mu_i = \mu_i^0 + R \cdot T \cdot \ln \frac{p_i}{p^0} \tag{2.109}$$

Here $\mu_i^0$ is the chemical standard potential for the component $i$ of the ideal gas at the standard pressure of $p^0 = 1$ bar; $\mu_i^0$ itself is not pressure dependent – alike $\mu^0$.

In analogy to definition (2.90), we define the molar fraction $x_i^g$ for the component $i$ of the gas mixture:

$$x_i^g = \frac{n_i^g}{n} \tag{2.110}$$

The superscript g indicates that we are dealing with a gas phase. The partial pressure $p_i$ of the component $i$ is related to $x_i^g$ and the total pressure $p_{tot}$:

$$p_i = x_i^g \cdot p_{tot} \tag{2.111}$$

We insert this relation into eq. (2.109), and we obtain

$$\mu_i = \mu_i^0 + R \cdot T \cdot \ln \frac{p_{tot}}{p^0} + R \cdot T \cdot \ln x_i^g \tag{2.112}$$

On the right-hand side of eq. (2.112), the first two summands are standing for the chemical potential of the component $i$ in the pure state – that is, the gas $i$. The third summand reflects the situation, where the gas $i$ is a component of the mixture. The chemical potential of the pure state of the gas or component $i$ is denoted $\mu_i^*$:

$$\mu_i^* = \mu_i^0 + R \cdot T \cdot \ln \frac{p_{tot}}{p^0} \tag{2.113}$$

If we insert eq. (2.113) into eq. (2.112), we get for chemical potential $\mu_i$ of the component $i$ of the gas mixture:

$$\mu_i = \mu_i^* + R \cdot T \cdot \ln x_i^g \tag{2.114}$$

In the same way, the chemical potential $\mu_i$ of the component $i$ of a liquid mixture is

$$\mu_i = \mu_i^* + R \cdot T \cdot \ln x_i^l \tag{2.115}$$

Here the superscript l stands for the liquid state. Equation (2.115) is valid for ideal mixtures and/or components with high concentrations. This annotation also holds for eq. (2.114) in case of the gaseous state.

Now we very briefly look at real gases and real solutions, respectively. The chemical potential of a real gas can be formulated in analogy to the ideal gas case. However, instead of using the pressure $p$ of the gas one has to use a modification of the pressure:

$$\phi = \gamma \cdot p \tag{2.116}$$

where $\phi$ is called the fugacity and $\gamma$ is the dimensionless fugacity coefficient. The fugacity is an intensive state variable; and its physical unit is equal to the unit of the pressure. The fugacity describes the deviation of the behavior of a real gas compared to the ideal gas. For ideal gases, the interactions and the volumes of the gas particles are neglected. Therefore, the ideal and the real effective pressures are only similar, if the pressure is very low – in this case, the fugacity coefficient is one or close to one. The fugacity is a similar physical or chemical measure than the activity $a$, which describes the effective concentration of a species in a solution or mixture. In this sense, the chemical potential depends on the activity of a real solution in the similar way than it depends on the concentration of the ideal solution – the same holds for a mixture:

$$a_i = f_i \cdot x_i \tag{2.117}$$

where $a_i$ is the activity of the component in the solution or mixture, $f_i$ is the activation coefficient and $x_i$ is the molar fraction as defined by eq. (2.90). Only if solutions or mixtures are strongly diluted, the activation coefficient is one and $a_i \approx x_i$ (or $a_i \propto n_i$).

Using the fugacity $\phi$ as defined in eq. (2.116), from eq. (2.108) for the ideal gas, the corresponding equation for the chemical potential of the real gas can be deduced:

$$\mu = \mu^0 + R \cdot T \cdot \ln \frac{\phi}{p^0} \tag{2.118}$$

The fugacity coefficient $\gamma$ can be derived from the following equation:

$$\ln \gamma = \int_0^p \frac{1}{p} \cdot \left( \frac{p \cdot V}{n \cdot R \cdot T} - 1 \right) dp \tag{2.119}$$

In eq. (2.119), we can define and calculate the so-called compressibility factor $Z$ as follows:

$$Z = \frac{p \cdot V}{n \cdot R \cdot T} = 1 + B_p(T) \cdot p + C_p(T) \cdot p^2 + \cdots \tag{2.120}$$

The temperature-dependent coefficients $B_p(T)$, $C_p(T)$ and so on are called the second, third and so on virial coefficients; and eq. (2.120) is called virial equation. The virial coefficients can be derived from experiments. In case of the ideal gas law – $p \cdot V = n \cdot R \cdot T$ – the compressibility factor $Z = 1$, and it is independent on the pressure.

For real gases, $Z \neq 1$ and it depends on pressure. Especially for high-density gases, the situation significantly deviates from the ideal situation. The reason is that intermolecular attraction forces and particle diameters cannot be neglected if the distances between gas molecules are becoming shorter at higher pressures and/or higher densities. This is shown in Figure 2.7, which illustrates the dependence of the compressibility on the pressure. For instance, at high pressures and/or high temperatures

$Z > 1$, $\gamma > 1$, and therefore, $\varphi > p$ since repulsive forces between the gas molecules are dominant. At the so-called Boyle temperature, the compressibility factor approaches 1, if the pressure approaches 0, that is, $Z \rightarrow 1$ for $p \rightarrow 0$.

The mixtures of real gases are similarly described as shown in eq. (2.114). However, instead of molar fraction $x_i^g$ for the component $i$ of the gas mixture and in the style of eq. (2.117), the activity $a_i^g$ is used:

$$a_i^g = f_i \cdot x_i^g \tag{2.121}$$

The activation coefficient $f_i$ is depending on the concentration $n_i$; and $f_i$ is approaching one ($f_i \rightarrow 1$), if $x_i^g$ is approaching 1 (i.e., $n_i \rightarrow n$, see eq. (2.90)). With the activity $a_i^g$, the chemical potential of the component $i$ of the real gas mixture can be written as follows:

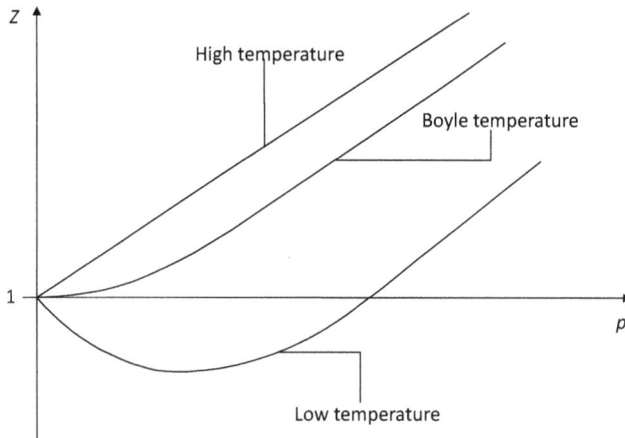

**Figure 2.7:** Compressibility factor in dependence on the pressure for various temperatures. At the so-called Boyle temperature, the compressibility factor approaches 1 if the pressure approaches 0.

$$\mu_i = \mu_i^* + R \cdot T \cdot \ln a_i^g \tag{2.122}$$

In an analogous way, the mixtures of real liquids or solutions can be described, that is,

$$\mu_i = \mu_i^* + R \cdot T \cdot \ln a_i^l \tag{2.123}$$

Here $a_i^l$ is the activity of the component $i$ in the real liquid mixture or solution.

# 3 Basics of electrochemistry

Electrochemistry is an important part of the major scientific field of physical chemistry. Electrochemical reactions deal with the chemical reactions of charged particles, that is, ionized atoms or molecules, and in particular with ionic conductors, that is, electrolytes. For instance, ionized atoms or molecules can arise from salts. Salts are formed by a chemical reaction between a base and an acid or between a metal and an acid. A typical representative is ammonium chloride ($NH_4Cl$), which is formed by the reaction of ammonia ($NH_3$ – the base) and hydrogen chloride (HCl – the acid). If salt molecules are dissolved in an aqueous solvent, they dissociate into ionized particles that are positively and negatively charged. The total charge of all ions remains constant. Dissociation can occur completely, that is, all initial salt molecules are dissociated into ions, or partly, if only fractions of the initial molecules dissociate. In a solvent (e.g., water) ammonium chloride dissociates into positively charged ammonium ions ($NH_4^+$) and negatively charged $Cl^-$ ions. Another example is the compound hydrogen chloride (HCl) that will be described more in detail in Section 3.1.2. At room temperature, HCl is a colorless gas. In water, HCl dissociates and reacts with the water molecules and finally forms positively charged hydronium ions ($H_3O^+$) and negatively charged $Cl^-$ ions. The aqueous solution of hydrogen chloride is named hydrogen chloride acid, and it is represented by the formula HCl, too. Aqueous solutions containing, for instance, $NH_4^+$ and $Cl^-$ ions or $H_3O^+$ and $Cl^-$ ions are electrolytes.

Moreover, one can state that electrochemistry deals with chemical reactions that occur at the interface of an electrode and an ionic conductor, that is, an electrolyte. The electrochemical reactions are closely correlated with the migration of electrical charges – that is, charge carriers like electrons or ions – between the electrodes and the electrolyte. Such reactions are named reduction–oxidation reactions, which are also well known as "redox" reactions. Those redox reactions always occur together and imply the change of the so-called oxidation states of the involved substances or elements. By the way, in the first instance, the oxidation state has nothing to do with the element oxygen. The element oxygen can be involved, but it must not necessarily be involved in redox reactions. If a substance or element gains an additional electron, its oxidation state is reduced and the corresponding reaction is named reduction. On the other hand, if a substance or element gives an electron away – or donates an electron – its oxidation state is enhanced and the corresponding reaction is named oxidation.

In the following chapters, the main players of electrochemical energy storage and conversion, which are important for the focus of this book, will be briefly described and discussed.

https://doi.org/10.1515/9783111618531-003

## 3.1 Electrolytes

Electrolytes are special types of electroconductive solutions. They play an important role for electrochemical processes, especially concerning electrochemical energy storage in batteries. Before we discuss the characteristics of electrolytes, we begin with a brief description of solutions.

### 3.1.1 Chemical solutions

A chemical solution is a homogenous mixture of two or more substances or compounds. Solutions are composed of the solute (or the solutes) which is (are) dissolved in the solvent. At first glance, solutions are characterized by the following properties. First, they are stable. The dissolved atoms, ions, molecules or small particles are homogenously dispersed in the solvent. Therefore, they are not instantly recognizable as solutions since they consist of a single homogenous phase and the ingredients of a solution are not visible.

The expression solubility describes the ability of a chemical substance or compound to dissolve in another chemical compound. Two substances or compounds are miscible, if one of them – the solute – can be completely dissolved in the other one – the solvent. In case of the contrary situation, where two chemical substances or compounds are not dissolvable at all, we say that they are immiscible.

Solutes can appear in gaseous, liquid or solid states. Solvents are usually supposed to be in the liquid state; however, solid solvents or gaseous solvents are also possible. Some common examples are now briefly addressed.

First, we have to state that in case of gaseous solvents only gases can be the solutes. The most prominent example of a gaseous solution is air, which is composed of nitrogen ($N_2$, 78.084%), oxygen ($O_2$, 20.946%), argon (Ar, 0.934%) and carbon dioxide ($CO_2$, actually ca. 0.041% with increasing tendency, the preindustrial value was 0.028%) and traces of other gases – mainly noble gases. Hence, we can state that air is a solution of various gases dissolved in nitrogen, that is, nitrogen is the solvent.

Many gases, liquids and solids can be dissolved in liquid solvents; in fact this holds for almost all gases, liquids and solids. Examples of gases that are dissolved in a liquid solvent can be, for instance, nitrogen ($N_2$) or oxygen ($O_2$) in water ($H_2O$). Ethanol ($C_2H_6O$) is an example for a liquid solute that can be dissolved in water. By the way, this solution is commonly known as an alcoholic beverage. Examples for solid compounds that can be dissolved in water are sodium chloride (NaCl), that is, table salt or calcium hydrogen carbonate $Ca(HCO_3)_2$.

If the solvent is a solid, the solutes can be in gaseous as well as in liquid and solid states – similar as in the case of liquid solvents. For instance, hydrogen gas ($H_2$) can be dissolved in the transition metal palladium (Pd). In fact, the solubility of hydrogen in palladium is very important for hydrogen storage. Liquids can also be dissolved in solid metals, for example, the liquid transition metal mercury (Hg) in the solid gold (Au) forms an alloy, that is, a so-called amalgam. Another example is the solution of water in solid salt or rock salt (halite, NaCl). Finally, steel is also a solution composed of carbon atoms (C), which are homogenously dispersed in a matrix of crystalline iron (Fe).

We now regard the solutes and the solvent as a closed system. In this case, we can say that if the solutes are dissolved in a solvent, that is, if a solution is formed, the entropy of the system is increased. This property is based upon the laws of thermodynamics, which predict that particles or chemical compounds exhibit the tendency to distribute homogenously inside a given space, whereupon the entropy is increased. Overall, we can say that solutions have positive entropies of mixing.

If water is the solvent, then we speak about aqueous solutions. The water molecule has polar structure (Figure 3.1) that makes it an effective solvent for polar substances in chemistry. Water is a protic solvent that dissociates into a proton ($H^+$) and a hydroxide ion ($OH^-$). Protic solvents are water-dissolvable salts (e.g., NaCl) or other polar substances (e.g., HCl). Polar substances are composed of polar molecules that exhibit a permanent dipole moment (Figure 3.2). Polar molecules stick together since the involved atoms carry different electrical charges due to the different electronegativity of the atoms. The electronegativity $\chi$ is a measure of the ability of an atom to attract electron pairs in case of a chemical bonding; it depends on the charge of the nucleus of the atom and the atomic radius. The larger the difference of the electronegativities of the involved atoms, the more pronounced is the polar character of the molecule. However, the difference of the electronegativities of the involved atoms in polar molecular bonding is not as high as in case of ionic bonding, where the electron of one atom is completely transferred to the other atom (this is the case, for instance, for NaCl that also exhibits a strong dipole moment).

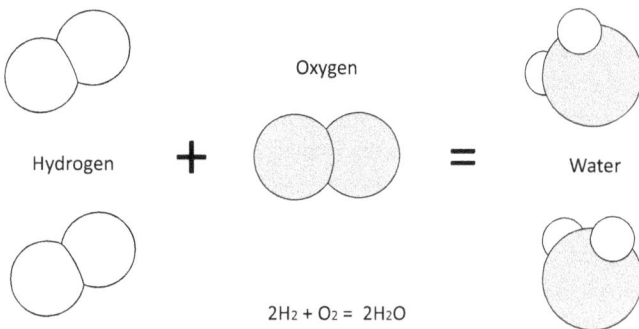

**Figure 3.1:** Two hydrogen gas molecules ($H_2$) combine with an oxygen gas molecule ($O_2$) and form two water molecules ($H_2O$). Water is a polar molecule with a double negative charge toward the oxygen side and two positive charges toward the hydrogen side (Figure 3.2).

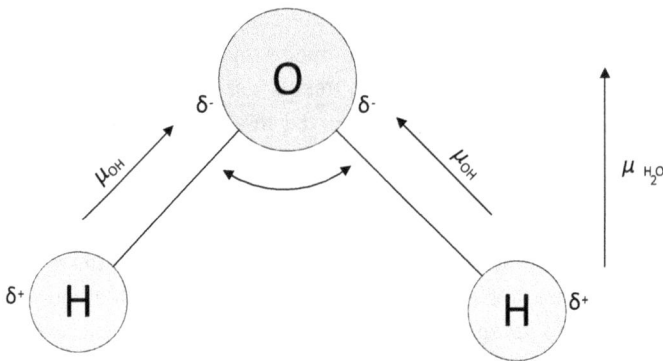

**Figure 3.2:** Polarity of the water molecule ($H_2O$); $\delta^+$ and $\delta^-$ are the positive and negative partial charges, and $\mu$ is the dipole moment.

### 3.1.2 Formation of electrolytes

An electrolyte is a substance that is electrically conductive if a voltage is applied. The carriers of this current are ions – not electrons as, for instance, in metals. Electrolytes are solutions, where – at least partly – the solute is dissociated in positive and negative ions. An electrolyte arises, if the solute is dissolved in a polar solvent – for instance, in water, that is, $H_2O$. The polar water molecules break up the crystal structure of the substance into ions that are separated from its adjacent neighbors in the crystal and are surrounded by $H_2O$ molecules – this process is called solvation and will be briefly described in the next section. In the polar solvent, the dissolution of the solute results into its dissociation into electrically charged ions. Hence, the solution is electrically conductive. The negatively charged ions are called anions and the positively charged are ions cations. Solutes that form electrolytes can be soluble salts, acids and bases. In addition, some gases, for example, hydrogen chloride (HCl), can be dissolved in polar solvents and become electrolytic. Furthermore, electrolytes can also arise, if polymers are dissolved in a polar solvent; this holds either for biological polymers, for instance, DNA, or for synthetic polymers. Those polymers contain charged functional groups that upon dissociation become the carriers that bear the electrical current. Anions and cations disperse uniformly through the solvent, and overall, the solution stays electrically neutral. However, the charged ions can move under the influence of an applied electrical field.

More generally, one can state that an electrolyte is a chemical compound that in a liquid state or in a solid state is dissociated into ions that migrate, if an electrical field is applied. In the broadest sense, electrolytes are substances, which at least partially exist as separated ions. To be more precise, an electrolyte consists of the electrolyte matrix – for instance, the polar solvent – and the dissociated ions of the dissolved substance.

If a solute completely or nearly completely dissociates within the polar solvent to form ions, then ion concentrations in the electrolytic solution are high, and we speak about a concentrated or strong electrolyte. On the other hand, if only a small fraction of the solute dissociates within the polar solvent, then the ion concentration is low, and we speak about a weak electrolyte (for weak electrolytes, see also Section 3.1.7).

A very prominent example for an electrolyte is sodium chloride, NaCl, which is dissolved in water. NaCl has a simple cubic lattice structure (Figure 3.3), where the lattice sites are located at the corners of a cube and are alternately occupied by sodium (Na) and chlorine (Cl) ions. In this cubic lattice, each sodium ion has six chlorine ions as direct neighbors, and vice versa, and each chlorine ion has six sodium ions as direct neighbors. In the periodic table of elements, sodium is an alkali metal, and therefore, a member of the first main group (period 3 and atomic number 11). Chlorine is a halogen and belongs to the seventh main group (period 3 and atomic number 17). Sodium chloride is an ionic compound, where sodium donates its outer electron to the chlorine atom resulting in the formation of $Na^+$ and $Cl^-$ ions. Then in the NaCl crystal, positive $Na^+$ ions are surrounded by negative $Cl^-$ ions and vice versa. The solid sticks together due to the electrostatic forces between the positively and negatively charged ions. The Coulomb law describes the electrostatic force:

$$F = \frac{1}{4\pi\varepsilon_0} \cdot \frac{q_1 \cdot q_2}{r^2} \tag{3.1}$$

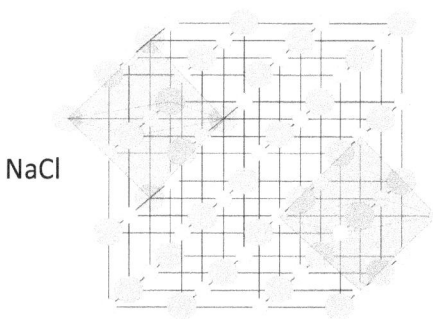

NaCl

**Figure 3.3:** Crystal structure of sodium chloride (NaCl); white balls: sodium ions and gray balls: chloride ions.

$F$ is the Coulomb force, $q_1$ and $q_2$ are the charges of the ions (for instance, for the $Na^+$ ion we get $q_1 = +1.602 \times 10^{-19}$ C and for the $Cl^-$ ion we get $q_2 = -1.602 \times 10^{-19}$ C), $r$ is the distance between the charges and $\varepsilon_0$ the permittivity of the vacuum. Since the interatomic distances within the NaCl crystal are very small ($r = 0.564$ nm), the Coulomb force is very large. Therefore, breaking the crystal structure requires large energy, and the melting temperature of NaCl is rather large (801 °C). However, if the solid NaCl crystals are poured into a solvent, for instance, water, the Coulomb force of attraction between the neighboring $Na^+$ and $Cl^-$ ions is significantly reduced; since in eq. (3.1) instead of $\varepsilon_0$ we have used $\varepsilon = \varepsilon_r \cdot \varepsilon_0$. In vacuum $\varepsilon_r = 1$ and in air $\varepsilon_r \approx 1$. However, in water $\varepsilon_r = 78.3$ (at 25 °C in a homogenous phase), and therefore, water can break off – dissolve – the NaCl crystal quite easily, even at room temperature since in water the Coulomb force is strongly reduced as compared to ambient air or in vacuum.

NaCl is a salt, that is, it is an ionic compound that arises if, an acid and a base are neutralized, that is, it occurs because of a neutralization reaction. Consider hydrochloric acid (HCl) and sodium hydroxide (NaOH), which is a strong base. If the acid hydrochloride and the base sodium hydroxide react, the $OH^-$ ions from the base and the $H^+$ ions from the acid are combined to form water, $H_2O$. Moreover, the $Cl^-$ ions from the acid and the $Na^+$ ions from the base are combined to form sodium chloride, that is, NaCl:

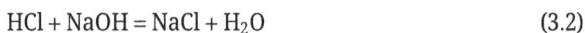

$$HCl + NaOH = NaCl + H_2O \qquad (3.2)$$

This is a neutralization reaction since the acidic and basic properties of the educts HCl and NaOH are vanished in the resulting products, that is, sodium chloride salt and water. In fact, using this simple chemical formula, the situation is only superficially described. To be more precise, we consider, for instance, 0.1 M solutions of HCl and NaOH in water, that is, 0.1 mol HCl or NaOH per liter $H_2O$. In water, HCl dissociates into a negatively charged $Cl^-$ ion and a positively charged hydronium ion, $H_3O^+$:

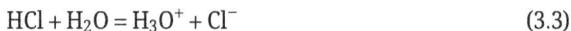

$$HCl + H_2O = H_3O^+ + Cl^- \qquad (3.3)$$

Furthermore, in water NaOH is dissociated into $Na^+$ and $OH^-$ ions. Thus, the neutralization reaction now can be more precisely written as

$$H_3O^+ + Cl^- + Na^+ + OH^- \rightarrow Na^+ + Cl^- + 2H_2O \qquad (3.4)$$

In this case, the net reaction for the neutralization is given by

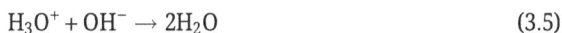

$$H_3O^+ + OH^- \rightarrow 2H_2O \qquad (3.5)$$

According to formula (3.4), the product is not a mixture of an acid, a base and water. But it consists of $Na^+$ and $Cl^-$ ions which are dissolved in water, that is, the result is salty water, and this is an electrolyte. The same holds, if NaCl salt is dissolved in water.

Thus, we can say that an electrolyte can be established, when a salt is poured into a solvent. The salt is dissolved in the solution and dissociates into its ionic components. We stay with the example sodium chloride, which is a solid, and therefore, labeled by the subscript (s), that is, $NaCl_{(s)}$. If the sodium chloride is dissolved in water, it dissociates into its ionic components $Na^+_{(aq)}$ and $Cl^-_{(aq)}$, which are dispersed in the liquid water, and therefore, labeled by the subscript (aq) – from the Latin word aqua. Then we get the dissociation reaction:

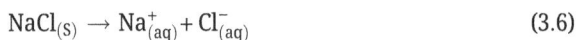

$$NaCl_{(s)} \rightarrow Na^+_{(aq)} + Cl^-_{(aq)} \qquad (3.6)$$

The situation might get a little bit more complicated, if a substance reacts with the solvent resulting in the production of ions. As an example, we consider gaseous carbon dioxide, $CO_2$, which dissolves in water. Together with the water, $CO_2$ forms carbonic acid ($H_2CO_3$), sometimes also written in the form $OC(OH)_2$. Furthermore, carbonic acid forms two kinds of salts, named carbonates and bicarbonates, which either are characterized by the carbonate ions $CO_3^{2-}$ with a double negative charge or by singly charged bicarbonate ions $HCO_3^-$; the latter ones are also called hydrogen carbonate ions. Within water, the following reactions occur:

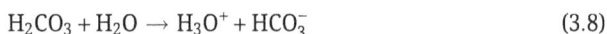

$$CO_2 + H_2O \rightarrow H_2CO_3 \tag{3.7}$$

$$H_2CO_3 + H_2O \rightarrow H_3O^+ + HCO_3^- \tag{3.8}$$

$$HCO_3^- + H_2O \rightarrow CO_3^{2-} + H_3O^+ \tag{3.9}$$

By these reactions, hydronium ions, carbonate ions and hydrogen carbonate ions are provided, which are dissolved in water and form the electrolyte.

### 3.1.3 Solvation

In liquid solutions, a process called solvation occurs, where the molecules of the solvent are attracted by ions of the dissociated solute. For instance, $Na^+$ ions of the dissociated NaCl molecules that are dissolved in water attract water molecules (Figure 3.4). The charged ion ($Na^+$) and the attracted $H_2O$ molecules form an associate. This is done in a way that the negative part of the polar water molecules points toward the $Na^+$ ion, and the negative side of the water dipole points into the opposite direction – off the $Na^+$ ion. Water molecules surround the $Na^+$ ion and they form a so-called solvation or hydration shell, where the latter notation is only used, if the solvent is water. The same occurs with the negative $Cl^-$ ions. The $Cl^-$ ions also attract the polar water molecules. The water molecules surround the $Cl^-$ ion, but now the positive side of water dipoles point toward the $Cl^-$ ions. The solvation process supports the even distribution of the positive and negative ions of the dissociated molecules of the dissolved solute within the solvent. This finally leads to a stabilization of the dissociated solute within the solvent since the solvation shell shields the positive and the negative ions; hence, they cannot recombine anymore.

Without extraneous influence or disturbances, for instance, if no electrical field is applied, the ion is symmetrically surrounded by the dipoles of the solvent (Figure 3.4). In case an electrical field is applied, the ion migrates through the electrolyte toward the corresponding electrode. However, in the latter case, the solvation shell is not symmetrically adjusted around the ion anymore due to inertia (Figure 3.5). This is a dissipative effect. Hence, the migration of the ion is retarded due to the impact of the corresponding electrostatic counter forces from the solvation shell.

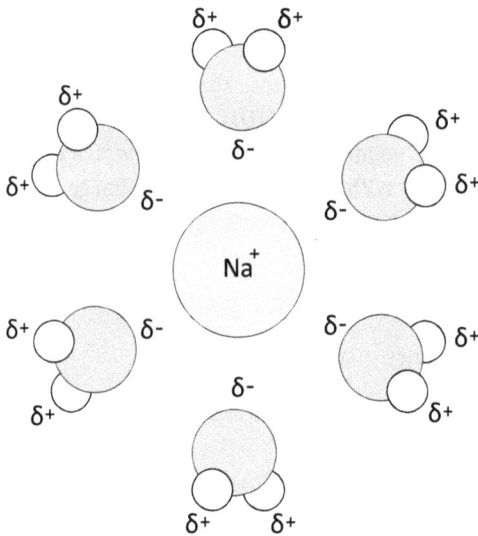

**Figure 3.4:** Sodium ions (Na$^+$) in an aqueous solution attract water molecules (H$_2$O). The negatively charged oxygen side of the water molecule (dipole) points toward the Na$^+$ ion. The water molecules form a hydration shell around the Na$^+$ ion.

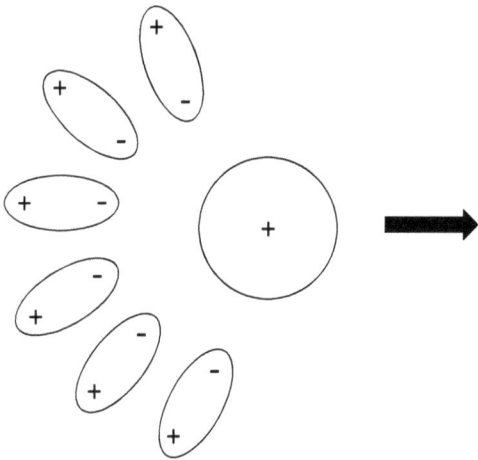

**Figure 3.5:** In an electric field, a sodium ion (Na$^+$) with a hydration shell migrates through an aqueous solution. Due to inertia, the hydration shell around the Na$^+$ ion is not symmetrically adjusted.

Furthermore, one can point here also to frictional effects. In the electrolyte, positively and negatively charged ions occur in parallel, and within an electrical field, they migrate in opposite direction. Of course, a solvation shell surrounds both types of ions. Hence, upon migration in opposite directions, the hydration shells of the positive and negative ions interact with each other and cause a retardation of the ion mobility. In chemistry, this effect is also known as electrophoresis.

In the next section, we look a little more in detail at the mobility of ions in electrolytes.

### 3.1.4 Ionic mobility and conductivity in electrolytes

If charged ions in an electrolyte are homogenously distributed, they statistically move without any favored direction according to the "Brownian motion" (Figure 3.6). Brownian motion is a random motion of the small particles that are suspended in a liquid or gas. This random motion results from collisions between the particles and the atoms or molecules of the liquid or gas. The atoms or molecules of the liquid or gas rapidly move in permanently changing random directions due to thermal activation. However, in the electrolytes, the ions also can migrate in certain well-defined directions, and the directed migration is superposed to the Brownian motion. There exist two possible reasons for such directed migrations of ions in electrolytes. First, a directed migration can be caused by the application of an electric field, and second, the directed migration can be forced by a concentration pressure according to a nonhomogeneous distribution of the ions with the solution, that is, by a diffusion mechanism. We now concentrate on the description of the first case, that is, the migration of ions in an electrolyte under the influence of an applied electric field. The diffusion of ions within an electrolyte will be discussed in Section 3.1.6.

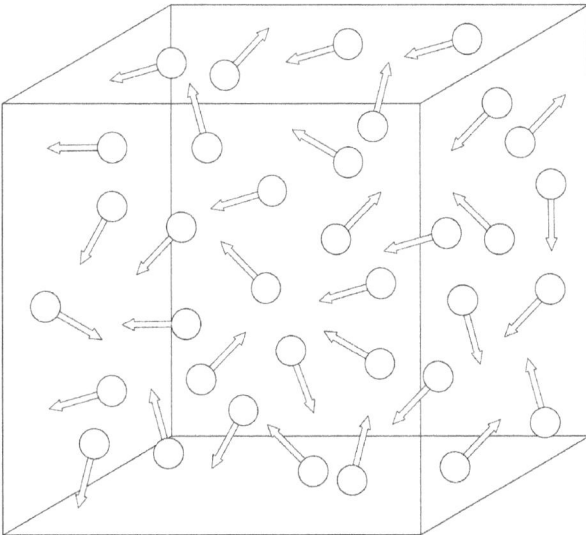

**Figure 3.6:** Brownian motion.

Consider, for instance, the dissolution of NaCl salt in water that is described by the dissociation reaction is already shown in formula (3.6). As shown in this formula, positively charged sodium ions and negatively charged chloride ions are formed upon dissolution of NaCl in water. If an electric field is applied, for instance, by introducing platinum (Pt) electrodes into the electrolyte, the charged ions within the electrolyte migrate to the according electrodes. If we continue with the example of the aqueous NaCl solution, then the positive sodium ions ($Na^+$) migrate toward the negative Pt cathode and the negative chloride ions ($Cl^-$) toward the positive Pt anode. Hence, an ionic current occurs within the electrolyte that can be measured. The ionic current in the electrolyte follows Ohm's law:

$$I = \frac{U}{R} \tag{3.10}$$

where $I$ is the ionic current, $U$ is the applied voltage at the electrodes and $R$ is the resistance. This is empirically well proved for electrolytes.

The ohmic law describes a certain electrolyte and the voltage-independent parameters that stand for the characteristic information about the electrolyte are summed up in the resistance – or in the conductance, that is, the reciprocal resistance $1/R$.

The existence of ions and their concentration within an aqueous solution or an electrolyte can be proved and quantified by measuring the resistance or the conductance, respectively – for instance, with a Wheatstone bridge (Figure 3.7). However, the measurement of the real resistance or conductance of an electrolyte is not trivial. If only a simple DC measurement is employed, where a constant DC voltage is applied at the electrodes, a counter voltage appears at the electrodes, which is directed in opposition to the initially applied voltage and the initially applied electric field. The reason is that at the interface of the electrolyte and the electrode, electrochemical reactions occur that are responsible for the counter voltage. The effect is described as electrolytic polarization. In addition, the ions that migrate toward the corresponding electrode cause a shielding of the applied electric field so that the effective electric field is significantly reduced. Therefore, the resistance that is characteristic for the material parameters of the electrolyte is altered, and hence, not correct. Such problems can be avoided by high-frequency AC measurements in the kHz range, that is, an AC voltage has to be applied at the electrodes, then a corresponding AC current can be measured, and by this, the resistance or the conductance of the electrolyte can be deduced much more precisely. Since the ionic current through the electrode–electrolyte interfaces periodically changes its algebraic sign, the chemical interface reactions are permanently inverted and the electrochemical polarization can be avoided. In addition, also the electric shielding of the electrodes by the formation of ion clouds close to them does not occur anymore.

In general, the resistance depends on two parameters, that is on the specific resistance – or in other words the resistivity $\rho$ – which is a material parameter, and on a

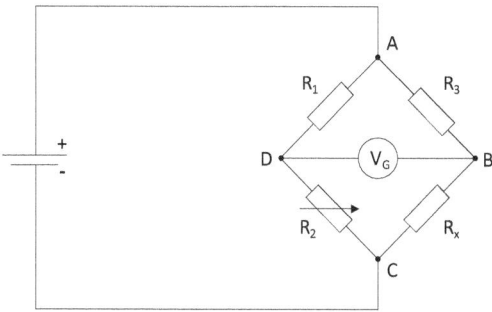

**Figure 3.7:** Wheatstone bridge.

geometric factor $C$ that describes the geometry of the electrolytic measurement cell, in which the resistivity measurements of the electrolyte are employed (Figure 3.8), that is,

$$R = \rho \cdot C \qquad (3.11)$$

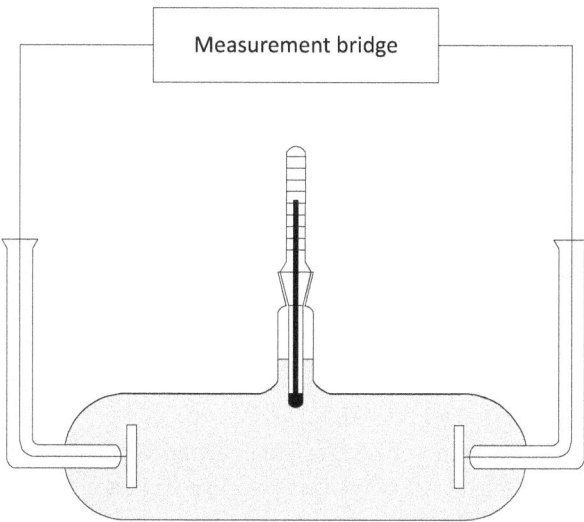

**Figure 3.8:** Typical setup of an electrolytic measurement cell with two electrodes and a thermometer.

The resistivity is temperature-dependent; its physical unit is $\Omega$ m. Equation (3.11) holds by the way for any conductor, both electronic as well as ionic conductors. In the ideal case, where the current flows homogenously through a conductor – the carriers could be either electrons or ions in this case – the geometric factor $C$ is given by

$$C = \frac{l}{A} \qquad (3.12)$$

where $l$ is the length of the conductor or – more precisely for our considerations – the distance of the electrodes in the electrolytic measurement cell; $A$ is the cross-sectional area of the electrodes, where the ionic current meets the electrodes. From eqs. (3.11) and (3.12), we directly get the resistivity $\rho$:

$$\rho = \frac{R}{C} = \frac{R \cdot A}{l} \tag{3.13}$$

Often the conductivity $\kappa$ is used instead of the resistivity and $\kappa$ is given by

$$\kappa = \frac{1}{\rho} = \frac{l}{R \cdot A} \tag{3.14}$$

The conductivity is the characteristic parameter that describes the electrical properties of an electrolyte. In contradiction to the electronic conductivity in metals and semiconductors usually expressed by the Greek letter $\sigma$, for the ionic conductivity of electrolytes, we use the Greek letter $\kappa$ to distinguish the latter case from the former one. Anyway, the physical unit in all cases, that is, for $\sigma$ and $\kappa$, is S/m $= \Omega^{-1}\,m^{-1}$ (where $S$ stands for Siemens, and $S = \Omega^{-1}$).

If an electric field $E$ is applied to the electrolyte, the ions move according to their positive or negative charge state, that is, either in field direction or opposite to it. Hence, an ionic current is provoked. The ionic current density $j$ is defined by

$$\vec{j} = \kappa \cdot \vec{E} \tag{3.15}$$

The ionic current density depends first on the applied electric field $E$, and therefore, via the geometric factor $C$ – eq. (3.11) – on the voltage that is applied to the electrodes of the electrolytic measurement cell. Because of the physical parameters that are included in the ion conductivity $\kappa$, the current density also depends on the ion concentration $n_i$ in the electrolyte and the charge of the ions that is related to their valence $z_i$. Finally, $j$ also depends on the drift velocity $v_i$ of the migrating specific ions within the electrolyte – in and opposite to the field direction.

The electrical conductivity of electrolytes is based upon the migration of ions in an electric field. To be more precise, the migrating species that are moving under the influence of the electric field are not just the ions but rather solvated ions, that is, ions surrounded by a cloud of polar molecules of the solvent (Figures 3.4 and 3.5). However, to keep our descriptions less complicated, in the following discussions we simply speak about ions keeping in mind that these ions are solvated within the electrolyte.

The electric force $F_e$ that acts on the charged ions cause the transport of the ions through the electrolyte, that is, the ionic current. In general, according to their valence the ions exhibit a charge of $z_i \cdot e$, where $e$ is the elementary charge (i.e., the charge of an electron), and therefore, the electric force $F_e$ is determined by

$$\vec{F}_e = z_i \cdot e \cdot \vec{E} \tag{3.16}$$

The ions – solvated ions – in the electrolyte are accelerated by the electric field or rather the electric force. However, in addition they also experience a frictional force that becomes a fortiori stronger the higher the velocity of the ions gets. Hence, the ion velocity is increasing because of acceleration by the electric force $F_e$. However, this holds only for a very short period of about $10^{-13}$ s. After this period, the average velocity of the ions approaches a residual speed and becomes constant because the frictional force works against the electric force. The frictional force – also called hydrodynamic friction or Stokes' drag – is proportional to the ion velocity $v_i$; it is described by the well-known Stokes' law:

$$\vec{F_d} = 6 \cdot \pi \cdot \eta \cdot r_i \cdot \vec{v}_i \tag{3.17}$$

$F_d$ names the frictional force or Stokes' drag, $\eta$ is the dynamic viscosity, $r_i$ stands for the radius of the migrating species, that is, of the solvated ions, and $v_i$ is their velocity.

For the stationary flow of the ions with constant final speed $v_{i-max}$, the electric and the frictional forces are equal (but of course opposite in sign):

$$\vec{F_e} = -\vec{F_d} \tag{3.18}$$

According to eqs. (3.16) and (3.17) the following relation holds:

$$z_i \cdot e \cdot \vec{E} = 6 \cdot \pi \cdot \eta \cdot r_i \cdot \vec{v}_{i-max} \tag{3.19}$$

Hence, we get for the final speed of the ions:

$$\vec{v}_{i-max} = \frac{z_i \cdot e \cdot \vec{E}}{6 \cdot \pi \cdot \eta \cdot r_i} \tag{3.20}$$

For a particular electrolyte with a dedicated dynamic viscosity $\eta$ and a given applied electric field $E$, every ion species that is contained in the electrolyte has a drift velocity that is dependent on the ion charge ($z \cdot e$) and the radius of the solvated ion ($r_i$). The sign of the charge gives the direction of the drift velocity.

The drift velocity of the ions within an electric field can be expressed as follows:

$$\vec{v}_i = u_i \cdot \vec{F_e} \tag{3.21}$$

where $u_i$ is the ion mobility. According to eqs. (3.16) and (3.21), we get the following:

$$\vec{v}_i = u_i \cdot z_i \cdot e \cdot \vec{E} = u \cdot \vec{E} \tag{3.22}$$

In this connection, $u$ is defined as

$$u = u_i \cdot z_i \cdot e \tag{3.23}$$

The parameter $u$ is the carrier mobility, as defined in electrodynamics, and stands for the relation of the drift velocity of carriers and the electric field. We can also call it as mobility parameter or simply mobility. Hence, both mobility parameters $u$ and $u_i$ are strongly related to each other; however, they are differently defined. At this point, one has to be careful not to be confused.

In general, the current density $j$ of free moving carriers is proportional to their velocity:

$$\vec{j} = e \cdot n \cdot \vec{v} \tag{3.24}$$

Here $n$ is the carrier density and $v$ the velocity of the carriers. However, in an electrolyte we are faced with positively and negatively charged ions (cations and anions) with valences $z_+$ and $z_-$, ion densities $n_+$ and $n_-$ and the oppositely directed velocities $v_+$ and $v_-$. Hence, we have to be a little more precise, and therefore, in frame of our considerations the total current density sums up as follows:

$$\vec{j} = \kappa \cdot \vec{E} = z_+ \cdot e \cdot n_+ \cdot \vec{v}_+ + z_- \cdot e \cdot n_- \cdot \vec{v}_- \tag{3.25}$$

The positive and negative subscripts refer to the positively and negatively charged ions. According to eq. (3.22), we obtain

$$\vec{j} = e \cdot (z_+ \cdot n_+ \cdot u_+ + z_- \cdot n_- \cdot u_-) \cdot \vec{E} \tag{3.26}$$

where $u_+$ and $u_-$ are used in the sense of definition (3.23). Finally, in accordance with eq. (3.15), we receive for the conductivity $\kappa$ the following equation:

$$\kappa = e \cdot (z_+ \cdot n_+ \cdot u_+ + z_- \cdot n_- \cdot u_-) \tag{3.27}$$

Equation (3.27) shows that the ion conductivity depends on the valences, the concentrations and the mobilities of the positive and negative ions within the electrolyte.

We also want to consider the case that the salt molecules are not completely dissociated in the solvent, that is, a part of the molecules is not dissociated within the electrolyte. Then the equation for the ion conductivity $\kappa$ is somewhat modified. We denominate $w_+$ and $w_-$ the number of positive and negative ions, into which a molecule dissociates and $n_m$ should be the concentration of molecules that are added to the solvent. Furthermore, we define the electrochemical valence $z_e$ as follows:

$$z_e = w_+ \cdot z_+ = w_- \cdot z_- \tag{3.28}$$

with the valences $z_+$ and $z_-$ of the positive and negative ions. For the corresponding ion concentrations, we get the following:

$$n_+ = w_+ \cdot n_m, n_- = w_- \cdot n_m \tag{3.29}$$

Finally, we receive for the ion conductivity

$$\kappa = e \cdot z_e \cdot n_m \cdot (u_+ + u_-) \tag{3.30}$$

Hence, the ion conductivity is proportional to the concentration of the dissociated molecules, the electrochemical valence and the mobilities (in the sense of eq. (3.23)) of the positive and negative ions.

The ion concentration determines the conductivity of an electrolyte, or vice versa, and the conductivity of an electrolyte is dependent on its ion concentration. To simplify matters, now we consider strong electrolytes that are completely dissociated, that is, the aqueous solutions of completely dissociated inorganic acids like HCl, $H_2SO_4$ or $HNO_3$ or salts like NaCl, where the acid or salt molecules are completely dissociated. For such electrolytes, a linear dependence of the conductivity on the electrolyte concentration or the respective concentration of the ions from the dissociated salt or acid molecules could be expected. However, this simple relation only holds for strongly diluted electrolytes, where only small concentrations of acids or salts are dissolved in the solvent, for example, water. Since the ions are positively and negatively charged within the electrolyte, a rising ion concentration provokes increasing interactions among themselves due to their electrostatic forces. If ion concentrations are rising within the electrolyte, the average distances between charged ions is reduced, and therefore, according to eq. (3.1), the electrostatic forces between the ions get stronger and impede the ion migration, and hence, the conductivity of the electrolyte as well (Figure 3.9).

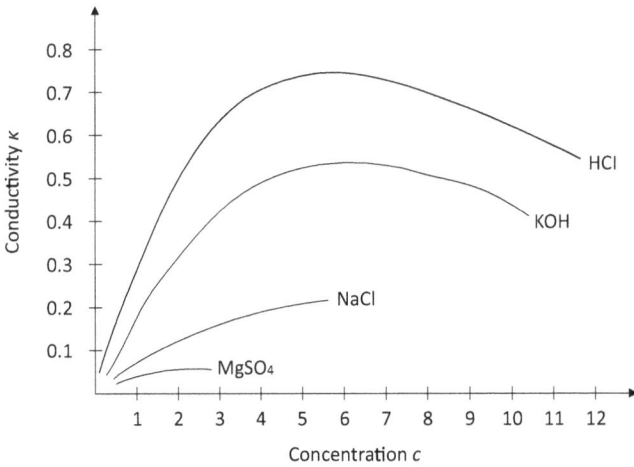

**Figure 3.9:** Ion conductivity $\kappa$ versus ion concentrations $c$ in the electrolyte for various electrolytes.

Furthermore, also the charge of state of the ions has an impact on the conductivity of the electrolyte since the electrostatic forces are also stronger, if the ion charge is increased. This can be also seen from eq. (3.1). For instance, if we consider NaCl that is dissolved in water, then the positive sodium ions ($Na^+$) and negative chloride ions ($Cl^-$) both are singly charged. On the other hand, if dissolved in water, magnesium sulfate ($MgSO_4$) dissociates into $Mg^{2+}$ and $SO_4^{2-}$ ions. Hence, those ions are doubly charged, and the electrostatic force is much stronger than in the case of dissociated NaCl molecules. The consequence is that the conductivity of the $MgSO_4$ electrolyte is significantly lower than the one of the NaCl electrolyte (Figure 3.9).

If the ion concentrations in the electrolyte get very high, then the conductivity can even become smaller (Figure 3.9). The reason is that the positive and negative ions can interact due to the Coulomb forces. The ions do not recombine to molecules again, but they cluster and form agglomerates that act as neutral particles. Hence, their migration in an electric field is prevented and for the conductivity of the electrolyte, they do not play a role anymore.

### 3.1.5 Molar conductivity

The conductivity $\kappa$ is the inverse of the resistivity $\rho$ and proportional to the path length of the ion current $l$ – that is, the distance between the electrodes – and inversely proportional to the resistance $R$ and the cross-sectional area $A$ that is flown through by the ionic current. This correlation was given already in eq. (3.13). The conductivity of an electrolyte depends on the number of the electric carriers, that is, the ion concentration. In this connection, the molar conductivity $\Lambda_m$ is defined by

$$\Lambda_m = \frac{\kappa}{c} \tag{3.31}$$

Here $c$ is the molar concentration – also called molarity – in $mol/m^3$ and hence, $\Lambda_m$ is given in $\Omega^{-1}\,m^2\,mol^{-1}$. One can mention that another often used unit for the molar concentration is mol per liter (mol/L), and furthermore, a solution with the concentration of 1 mol/L is called 1 molar (1 M).

The molar concentration $c$ is defined by the amount $n$ of the solute per unit volume $V$ of the solution. If $N$ denotes the number of molecules present in the volume $V$ of the solution ($V$ can be given for instance in liters), then the ratio $N/V$ defines the so-called number concentration $C$. Using Avogadro's constant ($N_A \approx 6.0221 \times 10^{23}\,mol^{-1}$), we obtain

$$c = \frac{n}{V} = \frac{N}{N_A \cdot V} = \frac{C}{N_A} \tag{3.32}$$

Anyway, the conductivity not only depends on the ion concentration but also on the charge state of the ions. For instance, if the ions are twofold positively and/or negatively charged, the conductivity is higher than for singly positively and/or negatively charged ions. Hence, the higher the charge state of the ions, the higher the conductivity of the electrolyte. Therefore, if we use the charge number $z$, we can define the equivalent molar concentration by the following equation:

$$c_{eq} = z \cdot c = z \cdot \frac{n}{V} \qquad (3.33)$$

The charge number $z$ multiplied with the elementary charge $e$ gives the charge $Q$ of a particle or a system. Using eq. (3.33), we define

$$\Lambda_{eq} = \frac{\kappa}{c_{eq}} \qquad (3.34)$$

$\Lambda_{eq}$ can be called equivalent molar conductivity. For completely dissociated electrolytes, $\Lambda_{eq}$ exhibits values in the order of $10^2 \ \Omega^{-1} \ cm^2 \ mol^{-1}$ (or $10^{-2} \ \Omega^{-1} \ m^2 \ mol^{-1}$). The equivalent molar concentration $c_{eq}$ can be related to the concentrations of positive or negative charge carriers, that is, ions. Hence, more precisely than in eq. (3.33), we can write

$$c_{eq} = c \cdot z_+ \cdot v_+ = c \cdot |z_-| \cdot v_- \qquad (3.35)$$

Here $z_+$ and $z_-$ are the signed charge numbers and $v_+$ and $v_-$ are the stoichiometric coefficients of the cations and anions, respectively. The stoichiometric coefficients give the ratio of the amount of the positive and negative particles or ions.

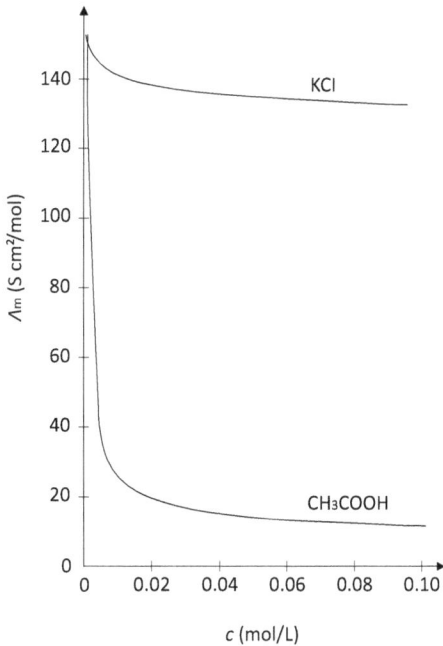

**Figure 3.10:** Dependence of the molar conductivity on the concentration. Two examples are shown: a typical strong electrolyte (potassium chloride, KCl) and a typical weak electrolyte (acetic acid, $CH_3COOH$). Strong electrolytes exhibit only weak concentration dependence of the molar conductivity.

Although in definition (3.31) the molar conductivity $\Lambda_m$ is normalized to the molar concentration $c$, it still depends on $c$ (Figure 3.10). The reason is that interactions of the ions in the electrolyte occur, and the dissociation depends on $c$. In case of strong electrolytes (i.e., salts, strong acids or strong bases), where the dissociation is complete, the molar conductivity depends only weakly on the molar concentration, as follows:

$$\Lambda_m = \Lambda_m^0 - K \cdot \sqrt{c} \tag{3.36}$$

This is Kohlrausch's nonlinear law for strong electrolytes that is also called Kohlrausch's square-root law. $\Lambda_m^0$ is the so-called limiting molar conductivity that is the molar conductivity at infinite dilution, that is, for the solution itself in the limit of zero concentration. It can be deduced by linear extrapolation toward the molar concentration value $c = 0$, if $\Lambda_m$ is drawn as a function of the square root of $c$ (Figure 3.11). $K$ is an empirical constant and is called Kohlrausch coefficient; it depends mainly on the stoichiometry of the solute, for instance, a salt. Kohlrausch' nonlinear or square-root law only holds for low electrolyte concentrations.

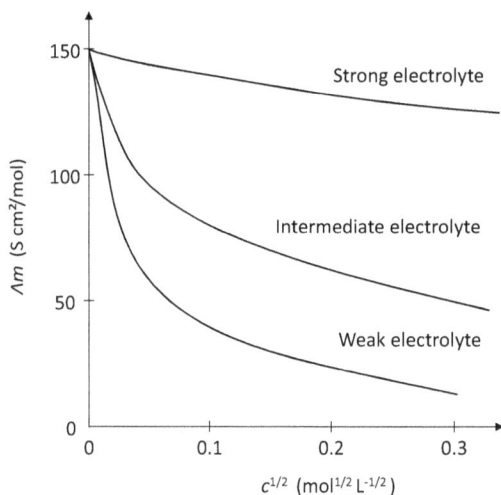

**Figure 3.11:** Kohlrausch's square-root law for strong electrolytes (upper curve). Intermediate and weak electrolytes do not follow Kohlrausch's square-root law.

Moreover, the molar conductivity of a strong electrolyte even becomes directly proportional to the molar concentration, if $c$ becomes sufficiently low, as shown, for instance, from a Taylor series expansion of the square root of $c$. If $\Lambda_m$ is plotted in dependence on the square root of $c$, the limiting molar conductivity $\Lambda_m{}^0$ can be obtained for strong electrolytes by linear extrapolation toward zero (Figure 3.11).

For an ideally diluted solution without interionic interactions, both positively as well as negatively charged ions exhibit an independent migration velocity irrespectively on the other oppositely charged ion. This is the so-called law of independent ion migration. Hence, the limiting molar conductivity $\Lambda_m{}^0$ is additively composed of the limiting molar conductivities of positively and negatively charged ions itself:

$$\Lambda_m^0 = v_+ \cdot \lambda_{m+}^0 + v_- \cdot \lambda_{m-}^0 \tag{3.37}$$

where $\lambda_{m+}^0$ and $\lambda_{m-}^0$ are the limiting molar conductivities of the positively charged cations and the negatively charged anions; $v_+$ and $v_-$ are again the stoichiometric coefficients of these cations and anions, respectively. In general, a molar conductivity $\lambda$ is related to the ion mobility $u$ as follows:

$$\lambda = z \cdot u \cdot F \tag{3.38}$$

$F$ is the Faraday constant ($F = N_A \cdot e$, where $N_A \approx 6.0221 \times 10^{23}$ mol$^{-1}$ is the Avogadro constant as mentioned earlier); $z$ is the valence. Hence, with relation (3.38), eq. (3.37) – for the case of the independent ion migration – can be modified as follows:

$$\Lambda_m^0 = (z_+ \cdot u_+ \cdot v_+ + z_- \cdot u_- \cdot v_-) \cdot F \tag{3.39}$$

For the so-called symmetrical $z$: $z$ electrolytes, where $z_+ = z_- = z$ and $v_+ = v_- = 1$, eq. (3.39) gets more simple:

$$\Lambda_m^0 = z \cdot (u_+ + u_-) \cdot F \tag{3.40}$$

Examples for symmetrical electrolytes are sodium chloride ($z = 1$) that dissociates in water according to $NaCl_{(s)} \rightarrow Na^+{}_{(aq)} + Cl^-{}_{(aq)}$ and copper(II) sulfate ($z = 2$) that dissociates according to $CuSO_{4(s)} \rightarrow Cu^{2+}{}_{(aq)} + SO_4{}^{2-}{}_{(aq)}$.

The Kohlrausch's nonlinear law for strong electrolytes was experimentally found. A microscopic theoretical description was later given by Debye–Hückel–Onsager model that provides quantitative results for the conductivity of ions in dilute solutions of strong electrolytes, and which finally delivers Kohlrausch's equation (3.36). The model is based on two effects, which are the reason for the dependence of the ion conductivity on the concentration. According to the interpretation in frame of the Debye–Hückel–Onsager model, the ions cannot freely move. In fact, far-reaching electrostatic forces cause a mutual interference of the migrating ions. On average, the ions are surrounded by more ions that are oppositely charged than by ions of an equal charge – due to the electrostatic action. That is, around the ions an ion cloud of opposite charge is formed. This ion cloud agglomerates, if the concentration is increased. Two implicit effects get stronger with increasing concentrations as follows:

The first effect is a relaxation effect. If the electric field is zero, the ions are symmetrically surrounded by oppositely charged ions (and/or by a hydration shell as explained in Section 3.1.3). However, if an electrical field is applied, the ion moves, and therefore, at each location of the ion, a new ion cloud is built up. This process occurs continuously and has the effect that the ion hurries ahead its ion cloud. Hence, if an electrical field is applied, the moving ion is surrounded by a dissymmetrical ion cloud, that is, the ion has a stronger inertance than the ion upon migration (Figure 3.12). By this effect, the ion migration is retarded.

The second effect is the electrophoretic effect. If an electric field is applied, positively and negatively charged ions surrounded by an oppositely charged ion cloud (or by a hydration shell) are moving in opposite directions. The migration of the ions is retarded by the friction of the ion clouds (or the hydration shells) of the ions. In other words, the ion cloud moves in opposite direction to its central ion, since the central ion and the surrounding cloud are oppositely charged. Hence, the central ion is swimming against the stream of ions of the ion cloud. This causes friction.

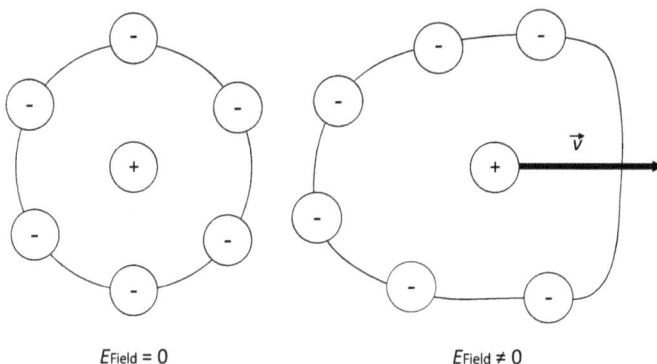

$E_{Field} = 0$                              $E_{Field} \neq 0$

**Figure 3.12:** Positive ion surrounded by an oppositely charged ion cloud. The oppositely charged ion cloud is symmetrical in case of an absent electrical field and asymmetrical in case of an applied electric field that makes the ion moving. The moving ion hurries ahead its surrounding ion cloud.

The Debye–Hückel–Onsager model leads to an expression for the Kohlrausch coefficient $K$:

$$K = A + B \cdot \Lambda_m^0 \tag{3.41}$$

where $A$ and $B$ are given by

$$A = \frac{z^2 \cdot e \cdot F^2}{3 \cdot \pi \cdot \eta} \cdot \sqrt{\frac{2}{\varepsilon \cdot R \cdot T}} \tag{3.42}$$

and

$$B = q \cdot \frac{z^3 \cdot e \cdot F}{24 \cdot \pi \cdot \varepsilon \cdot R \cdot T} \cdot \sqrt{\frac{2}{\varepsilon \cdot R \cdot T}} \tag{3.43}$$

In eqs. (3.42) and (3.43), $\varepsilon$ is the permittivity of the solvent ($\varepsilon = \varepsilon_r \cdot \varepsilon_0$, $\varepsilon_r$ and $\varepsilon_0$ are the relative permittivity and the permittivity of the vacuum, respectively), $e$ is the elementary charge, $z$ is the valence of an ion, $F$ is the Faraday constant, $\eta$ is the viscosity of the solvent, $R$ is the universal gas constant ($R \approx 8.3145$ J/mol K, $R = N_A \cdot k_B$ and $k_B \approx 1.3806 \times 10^{-23}$ J/K is the Boltzmann constant) and $q$ is a number that varies for different types of electrolytes.

For the symmetrical so-called 1:1 electrolytes such as sodium chloride, where each ion is singly charged, $q = 0.586$.

Finally, this leads to an expression that looks quite similar to the square-root law of Kohlrausch, that is, similar to experimentally deduced eq. (3.36):

$$\Lambda_m = \Lambda_m^0 - \left(A + B \cdot \Lambda_m^0\right) \cdot \sqrt{c} \tag{3.44}$$

However, this relation holds only for low molar concentrations; hence, the molar concentration should usually be less than about $10^{-3}$ mol/L, that is, <0.001 M, depending on the charge type. Under this premise, the agreement between theory and experiment is quite good.

## 3.1.6 Diffusion of ions in electrolytes

We briefly discussed in the last section the directed motion of ions in an electrolyte under the influence of an electric field; this implies that an electric force is acting on the ions (eq. (3.16)). But as mentioned earlier, ion migration in an electrolyte can also occur without external force forming an applied electrical field; that is, it can originate from a concentration pressure according to a nonhomogeneous distribution of small particles within a solution, or in case of our considerations, a nonhomogeneous distribution of ions in an electrolyte. Thus, local differences in ion concentrations cause a directed motion of the ions from regions with high ion concentrations toward regions with lower ion concentrations (Figure 3.13). This effect is commonly known as diffusion.

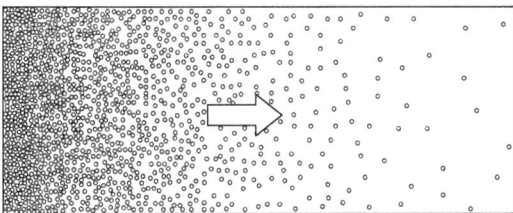

**Figure 3.13:** Diffusion of particles from a high-density region toward a region with lower particle density.

The directed migration according to a concentration gradient is superposed to the Brownian motion (Figure 3.14). Hence, more precisely defined, diffusion is a net movement – here of ions – from a region with higher to a region with lower concentrations. We assume a constant temperature and pressure. No expansion occurs here, in this case, since temperature and pressure are constant. At constant temperature and pressure, the amount of work $\delta W$ that can be applied per mole on a substance is related to a change of the chemical potential $d\mu$ of the substance, if the substance moves from a position with the chemical potential $\mu$ to a position with the chemical potential $\mu + d\mu$, that is,

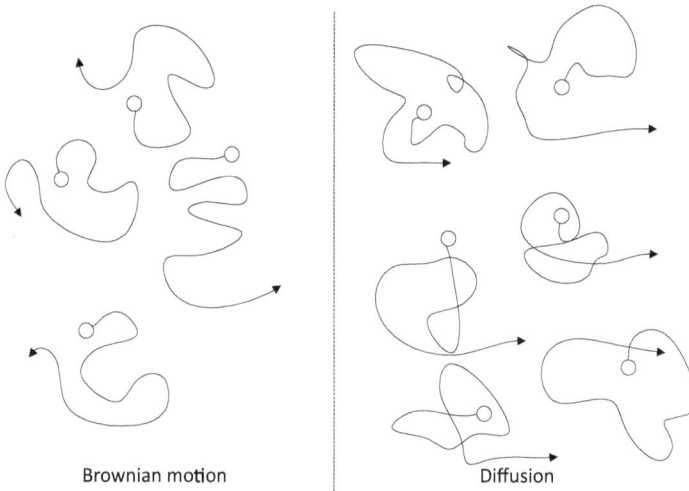

**Figure 3.14:** Brownian motion (left side) and diffusion (right side) of small particles. Diffusion of particles from the left to the right is superimposed to their Brownian motion.

$$\delta W = d\mu = \left(\frac{\partial \mu}{\partial x}\right)_{p,T} dx \tag{3.45}$$

For simplification, here we assumed that the migration of the substance only occurs in $x$-direction. In general, work can be expressed in terms of an opposing force, that is,

$$\delta W = -F \cdot dx \tag{3.46}$$

Hence, instead of speaking in case of diffusion about the directed net migration according to a concentration gradient, we analogously can say that the directed net migration occurs from regions with high chemical potentials toward regions with lower chemical potentials – this is the thermodynamics point of view. The thermodynamic driving force for the diffusion is the gradient of the chemical potential $\mu$. This gradient is negative in the direction of the diffusion:

$$F_{th} = -\left(\frac{\partial \mu}{\partial x}\right)_{p,T} \tag{3.47}$$

The migration of ions occurs as long as the thermodynamic equilibrium is reached, where the concentrations and/or chemical potentials are constant throughout the whole regions. If we divide $F_{th}$ by the Avogadro constant ($N_A \approx 6.0221 \times 10^{23}$ mol$^{-1}$), then we get the thermodynamic driving force that acts on a single particle or ion $j$:

$$F_{th-j} = -\frac{1}{N_A} \cdot \left(\frac{\partial \mu_j}{\partial x}\right)_{p,T} \tag{3.48}$$

Thus, the gradient of the chemical potential $\mu$ is driving the migration of the ion $j$, or respective the flow of material. In particular, the material flow density – that is, the flux $J$ – is proportional to the gradient of the chemical potential ($\mu_j$). If we assume again only a migration in $x$-direction, we get:

$$J_x = -C \cdot \left(\frac{\partial \mu_j}{\partial x}\right)_{p,T} \tag{3.49}$$

The flux $J_x$ is given in mol/m$^2 \cdot$ s, the chemical potential $\mu_j$ in J/mol and the length $x$ in m. The proportionality constant $C$ is given in mol$^2 \cdot$ s/kg $\cdot$ m$^3$. If the solution or the electrolyte is sufficiently diluted, the chemical potential $\mu_j$ can be expressed by

$$\mu_j = \mu^0 + R \cdot T \cdot \ln\left(\frac{c_j}{c^0}\right)_{p,T} \tag{3.50}$$

$R$ is again the universal gas constant ($R \approx 8.3145$ J/mol $\cdot$ K), $c_j$ is the molar concentration of the solute – that is, the molar concentration of the ions $j$ in the electrolyte – and $c^0$ is the molar concentration under standard pressure (101.325 kPa or 1.01325 bar). In the same manner, $\mu^0$ is the chemical potential under standard pressure. In general, $\mu^0$ and $c^0$ still can be temperature-dependent; however, here we keep the temperature also constant. Inserting the expression (3.50) for $\mu_j$ into eq. (3.49), and keeping in mind that the concentration is dependent on the position and the time, that is, $c_j = c_j(x, t)$, we get

$$J_x = -C \cdot \frac{\partial}{\partial x}\left(\mu^0 + R \cdot T \cdot \ln\left(\frac{c_j(x,t)}{c^0}\right)_{p,T}\right) \tag{3.51}$$

or rather

$$J_x = -C \cdot \frac{R \cdot T}{c_j(x,t)} \cdot \left(\frac{\partial c_j(x,t)}{\partial x}\right)_{p,T} \tag{3.52}$$

On the right side of eq. (3.52), we can combine the parameters in front of the partial derivative into the parameter $D_j$, that is,

$$D_j = -C \cdot \frac{R \cdot T}{c_j(x,t)} \tag{3.53}$$

or

$$C = -D_j \cdot \frac{c_j(x, t)}{R \cdot T} \tag{3.54}$$

$D_j$ is the diffusion coefficient – given in m$^2$/s – as used in Fick's first law of diffusion:

$$J_x = -D_j \cdot \left(\frac{\partial c_j(x, t)}{\partial x}\right)_{p,T} \tag{3.55}$$

The concentration gradient $\partial c/\partial x$ has the unit mol/m$^4$ – the unit of $c$ is mol/m$^3$. The general formulation of Fick's first law of diffusion for three dimensions is given by

$$\vec{J} = -D \cdot \nabla c(x, y, z, t) \tag{3.56}$$

$\nabla$ is a vector differential operator, that is, $\nabla = (\partial/\partial x, \partial/\partial y, \partial/\partial z)$. It is called del operator or nabla operator.

The material flow density or flux can be also described by the drift velocity $v_j$ of the diffusing species – here these are the ions in the electrolyte – and the concentration $c_j$. Then we get

$$J_x = v_j \cdot c_j(x, t) \tag{3.57}$$

Hence, from eqs. (3.55) and (3.57) we obtain the diffusion coefficient of the following expression:

$$D_j = -v_j \cdot c_j(x, t) \cdot \left(\frac{\partial c_j(x, t)}{\partial x}\right)_{p,T}^{-1} \tag{3.58}$$

If we go back to eq. (3.48) and keep in mind that for the gas constant the equation $R = N_A \cdot k_B$ holds ($k_B \approx 1.3806 \times 10^{-23}$ J/K is the Boltzmann constant), then we get the thermodynamic driving force that acts on a particle or an ion $j$:

$$F_{th-j} = -\frac{k_B}{R} \cdot \left(\frac{\partial c_j(x, t)}{\partial x}\right)_{p,T} \tag{3.59}$$

Now we can insert eq. (3.50) into eq. (3.59) and get

$$F_{th-j} = -\frac{k_B \cdot T}{c_j(x, t)} \cdot \left(\frac{\partial c_j(x, t)}{\partial x}\right)_{p,T} \tag{3.60}$$

Hence, by comparison of eqs. (3.58) and (3.60) we find the following expression for the diffusion coefficient of the particle or ion $j$:

$$D_j = \frac{v_j \cdot k_B \cdot T}{F_{th-j}} \tag{3.61}$$

From eq. (3.61), the diffusion coefficient can be calculated, if the drift velocity $v_j$ of the migrating particle or ion and the thermodynamic driving force $F_{th-j}$ are known.

Equation (3.57) shows that the flux is related to the drift velocity. As shown in Figure 3.15, within the time interval $\Delta t$, all particles or ions laying inside a distance $\Delta x = v_j \cdot \Delta t$ or rather inside the volume $\Delta V = v_j \cdot A \cdot \Delta t$ can pass through the cross-sectional area $A$. Therefore, the amount of substance that can pass through the cross-sectional area $A$ is $v_j \cdot A \cdot \Delta t \cdot c_j(x, t)$. From eq. (3.58), we get the following for the drift velocity:

$$v_j = -\frac{D_j}{c_j(x,t)} \cdot \left(\frac{\partial c_j(x,t)}{\partial x}\right)_{p,T} \tag{3.62}$$

$$\Delta V = v_j \cdot A \cdot \Delta t$$

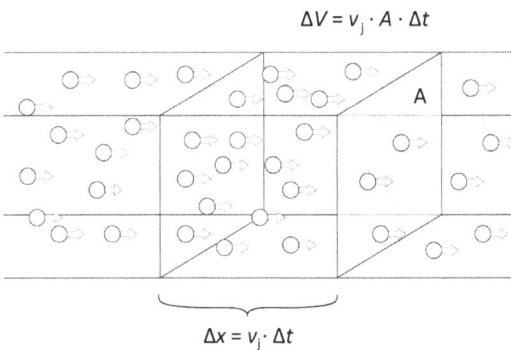

$$\Delta x = v_j \cdot \Delta t$$

**Figure 3.15:** The flux is related to the drift velocity. Within the time interval $\Delta t$, all particles that are lying inside the volume $\Delta V$ can pass the cross-sectional area $A$.

Comparing this expression for the drift velocity with eq. (3.60) and keeping in mind that for the gas constant $R = N_A \cdot k_B$ holds, we obtain the following expression for the drift velocity:

$$v_j = \frac{D_j \cdot F_{th-j}}{R \cdot T} \tag{3.63}$$

Hence, if we know the thermodynamic driving force $F_{th-j}$ and the diffusion coefficient $D_j$, we can calculate the drift velocity $v_j$ of the particles or ions.

Now we somewhat broaden our considerations. In general, one can say that a mass transport can occur by diffusion due to a concentration gradient as well as by migration due to the influence of an effective force. At this point, it is to mention that the driving force $F_{th-j}$ in eq. (3.63) can have any origin. Hence, if we come back to the example that an ion migrates in an electric field of the strength $E$, the effective driving force on the migrating ion has the form:

$$F_{th-j} \underset{(here)}{=} F_{th-j,E} = z_j \cdot e \cdot E \tag{3.64}$$

Again, $z_j$ is the valence of the ion and $e$ is the elementary charge. The mobility of the ion $j$ is given by $u_j$. If the ion in a solution faces the electric force $F_{\text{th}-j,E}$, then its drift velocity $v_j$ is given by

$$v_j = u_j \cdot E \tag{3.65}$$

Here $u_j$ is again the mobility of the ion $j$.

If we insert the expressions for electric driving force (3.64) and the drift velocity (3.65) into eq. (3.61) for the diffusion coefficient; finally, we obtain the famous Einstein relation that provides the correlation between the diffusion coefficient and the ion mobility:

$$D_j = \frac{u_j \cdot k_\text{B} \cdot T}{z_j \cdot e} \tag{3.66}$$

Using $R = N_\text{A} \cdot k_\text{B}$ and $F = N_\text{A} \cdot e$ for the gas constant and the Faraday constant, respectively, the Einstein relation can be written in another form that is often used:

$$D_j = \frac{u_j \cdot R \cdot T}{z_j \cdot F} \tag{3.67}$$

We now want to reference the connection between the diffusion coefficient and the limiting molar conductivity. If we remember eq. (3.38) for the molar conductivity $\lambda$ in dependence on the ion mobility $u$, then we can express the limiting molar conductivity of the ion $j$ in a solution as follows:

$$\lambda^0_{m,j} = z_j \cdot u^0_j \cdot F \tag{3.68}$$

resulting in

$$u^0_j = \frac{\lambda^0_{m,j}}{z_j \cdot F} \tag{3.69}$$

Here $u_j^0$ is the mobility of the ion $j$ for the ideally diluted solution in the limit of zero concentration, where no interionic interactions occur.

Expression (3.69) for $u_j^0$ can be inserted into the Einstein relation in eq. (3.67) and we obtain

$$D_j = \frac{\lambda^0_{m,j} \cdot R \cdot T}{z_j^2 \cdot F^2} \tag{3.70a}$$

or rather

$$\lambda^0_{m,j} = \frac{D_j \cdot z_j^2 \cdot F^2}{R \cdot T} \tag{3.70b}$$

Equation (3.70) can be inserted into eq. (3.37) separating the parameters for positively and negatively charged ions. Hence, based on the law of independent ion mobility, we finally obtain

$$\Lambda_m^0 = \left( v_+ \cdot D_{j+} \cdot z_+^2 + v_- \cdot D_{j-} \cdot z_-^2 \right) \cdot \frac{F^2}{R \cdot T} \tag{3.71}$$

This is the so-called Nernst–Einstein equation; $v_+$ and $v_-$ are the stoichiometric coefficients of the positive cations and the negative anions, and $z_+$ and $z_-$ are their signed charge numbers. In principle, with this equation one can calculate ionic diffusion coefficients from experimentally determined conductivities.

However, in general one cannot observe the diffusion of single ions because of the electroneutrality. Only with radioactive tracer ions, this can be achieved. For instance, by doping of NaCl salt with radioactive $^{24}Na^+$ ions (the stable sodium ion is $^{23}Na$), diffusion coefficients of sodium ions in aqueous electrolytes can be obtained.

In general, a salt can be denoted by $cat_{v+}^{z+} an_{v-}^{z-}$; $v_+$ and $v_-$ are again the stoichiometric coefficients of the cations ($cat_{v+}^{z+}$) and the anions ($an_{v-}^{z-}$) of the salt, respectively, $z_+$ and $z_-$ are their signed charge numbers. According to Nernst, the mean value of the diffusion coefficient of a salt can be expressed by

$$D = (v_+ + v_-) \cdot \frac{D_+ \cdot D_-}{(D_+ + D_-)} \tag{3.72}$$

The subscripts $+$ and $-$ identify the diffusion coefficients of the cations and anions.

Speaking about diffusion, similarly as briefly mentioned in Section 3.1.4, a frictional force works against the driving force. The frictional force is given by the Stokes' law (eq. (3.17)). Hence, the thermodynamic driving force can be expressed by the frictional force or Stokes' drag $F_d$:

$$F_{th-j} = F_d = 6 \cdot \pi \cdot \eta \cdot r_j \cdot v_j \tag{3.73}$$

As aforementioned, $\eta$ is the dynamic viscosity, $r_j$ stands for the radius of the migrating ion $j$, and $v_j$ stands for its velocity. Inserting eq. (3.73) into eq. (3.61), we obtain a relation between the diffusion coefficient and the dynamic viscosity $\eta$:

$$D_j = \frac{k_B \cdot T}{6 \cdot \pi \cdot \eta \cdot r_j} \tag{3.74}$$

Equation (3.74) is called Einstein–Stokes relation. It holds for spherical particles that are not too small. In particular, the Einstein–Stokes relation also holds for electrically neutral particles (or molecules) since it shows no dependence on the charge state of the diffusing species.

Up to now, we were speaking about strong electrolytes that are completely ionized. In the next section, we will briefly comment on weak electrolytes, especially concerning the molar conductivity of weak electrolytes.

### 3.1.7 Weak electrolytes

Weak electrolytes are not completely ionized. Examples for weak electrolytes are so-called Brønsted acids and bases, for instance, the acetic acid ($CH_3COOH$) or ammonia ($NH_3$) that is a Brønsted base.

Before continuing our brief considerations about weak electrolytes, we first have to clarify some definitions of acids and bases. According to Arrhenius, acids are defined as substances that dissociate in aqueous solution to give hydrogen ions ($H^+$). In this regard, bases are defined as substances that dissociate in aqueous solution to provide hydroxide ions ($OH^-$). The acid and base definitions of Brønsted and Lowry are a generalization of the Arrhenius definition of acids and bases. According to this latter theory, acids and bases are defined by the way, how they react with each other. The acid forms the so-called conjugate base, and the base forms its conjugate acid. This is realized by the exchange of a proton, that is, a hydrogen cation $H^+$. The acid gives the proton away to become a conjugate base, and the base accepts the proton to become a conjugate acid. Hence, the definition of acids and bases by Brønsted and Lowry can be expressed in terms of an equilibrium reaction:

$$\text{Acid} + \text{base} \leftrightarrow \text{conjugate base} + \text{concugate acid} \tag{3.75}$$

For any acid HA and any base B, we can write the equilibrium reaction (3.75) in the following form that makes the exchange of the proton $H^+$ clearer:

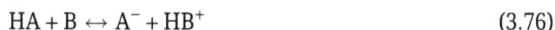

$$HA + B \leftrightarrow A^- + HB^+ \tag{3.76}$$

$A^-$ is the conjugate base and $HB^+$ is the conjugate acid, respectively. The equilibrium reaction for an aqueous solution is given by

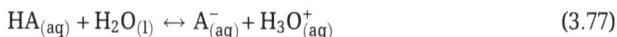

$$HA_{(aq)} + H_2O_{(l)} \leftrightarrow A^-_{(aq)} + H_3O^+_{(aq)} \tag{3.77}$$

The acid HA is dispersed within the liquid water. Therefore, it is labeled by the subscript (aq). The same holds for the conjugate base $A^-$ and the conjugate acid $HB^+$, that is, the hydronium ion $H_3O^+$. The liquid water is labelled by the subscript (l). For the example of the acetic acid ($CH_3COOH$), formula (3.77) can be written as follows:

$$CH_3COOH_{(aq)} + H_2O_{(l)} \leftrightarrow CH_3COO^-_{(aq)} + H_3O^+_{(aq)} \tag{3.78}$$

Acetic acid donates a proton to water and becomes the conjugate base, that is, the acetate ion $CH_3COO^-$. $H_2O$ is a base; it accepts the proton of $CH_3COOH$ and hence, becomes a conjugate acid, that is, a hydronium ion $H_3O^+$.

The dependence of the molar conductivity on the concentration in case of a weak electrolyte is much different from the concentration dependence of a strong electrolyte (Figures 3.10 and 3.11). As we have seen earlier, for strong electrolytes Kohlrausch's square-root law (3.36) holds for sufficient low concentrations. The molar conductivity of a weak electrolyte, that is, a partly dissociated substance, can be expressed by

$$\Lambda_m = a \cdot \Lambda_m'$$ (3.79)

where $\Lambda_m'$ is the molar conductivity for the complete dissociation, and $a$ is the degree of ionization. Hence, for weak electrolytes, the molar conductivity depends on the number of ions in the solution and on $a$. At equilibrium, the degree of ionization $a$ can be defined for any acid HA at a molar concentration $c$ by the following relations:

$$[HA] = (1 - a) \cdot c$$ (3.80a)

$$[A^-] = a \cdot c$$ (3.80b)

$$[H_3O^+] = a \cdot c$$ (3.80c)

The square brackets denote the concentrations of the acid HA of the conjugate base and of the conjugate acid, respectively. Using eqs. (3.80a)–(3.80c), the acidity constant $K_a$ can be defined by

$$K_a = \frac{[A^-] \cdot [H_3O^+]}{[HA]} = \frac{a^2 \cdot c}{1 - a}$$ (3.81)

Here, the concentration of $H_2O$ is assumed to be constant. Expression (3.81) is a quadratic equation for the degree of ionization $a$. Hence, we get

$$a = \frac{K_a}{2 \cdot c} \left[ \sqrt{1 + \frac{4 \cdot c}{K_a}} - 1 \right]$$ (3.82)

If concentrations are sufficiently low, $\Lambda_m'$ can be replaced by the limiting molar concentration $\Lambda_m^0$, that is,

$$\Lambda_m = a \cdot \Lambda_m^0$$ (3.83)

Inserting eq. (3.82) into (3.83) we obtain

$$\Lambda_m = \frac{K_a}{2 \cdot c} \left[ \sqrt{1 + \frac{4 \cdot c}{K_a}} - 1 \right] \cdot \Lambda_m^0$$ (3.84)

Hence, for a known acidity constant $K_a$, one can deduce the concentration dependence of the molar conductivity.

Equation (3.81) for the acidity constant $K_a$ can be rearranged into the following form:

$$\frac{1}{a} = 1 + \frac{a \cdot c}{K_a}$$ (3.85)

If we insert $a = \Lambda_m / \Lambda_m^0$ from eq. (3.83) into eq. (3.85), we finally obtain:

$$\frac{1}{\Lambda_m} = \frac{1}{\Lambda_m^0} + \frac{\Lambda_m \cdot c}{K_a \cdot \left( \Lambda_m^0 \right)^2}$$ (3.86)

Equation (3.86) is the so-called Ostwald's dilution law that describes the dependence of the molar conductivity on the concentration for weak electrolytes. Since, for weak electrolytes the degree of ionization is low, the interionic interactions are not so relevant than for the case of strong electrolytes. $\Lambda_m$ can be obtained from experiments. If $\Lambda_m^{-1}$ is plotted in dependence on $\Lambda_m \cdot c$, the limiting molar conductivity can be obtained by extrapolation toward the interception (Figure 3.16). Using this extrapolated value of $\Lambda_m^0$, the degree of ionization $\alpha$ can be calculated with eq. (3.83).

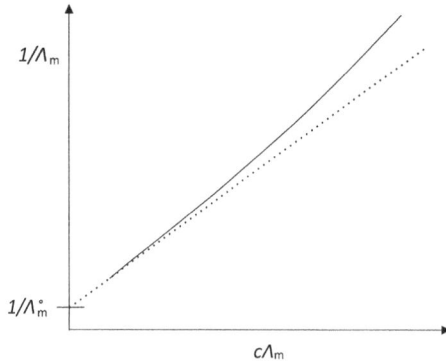

**Figure 3.16:** Determination of the limiting molar conductivity $\Lambda_m^0$ of a solution by linear extrapolation to the $c\Lambda_m = 0$.

## 3.2 Electrochemical electrodes

Colloquially described, an electrode is an electrical conductor that provides an electrical contact to a nonmetallic material or medium, for instance, to a semiconductor, to air or even to vacuum. The charge carriers in the electrode are electrons. Electrodes arise as a pair – the electrode and the counter electrode. They are typically made of metals or can also consist of graphite. In many cases, the electrode and its counter electrode are made of different materials. Together with a counter electrode, the electrode interacts with the nonmetallic medium that is located between them.

We speak about electrochemical electrodes, if such electrodes are in contact with an electrolyte, that is, an ionic conductor. For simplicity, we briefly speak later about electrodes and mean electrochemical electrodes. For the discussion of the properties of such electrodes, one always has to imply the interface to the electrolyte. With this regard, an electrode is the combined system of a piece of metal or graphite, for instance, rod-shaped, and the interface region to the surrounding electrolyte.

At the interface between the electrode and an ionic conductor, electrochemical reactions occur, which come along with an electrical current. Such electrode reactions can be ascribed to net oxidation or reduction processes that occur at the electrode–electrolyte interface. One or more electrons can be involved in these reactions and the reactions can occur either in a single electron transfer step or in successive two or more electron transfer steps. The substances that lose (i.e., donate) or receive (i.e., accept) one or more electrons, that is, the substance that are oxidized or reduced, respectively, are called electroactive species.

The electron transfer takes place within a very thin interface region at the surface of the electrode. The underlying physical process implies the quantum-mechanical tunneling of electrons between the electroactive species and the electrode. The activation energy of such processes is significantly constituted by the work that is required to replace the hydration shells of the ions, that is, the $H_2O$ molecules in the electrolyte that are attracted by the charged ions.

Depending on the medium that is surrounding the electrode, different kinds of electrode reactions and/or different interactions between the electrode and the surrounding medium can be observed.

## 3.2.1 Interface region between electrodes and electrolytes

The phase boundary characterizes the interface region, and it is crucial for the properties and the operation of electrodes. Likewise, phase boundaries are termed interfaces. Considering the solid state and the liquid state that form phase boundaries separating one phase from the other, we can find solid–liquid, liquid–gaseous and solid–gaseous phase boundaries or interface regions. The third state of matter – that is, the gas state – does not form a phase boundary on its own; hence, we do not have to consider it here. The same holds by the way for a plasma that is the fourth state of matter. The interface or phase boundary between a solid and a gas is usually termed surface. The distinction between phase boundaries or interfaces and surfaces has historical reasons and is often arbitrary. In the following explanations, depending on the context, we speak about surfaces, phase boundaries and interface regions; the latter one incorporates also regions that are a little farther from the actual phase boundary.

On the macroscopic scale, a phase boundary can be clearly seen – for instance, the phase boundaries between water and air or between a metal and water. At the surface of a phase, the bond forces acting on the atoms are directed toward the interior of the phase, thus a surface energy appears. This holds for a regular arrangement of the atoms at a phase surface (e.g., the surface of a crystal) as well as for irregularly arranged surface atoms (e.g., in case of amorphous substances or liquids). At first glance, a phase boundary has a well-defined surface. However, this is a very misleading impression since on the microscopic scale, phase boundaries look very complex, and furthermore, even rather dynamic processes can occur. Hence, we will see soon that the phase boundary is not really a strict and abrupt boundary.

If we look at the phase boundary on the microscopic scale, it becomes evident that it is not so easy to say precisely what is the boundary or the surface of phase and where exactly it is located. For instance, the surface of undisturbed water, that is, the phase boundary between water and the ambient atmosphere (e.g., air), looks very smooth, and the surface roughness is very low. However, water exhibits a vapor pressure.

Hence, at the phase boundary between the liquid and the ambient air atmosphere above the water, a statistical equilibrium of leaving and returning water molecules is established (Figure 3.17). In fact, this is a rather dynamic process since per second roughly $10^{22}$ molecules take part in the transit between the surface of the water and the air atmosphere and vice versa. Moreover, directly at the surface of water, only about $10^{15}$ water molecules can find place on the area of a square centimeter – that is, the surface density of the water molecules is about $10^{15}$ cm$^{-2}$. Hence, each molecule rests about $10^{-7}$ s at the surface of water before it leaves again. The mean velocity of the molecule when leaving is about 700 m/s. The molecules do not leave the surface in a straight orthogonal direction without any disturbance, but face countless collisions with each other or with the air molecules, that is, mainly $N_2$ and $O_2$ molecules. Hence, we can conclude that at the microscopic scale, the pretended smooth surface of water is a highly dynamic and chaotic thing.

The same holds for solid surfaces or phase boundaries, too, even if the ambient is a liquid. Solids also exhibit a vapor pressure, that is, atoms from the solid can leave the solid surface and enter the ambient gas or liquid and vice versa.

Now we look at a scenario, where a metal electrode – for example, a zinc (Zn) or copper (Cu) metal strip – is dipped into a liquid that can be, for instance, water or an aqueous solution (Figure 3.18). Aqueous solutions that conduct the electrical current contain electrolytes, for instance, dissolved salts in water. At the surface of the metal, the metal ions quit into the liquid leaving one or several electrons behind in the metal electrode. Since the metal ions are solved in the liquid, and since the electrons remain in the metal electrode, positive and negative charges are separated at the metal–liquid interface. From the examples, Zn and Cu mentioned earlier, we find the following reactions:

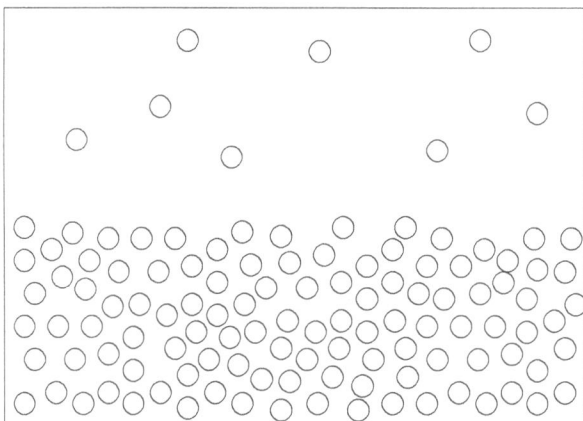

**Figure 3.17:** Phase boundary between a liquid (e.g., water) and the ambient air atmosphere: At the interface, a statistical equilibrium of leaving and returning molecules of the liquid is established.

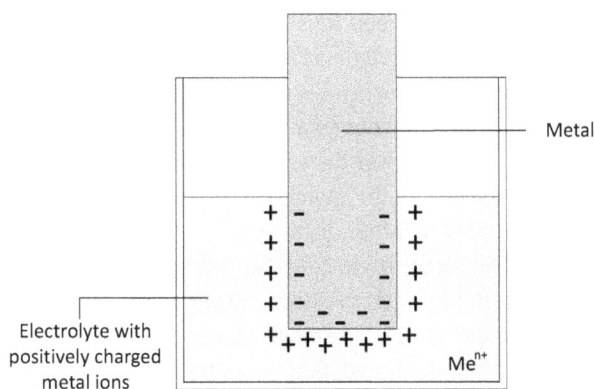

**Figure 3.18:** Metal electrode dipped into a liquid (e.g., an aqueous solution or electrolyte). Metal ions are entering the solution leaving one or several electrons behind in the metal electrode.

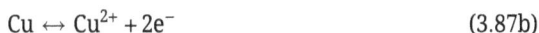

$$Zn \leftrightarrow Zn^{2+} + 2e^- \qquad (3.87a)$$

$$Cu \leftrightarrow Cu^{2+} + 2e^- \qquad (3.87b)$$

That is, both metals leave two electrons per solved ion behind in the metal electrode. The reactions are reversible; therefore, the reaction arrows point in both directions. The driving force for such a behavior is the solution pressure or solution tension. This is a measure of the tendency of the atoms or molecules of a solid to enter a solution, that is, to cross the interface between the solid phase and the liquid phase. Such a tendency can be observed, for instance, for metals, hydrogen and for some nonmetallic materials. The solution pressure, that is, the force that is directed from the metal toward the solution and that drives the metal or the metal ions into the liquid, can also be qualitatively explained by the strong concentration gradient of metal atoms (or ions) between the metal strip and the liquid (Figure 3.18). Hence, the metal atoms or ions are diffusing from a high concentration area (the metal electrode) to an area with lower concentration (the liquid).

In case that a metal electrode is dipped into water or an aqueous solution, the polar water molecules break off the metals' ions from the surface of the solid, that is, the $Zn^{2+}$ or $Cu^{2+}$ ions mentioned earlier. In metals, electrons can freely move since the electrons from the outer atomic shell are not attached to the atomic core. Therefore, they stay in the metal electrode, if the metal ions are erased from the electrodes surface. Hence, we get an excess of electrons in the electrode, and the electrode becomes negatively charged. This implies that the positively charged metal ions in the liquid are attracted by the excess electrons in the electrode, and they cannot move far away from the metal surface. Of course, the overall charge remains neutral in the interface region between the solid and the liquid.

The metal ions leave the electrode surface and enter the aqueous solution (due to the solution pressure), and they return to the electrode surface (due to electrostatic attraction). In addition, within the liquid, the metal ions also have a proper motion due to thermal activation since the processes we look at are occurring at reasonable temperatures – that is, more or less around room temperature – and not at absolute zero. Hence, we are speaking about statistical phenomena. There is a higher probability that a metal ion in solution is located at a closer distance to the electrode surface than at a larger distance. Hence, the density of the metal ions in solution is higher, close to the electrode surface – as compared to larger distances.

The situation even gets more complicated, if we consider that metal atoms can also enter the aqueous solution due to the solution pressure; however, the density of metal atoms in the liquid rapidly becomes smaller at larger distance from the surface of the electrode. If there are metal atoms in the liquid, they stay very close to the electrodes surface. The probability that metal atoms enter the aqueous solution is lower than the probability that metal ions enter it since the polar water molecules actively erase the metal ions from the electrodes surface.

Similar processes also occur for the electrons. Due to detaching metal ions from the electrodes surface, excess electrons are located in the near surface region of the electrode. Hence, these electrons also face a solution pressure and enter the liquid, and therefore we have excess electrons in the near surface region of the electrode and in the aqueous solution close to electrodes surface. At larger distance from the surface of the electrode, the electrons attach to positive ions in the solution.

Figure 3.19 summarizes the whole – rather complicated – situation. For explanation, we start from the left part of Figure 3.19 and continue to the right side. In the left part of the Figure 3.19, we see the pure metallic region of the electrode. The figure illustrates that the surface of the electrode is not really well-defined. We rather deal with a phase boundary that has a volume dilatation. Toward the right side of the surface of the pure metallic electrode, the phase boundary attaches. This phase boundary extends from the region, where the density of metal ions is strongly decreasing (on the left side), to the location, where the density of the metal ions in the aqueous solution has its maximum. This range is the smeared-out surface of the electrode. The extended phase boundary contains neutral metal ions with a strongly decreasing density toward the right. It also contains positively charged metal ions with an increasing density toward the right. These metal ions already are disconnected from the crystal compound due to breaking off by polar water molecules. Finally, the phase boundary contains electrons with the maximum density laying close to the region, where the density of the metal atoms approaches the bulk value of the pure metal electrode, that is, at the side where most of the metal ions are broken off by the polar water molecules of the aqueous solution.

Toward the right side of the phase boundary, the Helmholtz layer follows up (Figure 3.19) – named after its discoverer. The Helmholtz layer is a narrow range in the aqueous solution that is located directly near the electrodes surface – or better near the phase boundary of the metal–liquid interface. It contains positive excess charges originating from the metal ions that exhibit a maximum density at the border

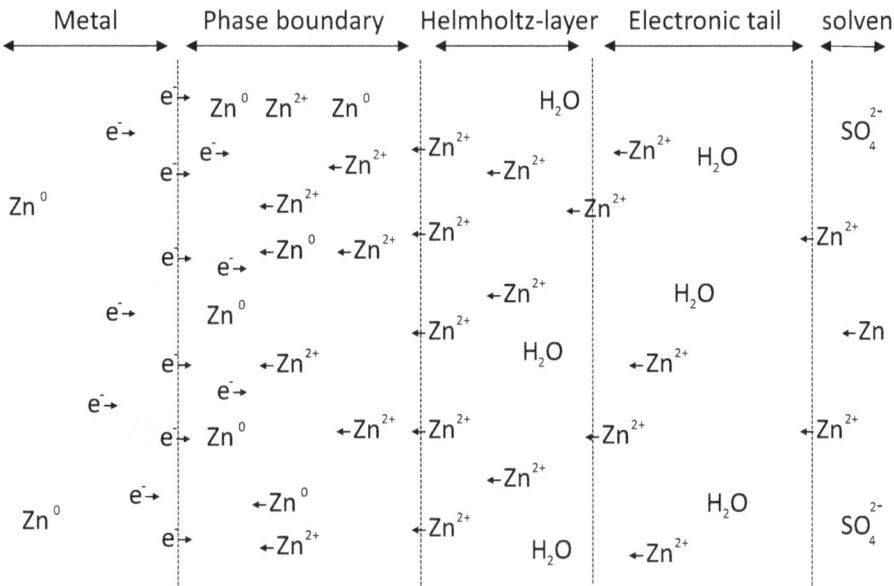

**Figure 3.19:** Interface region of a metal electrode (e.g., a zinc electrode) that is dipped into an aqueous electrolyte solution.

between the phase boundary and the Helmholtz layer (see Figure 3.19). As already explained, the positive metal ions accumulate in this region due to attraction by the excess electron charges in the phase boundary and the near surface region of the electrode.

According to a simple model understanding (Helmholtz model) that is represented in Figure 3.20, the Helmholtz layer is characterized by two planes, that is, the inner Helmholtz plane (IHP) at a distance $d_1$ from the phase boundary and the outer Helmholtz plane (OHP) at a distance $d_2$. Both planes are marked in Figure 3.20. The first one (IHP) marks water molecules that are directly attached to the electrodes surface, that is, to the phase boundary. Theses water molecules are adsorbed – or specifically adsorbed – by the electrode. The second one (OHP) marks the solvated metals ions with a hydration shell that are attached to the electrodes surface (respectively to the absorbed water molecules of the IHP), that is, to the phase boundary. In this later case, we speak about not specifically adsorbed metal ions. The polar water molecules surround the positive metal ions and form a hydration shell in a way that we already saw in Figure 3.4 for the case $Na^+$ ions (see Section 3.1.3). Both planes – IHP and OHP – are drawn through the center of the metal ions. For simplicity, in Figure 3.20 the phase boundary is drawn only as a line – without the detailed structure that was shown in Figure 3.19.

The right side of the Helmholtz layer can extend rather far into the aqueous solution. This part of the Helmholtz layer is called electrokinetic tail (Figure 3.19). There is a twofold reason (electrostatics and thermal activation) for the distribution of the in-

0 $d_1$ $d_2$

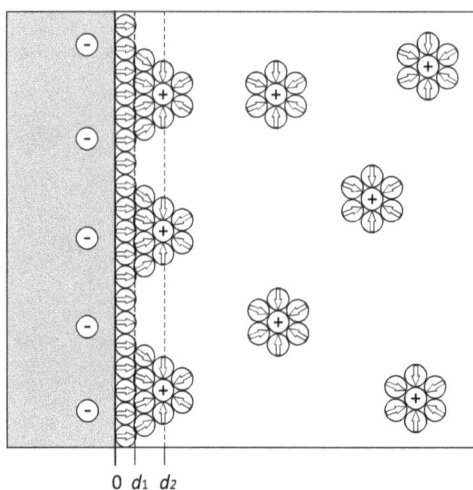

**Figure 3.20:** Helmholtz model: the phase boundary is marked by the line at $d = 0$; $d_1$ and $d_2$ mark the inner and outer Helmholtz planes, respectively.

volved metal ions in this local region of the aqueous solution. On one hand, the ions are still influenced by the negative layer of excess electrons in the near surface region of the electrode, and on the other hand, they are driven quite far away from the interface boundary due to thermally activated motion.

Far away from the electrode, the aqueous solution contains water molecules and ions of the electrolyte. For the example shown in Figure 3.19, the aqueous solution consists of an aqueous sulfuric acid solution ($H_2O$ mixed with $H_2SO_4$). In water, $H_2SO_4$ dissociates into two protons ($H^+$) and a sulfate ion ($SO_4^{2-}$), whereupon the protons attach water molecules and form hydronium ions ($H_3O^+$):

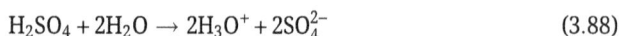

$$H_2SO_4 + 2H_2O \rightarrow 2H_3O^+ + 2SO_4^{2-} \tag{3.88}$$

The Helmholtz layer is only a part of the so-called Helmholtz double layer that is shown in Figure 3.21. The Helmholtz double layer consists of the Helmholtz layer and the layer with the excess electrons in the near surface region of the electrode – that is, it consists of a negatively charged layer (the excess electrons) and the positively charged Helmholtz layer. Hence, it looks somewhat like a capacitor with a negatively charged and a positively charged plane that are laying parallel to each other. In fact, such double layers can be used to construct very effective capacitors – so-called supercapacitors that are electrochemical capacitors with very high capacity. If a metal electrode is dipped into the aqueous solution, the Helmholtz double layer forms after a short period time in the millisecond to second range depending on the liquid and the metal. Helmholtz double layers always appear at the interface of solid and liquid phases – provided the solid phase does not solve in the liquid.

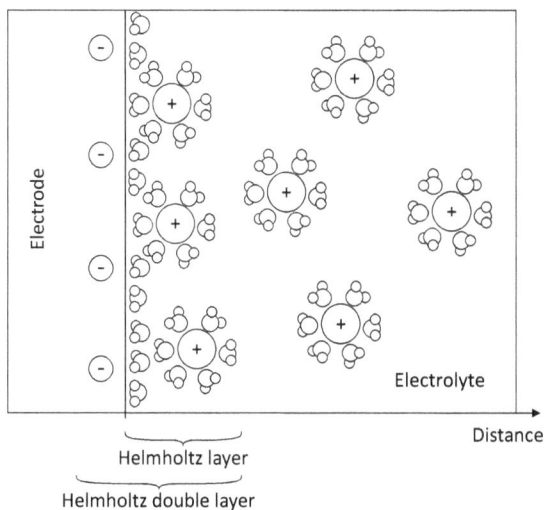

**Figure 3.21:** Region of the electrode–electrolyte interface according to the Helmholtz model showing the Helmholtz layer and the Helmholtz double layer. The solid line marks the phase boundary. The positive ions are surrounded by water molecules. $H_2O$ is a dipole; hence, the larger oxygen atoms point toward the positive ions since they represent the negative side of the dipoles.

The Helmholtz model – as shown in Figure 3.20 – is a simplification since it neglects thermal effects and thermal energy, respectively. Due to thermal energy, the OHP is dispersed; and positive charges, that is, positive solvated metal ions are migrating deeper into the liquid. The Helmholtz model neglects this disordering effect of thermally activated motion and emphasizes a rigidity of the Helmholtz layer that does not really reflect its nature. Another model for the solid–liquid interface region is the Gouy–Chapman model of a diffuse double layer. In this model, thermally activated motion of the metal ions is considered (Maxwell–Boltzmann statistics are applied) and the charge distribution of the ions is described as a function of the distance from the electrodes surface. Ions that have the opposite charge than the surface of the electrode cluster near the electrode, and ions of the same charge are repelled from the electrode (Figure 3.22). Hence, the electric potential exhibits an exponential decrease with growing distance from the electrodes surface.

Neither the Helmholtz model nor the Gouy–Chapman model is a good description of the double layer. The first one overemphasizes the rigidity of the double layer, and the latter one underemphasizes its structure. Other – more sophisticated – models were developed, but their description and discussion go beyond the scope of this book.

Because at the interface between the metal electrode and the aqueous solution, separation of charges occurs, at the phase boundary, an electrical potential or voltage appears. Before we focus on this point, we briefly discuss the physical terms potential and voltage a little more precisely. Both terms are sometimes intermixed that might cause imprecise explanations, and therefore, confusion.

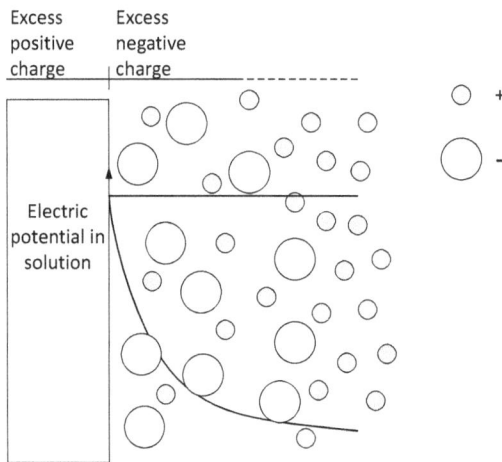

Excess positive charge

Excess negative charge

Electric potential in solution

+

−

**Figure 3.22:** Gouy–Chapman model: ions with opposite charge than the electrode surface (left side) cluster near the electrode; ions of the same charge are repelled from the electrode (right side).

An electrical potential is defined as the amount of work that has to be provided for the transportation of a unit charge from the charge free point of infinity to the location under consideration.

The electrical voltage is the difference between two potentials. For instance, in the place A, we have a potential $\Phi_A$; and in the place B the potential is $\Phi_B$; hence, the voltage is given by $V_{AB} = \Phi_A - \Phi_B$.

The important difference between electrical potential and voltage is that potentials are not measurable, only voltages are measurable.

Basically, if electrical charges are separated, a potential or voltage is established. For this charge separation, work has to be expended. In electrochemistry, the separation of charges occurs in consequence of the solution pressure. The metal ions enter the aqueous solution leaving electrons behind in the electrode, until the established potential or voltage is large enough to stop this one-way migration. The work $W$ that has to be employed to separate charges at the phase boundary of the electrode and the aqueous solution is given by

$$W = \int_{t_1}^{t_2} v \cdot i \cdot dt \tag{3.89}$$

where $v$ is the voltage that is built at the phase boundary, and $i$ is the current of the ions migrating from the electrode into the aqueous solution; $v$ and $i$ are dependent on the time $t$. The time interval $\Delta t = t_2 - t_1$ covers the duration until the maximum voltage is built and the steady state is achieved. After the steady state is achieved, the same amount of ions that migrate from the electrode into the aqueous solutions moves from the solution to the electrode (exchange current), that is, equal ion currents are flowing in opposite directions; therefore, no measurable charge transport can be observed – chemical and electrical energy are in equilibrium.

The voltage of the metal–liquid interface region of the electrode and the aqueous solution cannot be measured; if one would like to do so, one has to employ an electrical contact to the liquid, that is, a second electrode, to connect it to the measurement device. Hence, the second electrode would cause a second voltage. Therefore, the voltage of a single metal–liquid interface region or phase boundary cannot be measured. Only the voltage of an electrochemical cell that consists of two electrodes and an electrolyte in between is measureable. This will be discussed in Section 3.3.

The potential distribution with the Helmholtz layer can be estimated from electrostatics and the Poisson equation:

$$\frac{\partial^2 \Phi(x)}{\partial x^2} = -\frac{\rho}{\varepsilon} \tag{3.90}$$

$\Phi(x)$ is the potential, $x$ is the coordinate perpendicular to the electrodes surface, $\rho$ is the distribution of free charges and $\varepsilon$ is the permittivity ($\varepsilon = \varepsilon_r \cdot \varepsilon_0$, $\varepsilon_r$ and $\varepsilon_0$ are the relative permittivity and the permittivity of the vacuum, respectively). In our case, there are no charges located between the electrode surface and the IHP; hence, $\rho = 0$, and we get

$$\frac{\partial^2 \Phi(x)}{\partial x^2} = 0 \tag{3.91}$$

or

$$\frac{\partial \Phi(x)}{\partial x} = \text{const.} \tag{3.92}$$

Therefore, the potential exhibits a linear change between the surface and the IHP of the electrodes. Within the metal electrode, no free charges occur and $\rho = 0$. Also within the aqueous solution or the electrolyte, the average charge density equals zero. Hence, in the electrode and in the electrolyte, the change of the potential is constant, and even more, the change is zero $\partial \Phi / \partial x = 0$ and we have a constant potential $\Phi_{metal}$ within the electrode and a constant potential $\Phi_{electrolyte}$ within the electrolyte. If $d$ is the distance between the electrodes surface and the IHP, we get

$$\Phi(x) = \Phi_{metal}, x \leq 0 \tag{3.93a}$$

$$\Phi(x) = \left(\Phi_{electrolyte} - \Phi_{metal}\right) \cdot \frac{x}{d} + \Phi_{metal}, 0 < x \leq d \tag{3.93b}$$

$$\Phi(x) = \Phi_{electrolyte}, x \geq d \tag{3.93c}$$

Hence, in this first approximation, the distribution of the potential between the electrodes surface and the IHP – or better the potential of the interface region between the electrode and the electrolyte – can be drawn as shown in Figure 3.23.

The voltage of an electrode is dependent on the temperature and on the ion concentration in the acidic solution – therefore, the standard potential has to be measured under standard conditions to get comparable results.

### 3.2.2 The standard hydrogen electrode

We have already mentioned that the voltage of an electrode at the interface between the metal and the solution cannot be measured. The reason was that a second electrode within the solution would be necessary to provide electrical contact to a voltmeter. Hence, it is only possible to measure two electrodes against each other. Therefore, to get comparable experimental results, it is necessary to measure the voltage or the potential difference against a well-defined standard electrode with a fixed potential. Then the potential differences of other electrodes are assignable referring to the standard electrode. For practical reasons, the potential of the standard electrode is set to zero. Therefore, the potential differences of any electrode can be referred to this zero point.

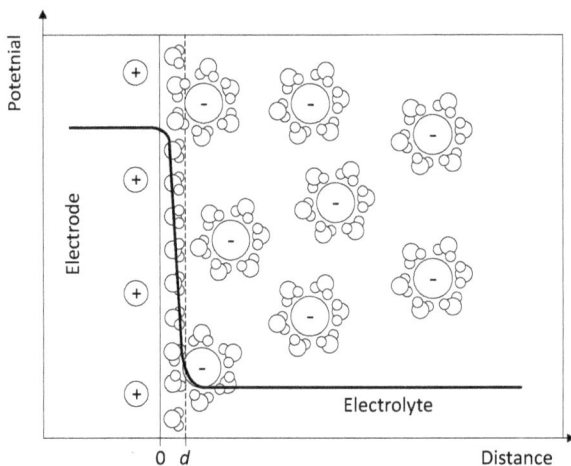

**Figure 3.23:** Potential distribution over the region at the electrode–electrolyte interface according to eqs. (3.93a)–(3.93c); $d$ is the distance from the surface of the electrode.

Per definition, the standard hydrogen electrode (SHE) is used as the standard for this purpose. The SHE is a redox electrode that is the base for the thermodynamic scale of redox potentials. The SHE is a redox half-cell. Pure hydrogen gas – $H_2$ molecules – bubbles through an acidic solution. At a platinum or platinized electrode, a redox reaction occurs. Here, due to catalytic effects at the platinum surface, the $H_2$ molecule splits into two hydrogen atoms that in turn donate their electrons to the platinum electrode; hence, protons ($H^+$) remain in the solution:

$$H_{2(g)} \leftrightarrow 2H^{+}{}_{(aq)} + 2e^{-} \qquad (3.94)$$

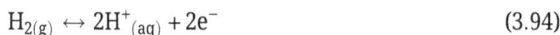

Formula (3.94) describes a redox reaction; it can occur in both directions. The gaseous hydrogen molecule ($H_2$) and the protons ($H^+$) form the redox pair or redox couple.

The drafted construction of a SHE is shown in Figure 3.24. Against this SHE, the potential of another electrode can be measured. However, to get a proper result, those measurements have to be carried out at standard conditions. Standard condition means that the measurements have to be executed at a temperature of 298.15 K (25 °C) and a pressure of 101.325 kPa (1.01315 bar); and furthermore, the acidic solution has to contain 1 mol of $H^+$ ions (protons) per liter, that is, $[H^+] = 1$ mol/L.

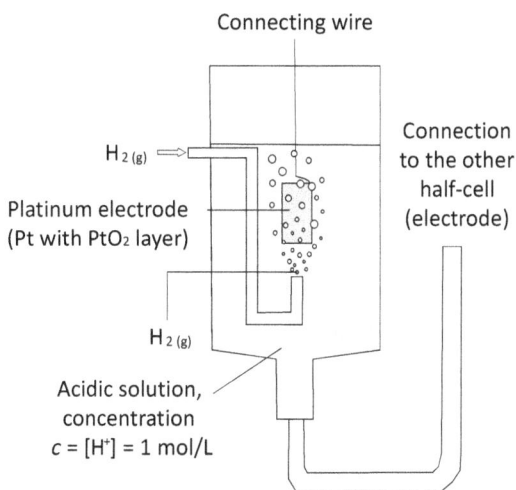

**Figure 3.24:** Typical setup of a standard hydrogen electrode (SHE).

The potentials of any metal electrodes have to be compared with that of the SHE. The absolute electrode potential of the SHE itself is $4.44 \pm 0.02$ V at 298.15 K. Under standard conditions, we get

$$E_{(abs)} = E_{(SHE)} + (4.44 \pm 0.02)V \qquad (3.95)$$

In eq. (3.95), $E_{(abs)}$ is the absolute potential of an arbitrary electrode and $E_{(SHE)}$ is the potential of this electrode measured against the SHE (V means volt). Hence, the absolute electrode potential is the potential measured against the SHE plus the absolute potential of the SHE itself. Usually, the potential difference – that is, the voltage – between two different metal electrodes is of interest. In this case, the absolute potential of the SHE cancels out when looking at the potential difference of the electrodes. Hence, in this case it does not matter that the potential of the SHE was set to zero.

The standard potentials or voltages of electrode materials can be obtained in galvanic cells (the general setup of galvanic cells will be described in the next section). These voltages have to be measured under the following conditions: one half-cell of the galvanic cell is the SHE, that is, the $H_2$–$H^+$ redox pair; the other half-cell is formed by the redox pair of the metal electrode that is of interest and that should be measured. In case of metals, the metal atom and the corresponding ion form the redox pair. The general formula can be written as follows:

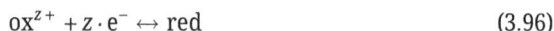

$$ox^{z+} + z \cdot e^- \leftrightarrow red \tag{3.96}$$

The abbreviations ox and red denote the oxidized and the reduced components, respectively, and z is the number of the transferred electrons.

For example, copper (Cu) shows a standard potential (voltage) of $E^0$ = +0.35 V against the SHE. This positive value indicates that copper is a noble metal. Zinc (Zn) has a standard potential of $E^0$ = −0.76 V and zinc is not a noble metal (i.e., a base metal) since $E^0$ is negative. In both cases $z = 2$, since two electrons are involved in each redox reaction step. Table 3.1 shows the standard potentials (voltages) of a selection of metals; $E^0$ is measured against the SHE.

**Table 3.1:** Standard potentials of various metals against the standard hydrogen electrode (SHE).

| Metal | Redox reaction | $E_0$ (V) | Metal | Redox reaction | $E_0$ (V) |
|---|---|---|---|---|---|
| Lithium (Li) | $Li^+ + e^- \leftrightarrow Li$ | −3.05 | Hydrogen ($H_2$) | $2H^+ + 2e^- \leftrightarrow H_2$ | 0 |
| Potassium (K) | $K^+ + e^- \leftrightarrow K$ | −2.92 | Copper (Cu) | $Cu^{2+} + 2e^- \leftrightarrow Cu$ | +0.35 |
| Sodium (Na) | $Na^+ + e^- \leftrightarrow Na$ | −2.71 | Silver (Ag) | $Ag^+ + e^- \leftrightarrow Ag$ | +0.81 |
| Manganese (Mn) | $Mn^{2+} + 2e^- \leftrightarrow Mn$ | −1.18 | Mercury (Hg) | $Hg^{2+} + 2e^- \leftrightarrow Hg$ | +0.85 |
| Zink (Zn) | $Zn^{2+} + 2e^- \leftrightarrow Zn$ | −0.76 | Platinum (Pt) | $Pt^{2+} + 2e^- \leftrightarrow Pt$ | +1.20 |
| Lead (Pb) | $Pb^{2+} + 2e^- \leftrightarrow Pb$ | −0.13 | Gold (Au) | $Au^{2+} + 2e^- \leftrightarrow Au$ | +1.50 |

Note: The standard potential of the standard hydrogen electrode is set to zero.

## 3.3 Electrochemical cells

In general, an electrochemical cell consists of two electrodes that are in contact with – or dipped into – an electrolyte or ion conductor. As mentioned earlier, the ionic conductor can be in a liquid or solid state. The electrode and the electrolyte are coherent and comprise an electrode compartment. Different configurations of electrochemical cells are possible, where various kinds of electrodes might be involved. An electrochemical cell is denoted in galvanic cell, if it can deliver an electrical current, as will be emphasized in the proceeding discussions.

For instance, two electrodes might share the same electrolyte, that is, they share one electrode compartment. In this case, the electrodes often – but not necessarily – are of the same type to avoid unwanted chemical reactions. An electrochemical cell with such a structure can be used, for instance, for electrolysis. Likewise, galvanic cells can be constructed with such a configuration; but in this case, the two electrodes that share the same electrode compartment have to be made of different materials.

However, in galvanic cells, that is, electrochemical cells that produce electricity by spontaneous endergonic electrochemical reaction processes, the two different electrodes that are employed mostly do not share the same electrode compartment. Each electrode has its own compartment; that is, both electrodes are located in two half-cells, and they are spatially separated from each other. Then heterogeneous chemical redox reactions can occur at these two different spatially separated electrodes.

An electrochemical cell that provides heterogeneous chemical redox reactions – that is, the galvanic cell – can be equipped with metal/metal–ion electrodes that were introduced already in the last section. The metal and the metal ion form a redox couple according to the following half-reaction (compare with eqs. (3.87a), (3.87b) or (3.96)):

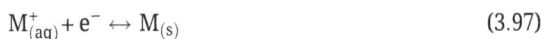

$$M^+_{(aq)} + e^- \leftrightarrow M_{(s)} \tag{3.97}$$

Half-reaction means the redox reaction at one electrode. If two different metal electrodes $M_1$ and $M_2$ are spatially separated from each other, the oxidation and reduction processes that are responsible for the overall redox reaction of the cell occur in the particular half-cells. The oxidation takes place in the electrode compartment with the less noble metal electrode $M_1$ (anode) and the corresponding electrolyte (the anolyte) and the reduction occurs in the half-cell with the more noble metal electrode $M_2$ (cathode) and the corresponding electrolyte (the catolyte). Hence, the first half-cell, where the oxidation occurs, is built up of the anode ($M_1$) and the anolyte; the solution contains positively charged ions (cations) released from the anode. The second half-cell, where the reduction occurs, is built up of the cathode ($M_2$) and the catolyte; the solution contains positively charged ions (cations) released from the cathode. The solutions in both half-cells contain negatively charged anions to balance the charge of the cations.

If the redox reactions spontaneously occur, electrical energy can be obtained from the electrochemical cell – this is just the case of the galvanic cell that consists of the combination of two redox pairs that are located in separated half-cells. These chemical processes and reactions will be discussed in detail in the next section with the help of the so-called Daniell element or Daniell cell that is one of the classical galvanic elements or batteries. The overall redox reaction can be executed, if an electrical circuit is closed, that is, if the half-cells are internally connected by a salt bridge (see Section 3.5.2) or by a semipermeable membrane and externally by electrical wiring (Figure 3.25). At the anode, the electrons are released:

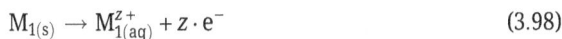

$$M_{1(s)} \rightarrow M^{z+}_{1(aq)} + z \cdot e^- \tag{3.98}$$

$M_{1(s)}$ is the (neutral) metal atom in the anode; $M_{1(s)}^{z+}$ is the metal ion that was released from the electrode into the solution of the half-cell; $z$ is the number of electrons that are involved in the oxidation process – that is, $z$ electrons remain in the anode and are the reason for its negative charge. Via the external electrical circuit, the electrons can move to the cathode $M_2$ in the other half-cell, where the reduction occurs:

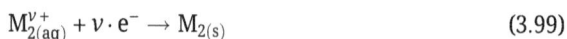

$$M^{\nu+}_{2(aq)} + \nu \cdot e^- \rightarrow M_{2(s)} \tag{3.99}$$

Here $M_{2(s)}$ is the (neutral) metal atom in the cathode, $M_{2(s)}{}^{\nu+}$ is the metal ion released from the cathode into the solution of the second half-cell and $\nu$ is the number of electrons that are involved in the reduction process.

Thus, in half-cells (often) two phases are employed for the half-cell potential, that is, a solution and a metal. Instead of a metal, also a gas can be employed, as we have seen earlier for the case of the SHE. The corresponding electrodes are called electrodes of the first kind. Typical examples are the various metal electrodes that are immersed in a solution that contains its own ions (Figure 3.26), that is, silver immersed in an aqueous silver nitride solution containing $Ag^+$ ions, zinc in an aqueous zinc(II) sulfate solution containing $Zn^{2+}$ ions or copper in an aqueous copper(II) sulfate solution containing $Cu^{2+}$ ions.

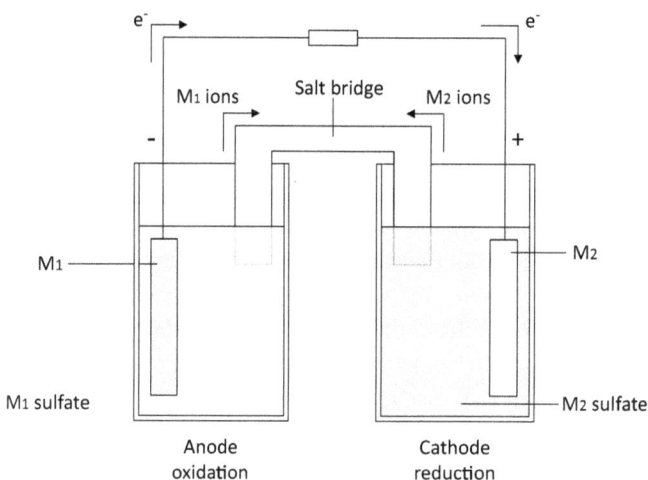

**Figure 3.25:** Galvanic cell with two separated half-cells that are externally connected by electrical wiring and internally by a salt bridge. The metal electrode $M_1$ is less noble than the metal electrode $M_2$. Hence, $M_1$ is the anode, where oxidation occurs, and $M_2$ is the cathode, where reduction occurs.

The chemical reactions in an electrochemical cell can be described in a shortened way that is called cell notation. In the cell notation (3.100), the separation of the half-cells (or electrode compartments) of the electrochemical cell are sketched by two vertical lines (i.e., a vertical double line). They represent a semipermeable membrane or a salt bridge. The anode is located at the left side of the vertical double line and the cathode on the right side. Within the half-cell, the individual phases – they can be solid, liquid or gaseous – are separated by a single vertical line. If the electrochemical cell is not divided into two half-cells, the vertical double line does not arise, and the different phases of the reacting species are separated by single vertical lines. Furthermore, the concentrations of dissolved species are written in parentheses, and the state of the phases can be annotated by the usual abbreviations – s (solid), l (liquid), g (gaseous), aq (aqueous solution) and the electrical charges of the protagonists of the cell reactions are marked by the superscript + or –. Uncharged species are either not marked or marked by the superscript 0. The following example presents the cell notation for the case where the electrochemical cell consists of two metal electrodes immersed into an aqueous media, for instance, diluted sulfuric acid:

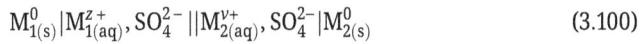

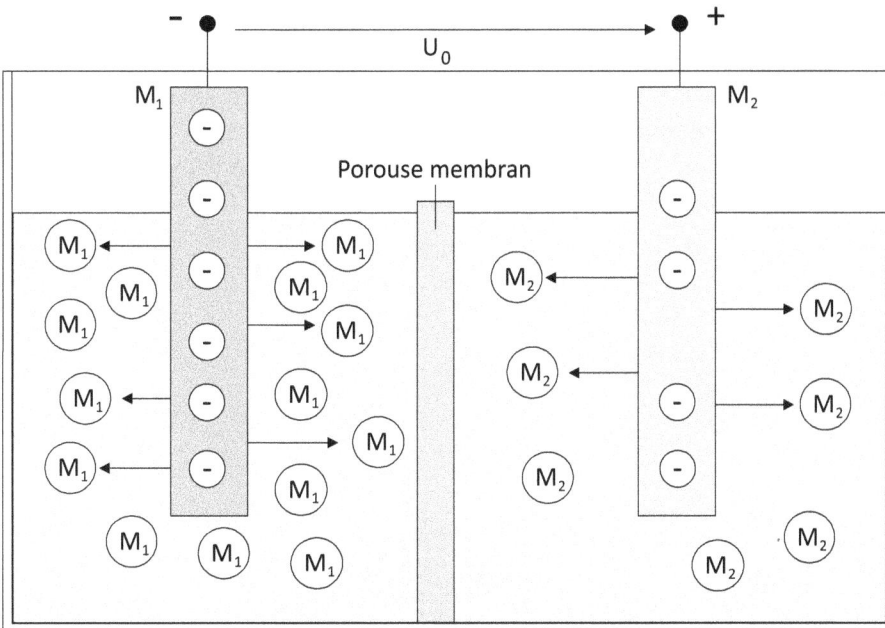

$$M^0_{1(s)}|M^{z+}_{1(aq)}, SO_4^{2-}||M^{v+}_{2(aq)}, SO_4^{2-}|M^0_{2(s)} \qquad (3.100)$$

**Figure 3.26:** Galvanic cell with two different electrodes that are immersed in solutions in the corresponding half-cells that contain its own ions. The electrode $M_1$ is less noble than $M_2$, that is, more positive $M_1$ ions enter the solution in the left half-cell than positive $M_2$ ions in the right half-cell, and therefore, more electrons are left behind in the left electrode than in the right electrode. Hence, the left electrode is the negative terminal and the right electrode is the positive terminal of the galvanic cell.

The electrochemical cell is separated into two half-cells. The left electrode is the anode; it is made of metal $M_1$ that is in the solid state and electrically neutral. The right electrode is made of metal $M_2$ that is also in the solid state and electrically neutral. Both electrodes are immersed into aqueous medium, where $M_1^{z+}$ and $M_2^{v-}$ ions are present, $z$ and $v$ are the number of electrons that are involved in the half-cell reactions. If we use an aqueous sulfuric acid solution ($H_2O$ mixed with $H_2SO_4$), the $H_2SO_4$ molecule dissociates according to formula (3.88), that is, $H_2SO_4$ dissociates into two protons ($H^+$) and a sulfate ion ($SO_4^{2-}$), whereupon the protons attach to water molecules and form hydronium ions ($H_3O^+$). The latter ones do not play an essential role for the cell reactions, and therefore, they are not mentioned in the cell notation (3.100).

As described in the beginning of this section, the electrode and the surrounding electrolytic solution are called half-cell or electrode compartment. Depending on the electrochemical reactions at the electrode in the half-cell, we distinguish various types of electrodes. For instance, in case of the gas electrode, a gas is involved in the electrode reactions. The gas flows around an inert metal electrode (Figure 3.27). A typical example is the SHE that enables the measurement of the standard potential of redox pairs. We introduced the SHE already in Section 3.2.2. Here the inert metal is made of platinum (Pt). However, the platinum is not directly involved in the redox reactions; rather it acts as a catalyst to provide atomic hydrogen. Furthermore, the platinum electrode acts a sink/source for electrons, that is, it acts as an electron acceptor or donor. The underlying redox reaction was already presented in formula (3.94), that is, $H_2(g) \leftrightarrow 2H^+(g) + 2e^-$. The reaction can occur in both directions. Using cell notation, the SHE can be described in the simplest form as follows:

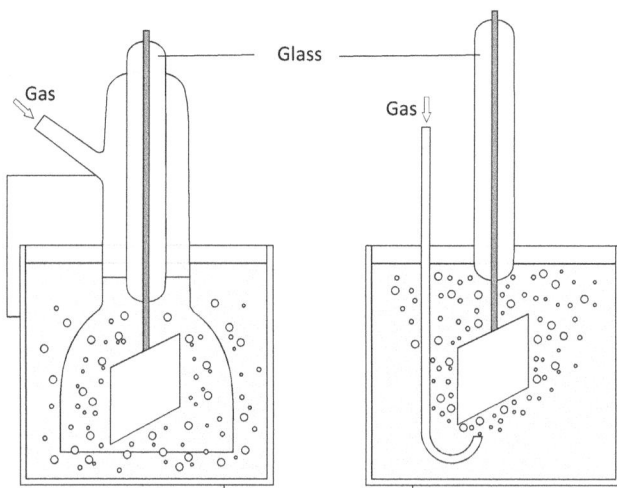

$$Pt|H_2, H^+ \tag{3.101}$$

**Figure 3.27:** Two versions of a gas electrode (with and without hood), where gas bubbles flow around the electrode. The standard hydrogen electrode (Figure 3.24) is at typical example of a gas electrode.

We can add the states of phases, and we get a more precise version of the cell notation:

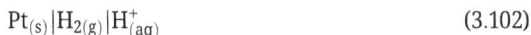

$$Pt_{(s)}|H_{2(g)}|H^+_{(aq)} \tag{3.102}$$

This is the cell notation of a half-cell. The electrochemical cell consists of an SHE (half-cell) and a redox pair of interest that is measured against the SHE. Then this redox pair of interest belongs to the second half-cell of the electrochemical cell.

We want to look at another example that can be expressed in cell notation (3.103). The electrochemical cell consists of a zinc electrode in one half-cell and a silver/silver chloride electrode in the other half-cell. In this case, the anode (i.e., the left electrode) is made of zinc, and the cathode (i.e., the right AG/AgCl electrode) is made from a silver wire that is covered by a silver chloride layer. The silver chloride layer is not soluble. The two electrodes are immersed into aqueous media, where zinc and chloride ions are present in the particular half-cells. The corresponding cell notation describes the complete situation:

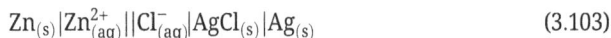

$$Zn_{(s)}|Zn^{2+}_{(aq)}||Cl^-_{(aq)}|AgCl_{(s)}|Ag_{(s)} \tag{3.103}$$

The zinc electrode, that is, the anode on the left in the cell notation (3.102) is a solid, and the same holds for the cathode that consist of an Ag wire covered with a porous layer of AgCl. The zinc cation ($Zn^{2+}$) and the chloride anion ($Cl^-$) are dissolved in the aqueous media. To provide more information, ion concentrations can be added to the cell notations. For instance, in case that all ion concentrations are just 1 mol/L, the cell notation (3.103) can be complemented as follows:

$$Zn_{(s)}|Zn^{2+}_{(aq, 1mol/L)}||Cl^-_{(aq, 1mol/L)}|AgCl_{(s)}|Ag_{(s)} \tag{3.104}$$

The right cathode in this example, that is, the Ag/AgCl electrode (Figure 3.28), is a special kind of electrode since the electrode potential only indirectly depends on the concentration of the electrolytic solution. Sometimes such electrodes are called electrodes of the second kind and they can be used as reference electrodes. Thus, electrodes of the second kind are electrodes, where beside the metal a second solid phase is involved. The second phase is a salt with a low solubility. Using the cell notation again, the Ag/AgCl electrode can be expressed by

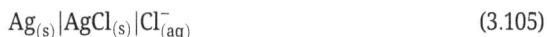

$$Ag_{(s)}|AgCl_{(s)}|Cl^-_{(aq)} \tag{3.105}$$

The salt $AgCl_{(s)}$ is in contact with the solution that contains the $Cl^-$ anion. The chloride concentration in the solution controls the silver cation concentration at the surface of the metallic silver, that is, at the Ag/AgCl interface. This occurs via the law of mass action. In equilibrium, this implies that for a chemical reaction the ratio of the concentrations of the educts and the products is constant, that is,

Ag(s)

AgCl(s)

KCl + AgCl(s)

Frit

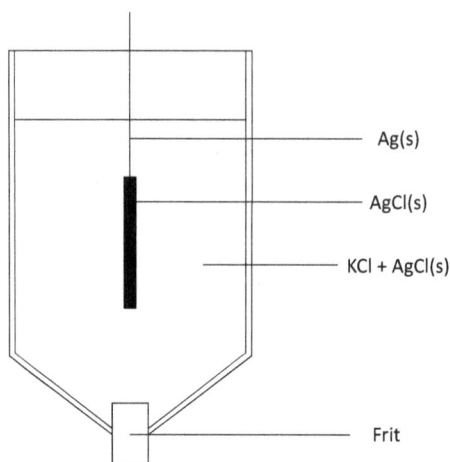

**Figure 3.28:** Ag/AgCl electrode – electrode of the second kind.

$$K = \frac{[Cl^-] \cdot [Ag^+]}{[AgCl]} \tag{3.106}$$

In eq. (3.106), the anion, cation and silver chloride concentrations are denoted by $[Cl^-]$, $[Ag^+]$ and $[AgCl]$, respectively. $K$ is called equilibrium constant. Overall, at the Ag/AgCl electrode the following reaction occurs:

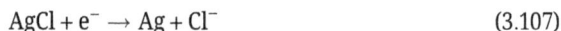

$$AgCl + e^- \rightarrow Ag + Cl^- \tag{3.107}$$

The AgCl layer takes an electron from the metal silver at the Ag/AgCl interface, and in return, AgCl releases a chloride anion to the aqueous solution.

For a better understanding, in the following section we will discuss thermodynamics of electrochemical cells and half-cells.

## 3.4 Thermodynamics of electrochemical cells

In this section, we briefly discuss some thermodynamic aspects of electrochemical and/or galvanic cells. In particular, we will deduce the Nernst equation that is one of the most important equations in electrochemistry. The Nernst equation can be derived from the changes in the Gibbs free energy $G$. In addition, from the thermodynamic point of view, we will also discuss special types of electrochemical or galvanic cells to deepen the understanding of their functionality. Before we start these discussions, we first have to look again on solutions, mixed phases and electrodes.

The chemical potential of the component $i$ of a mixed phase or solution corresponds to energy that is necessary to introduce an amount of 1 mol of this component from the outside (where $i$ does not interact with other phases or components) into the solution or mixed phase. The temperature dependence of the chemical potential $\mu_i$ of this component $i$ can be expressed by the following equation:

$$\mu_i = \mu_i^0 + R \cdot T \cdot \ln a_i \tag{3.108}$$

In this equation, $\mu_i^0$ is the chemical potential of the component $i$ for standard conditions, $a_i$ is the activity.

If two mixed phases or components A and B of a solution or mixed phase are in contact, and furthermore, if they are in equilibrium, their chemical potentials $\mu_{i(A)}$ and $\mu_{i(B)}$ are equal:

$$\mu_{i(A)} = \mu_{i(B)} \tag{3.109}$$

This holds for all components $i$. In particular, if we consider a metal electrode that we immerse in an aqueous solution, the corresponding equilibrium equation holds:

$$\mu_{\text{metal}^{z+}(aq)} = \mu_{\text{metal}^{z+}(s)} \tag{3.110}$$

For example, for the cases of divalent copper or zinc electrodes we get:

$$\mu_{\text{Cu}^{2+}(aq)} = \mu_{\text{Cu}^{2+}(s)} \tag{3.111a}$$

$$\mu_{\text{Zn}^{2+}(aq)} = \mu_{\text{Zn}^{2+}(s)} \tag{3.111b}$$

In this case, we considered that in the electrode – that is, in the solid(s) – the metal ions are located on defined lattice sites of the crystal, and hence, they are immobile, while the outer electrons are delocalized, and they are not bonded to the immobile ions and can move through the lattice.

The equilibrium equations (3.109) or (3.110) of the chemical potentials of electrode metals immersed in an aqueous solution are not valid in the very first moment of the immersion. Rather, immediately a balancing process starts. Depending on the energetic conditions, either a dissolution or a deposition process of the metal occurs. However, such reactions cannot run until the final equilibrium is achieved since just from the beginning of the process an electrical potential difference is formed.

If the chemical potential of the metal ions within the electrode – that is, in the solid state – is larger than the chemical potential of the metal ions in solution, the ions from the electrode enter the solution. Therefore, the region of the solution close to metal surface gets positively charged, and on the other hand, the electrode is negatively charged close to its surface since the dissolved metal ions left their delocalized outer electrons behind in the crystal lattice. Hence, an electrical double layer is formed. The dissolution of metal ions, that is, the migration of metal ions from the electrodes surface toward the solution comes to a stop, if the electrostatic attraction between the negatively charged surface of the electrode and the positively charged interface region of the aqueous solution equals the solution pressure of the metal ions.

If the chemical potential of the metal ions in solution is larger than the chemical potential of the metal ions within the electrode, the opposite processes and charge displacements occur. This results in a deposition of metal ions from the solution on the electrodes surface. In this case, the surface of the electrode is positively charged, and the interface region of the aqueous solution gets negatively charged. Again, this process comes to a stop, if the deposition pressure equals the electrostatic attraction within double layer.

Finally, an electrostatic potential difference between the electrode and the solution is established. This has to be considered, when – according to and in extension of eq. (3.109) – the equilibrium equations for the two phases of the component $i$ are formulated. Thus, we have to consider an electrochemical equilibrium condition.

We introduce 1 mol of the $z$-fold charged component $i$ into a mixed phase or solution with the electrostatic potential $\Phi$. Then in addition to the chemical potential $\mu_i$ we have to afford (or release – depending on the algebraic sign of $z_i$ and $\Phi$) electrical work $z_i \cdot F \cdot \Phi$, where $F$ is the Faraday constant. In this case, the electrochemical equilibrium condition – that is the in comparison to eq. (3.109) extended equilibrium condition – is given by the following equation:

$$\mu_{i(A)} + z_i \cdot F \cdot \Phi_{(A)} = \mu_{i(B)} + z_i \cdot F \cdot \Phi_{(B)} \tag{3.112}$$

In eq. (3.112), $\Phi_{(A)}$ and $\Phi_{(B)}$ are the electrostatic potentials within the phases A and B. They are called Galvani potentials or inner potentials in electrochemistry. They are the electric potential differences between two points in the bulk of two phases. In our considerations, the two phases are the electrode (A) and the solution (B), that is, a solid that is immersed into a liquid. However, these phases can also be two solids (e.g., metals) that are in contact. In general, Galvani potentials cannot be measured. This is only possible if the two phases have the same chemical composition.

The electrochemical potential $\mu_i^*$ is a thermodynamic measure of the chemical potential including the electrostatic energy contributions; it is expressed in the unit J/mol. The "normal" chemical potential $\mu_i$ omits the electrostatic energy contributions. Thus, $\mu_i^*$ is an extension of $\mu_i$. In other words, the electrochemical potential is the total potential that includes both the internal nonelectrical ("normal") chemical potential and the electric potential of an ion $i$ that is located within an electrical field or in an electrical potential $\Phi$. Hence, the electrochemical potential $\mu_i^*$ can be defined by the following equation:

$$\mu_i^* = \mu_i + z_i \cdot F \cdot \Phi \tag{3.113}$$

where $F$ is again Faraday's constant, $\Phi$ is the local electrical potential and $z_i$ is the valence (charge) of the ion $i$. For an uncharged neutral atom $z_i$ is zero, and therefore, the electrochemical potential equals the chemical potential, that is, $\mu_i^* = \mu_i$. Taking into account eq. (3.108), we get

$$\mu_i* = \mu_i^0 + R \cdot T \cdot \ln a_i + z_i \cdot F \cdot \Phi \tag{3.114}$$

In accordance with eq. (3.112), the electrochemical equilibrium for the components of the two phases of the component $i$ is defined by

$$\mu_{i(A)}{}^* = \mu_{i(B)}{}^* \tag{3.115}$$

For a constant temperature and pressure, and for a constant amount of substance of all other components of a system, the electrochemical potential $\mu_i^*$ defines the amount of work that is necessary to enhance the amount of the ions of type $i$ (in mole) from $n_1$ to $n_2$. Since this amount of work conforms to the change of the Gibbs free energy under these conditions, we get

$$\Delta G = \int_{n_1}^{n_2} \mu_i^* \cdot dn_i \tag{3.116}$$

The ability of any system to carry out work can be related to a potential difference. Therefore, chemical reactions under participation of ions are running until all electrochemical potentials of all components of a system were adapted.

The electrochemical potential should not be confused with the potential $E$ (or better the potential difference or even better the voltage $E$) of an electrode. The potential difference $E$ of an electrode corresponds to an energy per charge, that is, $E = dW/dQ$. On the other hand, the electrochemical potential corresponds to an energy per mole and is expressed in J/mol, as mentioned above, that is, $\mu_i^* = dG/dn_i$.

The electrochemical equilibrium can be regarded as a dynamic equilibrium and this holds by the way for any chemical equilibrium. This means that for the case of a metal electrode immersed in a solution, where oxidation or reduction reactions occur according to eqs. (3.98) and (3.99), the dissolution of the metal (oxidation) and the deposition of the metal (reduction) occur at equal reaction rates. If an external electrical circuit is employed, the electrical current $i$ is zero. Hence, the dynamic electrochemical equilibrium can be described by the following reaction:

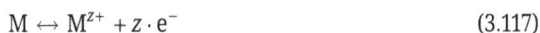

$$M \leftrightarrow M^{z+} + z \cdot e^- \tag{3.117}$$

The reactions from the right to the left side of this equation occur at the same reaction rate ("speed") as in opposite direction, that is, from the left to right side. For the examples of divalent copper or zinc electrodes immersed in aqueous solutions that contain the corresponding copper or zinc ions, respectively, the reactions for the (dynamic) electrochemical equilibrium can be written as follows:

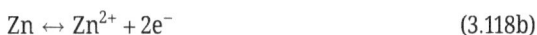

$$Cu \leftrightarrow Cu^{2+} + 2e^- \tag{3.118a}$$

$$Zn \leftrightarrow Zn^{2+} + 2e^- \tag{3.118b}$$

After these preceding considerations, in the next section we will deduce the important Nernst equation.

### 3.4.1 The Nernst equation

The dependence of the voltage of an electrochemical or galvanic cell on the temperature $T$ (in Kelvin) and on the activity $\alpha$ can be expressed with the Nernst equation:

$$E_{\text{electrode}} = E^0 + \frac{R \cdot T}{z \cdot F} \cdot \ln \frac{\alpha_{\text{ox}}}{\alpha_{\text{red}}} \tag{3.119}$$

The Nernst equation allows the determination of the potential or voltage – of a chemical cell under nonstandard conditions; more precisely, it describes the dependence of the electrode potential – or better the voltage – of a redox pair on the standard electrode potential $E^0$, on the temperature $T$ and on the specific activities $\alpha$ of the particular partners in the redox reaction. In eq. (3.119), $R$ is the universal gas constant, $F$ is the Faraday constant and $z$ is the number of electrons that are involved in the redox reaction. We introduced the activity $\alpha$ in frame of our discussions on the chemical potential in Section 2.3 – see, for instance, eq. (2.121).

Very often, the Nernst equation is expressed in terms of the logarithm with the base 10, at which the temperature is the standard temperature $T = 298.15$ K (25 °C). In this case, we merge the constants and get

$$E_{\text{electrode}} = E^0 + \frac{0.059\text{V}}{z} \cdot \log \frac{\alpha_{\text{ox}}}{\alpha_{\text{red}}} \tag{3.120}$$

The Nernst equation can be deduced as follows. We consider, for instance, a chemical reaction, at which four substances A, B, C and D with the corresponding stoichiometric coefficients $v_A$, $v_B$, $v_C$ and $v_D$ are involved, that is:

$$v_A \cdot A + v_B \cdot B = v_C \cdot C + v_D \cdot D \tag{3.121}$$

The (molar) Gibbs free energy of 1 mol of substance A is given by

$$G_{\text{m,A}} = G_A^0 + R \cdot T \cdot \ln \alpha_A \tag{3.122}$$

$G_A{}^0$ is the Gibbs free energy under standard conditions; $\alpha_A$ is the activity of the substance A. For an amount of $v_A$ mol of substance A, we get the following expression for the Gibbs free energy:

$$G_A = v_A \cdot G_{\text{m,A}} = v_A \cdot G_A^0 + v_A \cdot R \cdot T \cdot \ln \alpha_A \tag{3.123}$$

The logarithm transforms a factor $v_A$ into an exponent $v_A$, hence,

$$G_A = v_A \cdot G_A^0 + R \cdot T \cdot \ln \alpha_A{}^{v_A} \tag{3.124}$$

Analogous, the Gibbs free energies of the substances B, C and D are given:

$$G_B = v_B \cdot G_B^0 + R \cdot T \cdot \ln a_B{}^{v_B} \tag{3.125}$$

$$G_C = v_C \cdot G_C^0 + R \cdot T \cdot \ln a_C{}^{v_C} \tag{3.126}$$

$$G_D = v_D \cdot G_D^0 + R \cdot T \cdot \ln a_D{}^{v_D} \tag{3.127}$$

If we consider the chemical reaction that was described earlier by the chemical formula (3.121), the change of the Gibbs free energy upon such a reaction can be expressed as follows:

$$\Delta G = v_C \cdot G_{m,C} + v_D \cdot G_{m,D} - v_A \cdot G_{m,A} - v_B \cdot G_{m,B} \tag{3.128}$$

We insert eqs. (3.124)–(3.127) into eq. (3.128) and obtain

$$\Delta G = \Delta G^0 + R \cdot T \cdot \ln \left( \frac{a_C^{v_C} \cdot a_D^{v_D}}{a_A^{v_A} \cdot a_B^{v_B}} \right) \tag{3.129}$$

In eq. (3.129), we used $\Delta G_A{}^0$ as the expression for the change of the Gibbs free energy under standard conditions:

$$\Delta G^0 = v_C \cdot G_C^0 + v_D \cdot G_D^0 - v_A \cdot G_A^0 - v_B \cdot G_B^0 \tag{3.130}$$

Equation (3.129) can be generalized for any electrochemical redox reaction that has the form shown in eq. (3.96), that is, overall, the change $\Delta G$ of the Gibbs free energy can be expressed by the following equation:

$$\Delta G = \Delta G^0 + R \cdot T \cdot \ln \prod_{i=1}^{n} a_i^{v_i} \tag{3.131}$$

In eq. (3.131), $a_i$ and $v_i$ are the activity and the stoichiometric coefficient of the substance $i$, respectively. Again, $\Delta G^0$ is the change of the Gibbs free energy under standard conditions. Furthermore, eq. (3.131) holds for the case that $n$ substances $A_1$, $A_2$, ..., $A_n$ are involved in the reaction, that is,

$$v_1 A_1 + v_2 A_2 + \cdots + v_k A_k \rightarrow v_m A_m + \cdots + v_{n-1} A_{n-1} + v_n A_n \tag{3.132}$$

On the other hand, the work that is done by the reaction in an electrochemical or galvanic cell can by expressed by the electrical energy that is delivered under the actual conditions. This electrical energy is given by the potential difference (voltage) $E$ and the involved charge $Q$. The latter one is defined by the Faraday constant $F$ and the valence, that is, the number $z$ of electrons that are involved in the redox reaction. In summary, in electrochemistry, a spontaneous reaction at constant temperature and pressure leads to the relationship:

$$\Delta G = -Q \cdot E = -z \cdot F \cdot E \tag{3.133}$$

$E$ is the voltage of the electrochemical cell in equilibrium, that is, $E$ is identical to $E_{electrode}$ in eqs. (3.119) and (3.120). Hence, the potential difference or voltage $E$ that is associated with the redox reaction of the form (3.96) can be defined as the decrease in the Gibbs free energy per Coulomb of the charge that is transferred. The minus sign is established by convention. The voltage $E^0$ under standard conditions follows from

$$\Delta G^0 = -z \cdot F \cdot E^0 \tag{3.134}$$

or rather

$$E^0 = -\frac{\Delta G^0}{z \cdot F} \tag{3.135}$$

If $E^0 < 0$, then $\Delta G^0 > 0$, and the electrochemical reactions (3.128) or – more general – (3.132) do not occur spontaneously. Contrariwise, if $E^0 > 0$, then $\Delta G^0 < 0$, and those reactions occur spontaneously.

From reactions (3.129) and (3.133) or (3.134), respectively, we directly receive the Nernst equation:

$$E = -\frac{\Delta G}{z \cdot F} = E^0 + \frac{R \cdot T}{z \cdot F} \cdot \ln\left(\frac{a_A^{\nu_A} \cdot a_B^{\nu_B}}{a_C^{\nu_C} \cdot a_D^{\nu_D}}\right) \tag{3.136}$$

For the specific case of a metal electrode immersed in an aqueous solution, that is, where redox reactions like (3.97) or (3.116) occur, we get the simpler expressions (3.119) or (3.120) for the Nernst equation.

The change $\Delta G$ of the Gibbs free energy due to a chemical reaction is proportional to the number of the involved particles or elements (i.e., according to the syntax of the chemical formula). Hence, the activities are proportional to concentrations, for the case of strongly diluted mixtures or solution, as mentioned already in Section 2.3. This yields to a form of the Nernst equation, where the activities are approximated by concentrations. In fact, the original form of the Nernst equation was introduced in such a form using concentrations:

$$E_{electrode} = E^0 + \frac{R \cdot T}{z \cdot F} \cdot \ln\frac{c_{ox}}{c_{red}} \tag{3.137}$$

### 3.4.2 The galvanic series and the standard electrode potential

The galvanic series determines the nobility of metals (or semimetals like alkaline earth metals, graphite and others). Sometimes, the galvanic series is also called electrochemical series or electrode potential series. When two metals that are electrically connected via a wire are submerged in an electrolyte, the less noble one will experience galvanic corrosion. The extent of corrosion depends on the electrolyte and the difference in nobility of the two metals or semimetals. This difference can be measured as a difference in a voltage potential, that is, the less noble metal or semimetal has a lower (i.e., a more negative) electrode potential than the nobler one. On the other hand, the difference of the voltage potentials of the two different metal electrodes defines the voltage that a battery can deliver. In a galvanic cell that operates as a battery a spontaneous redox reaction occurs. This redox reaction drives the galvanic cell – that is, the battery – to produce an electric potential. For such a spontaneous reaction, the change of the Gibbs free energy has to be negative ($\Delta G^0{}_{cell} < 0$), in accordance with eq. (3.134), we get the following equation:

$$\Delta G^0_{cell} = -z \cdot F \cdot E^0_{cell} \tag{3.138}$$

Again, $z$ is the number of electrons that are exchanged during the electrochemical redox reaction, and $E^0{}_{cell}$ is given by

$$E^0_{cell} = E^0_{cathode} - E^0_{anode} \tag{3.139}$$

$E^0{}_{cathode}$ and $E^0{}_{anode}$ are the standard potentials at the cathode and the anode, respectively. These values are given in the table of standard electrode potentials (see, for instance, Table 3.1). As mentioned already earlier, $E^0{}_{cell}$ has to be positive to provide spontaneous reaction so that $\Delta G^0{}_{cell} < 0$. Instead of eq. (3.139), we also can use the following expression:

$$E^0_{cell} = E^0_{red} - E^0_{ox} \tag{3.140}$$

In this expression $E^0{}_{red}$ is related to the electrode, where the reduction occurs (i.e., at the cathode); and $E^0{}_{ox}$ is related to the electrode, where the oxidation occurs (i.e., at the anode). In a galvanic cell, the less noble metal operates as an anode, where oxidation occurs (half-reaction, where metal ions from the anode dissolve into the electrolyte leaving electrons behind in the anode (loss of electrons)). The nobler electrode is the cathode, where reduction occurs (gain of electrons). Hence, electricity is generated due to the electric potential difference between the two different metal electrodes of each half-cell, and the potential difference results from the difference between the individual potentials of the two metal electrodes with respect to the electrolyte.

If the Gibbs free energy is positive ($\Delta G^0{}_{cell} > 0$), we speak about an electrolytic cell that undergoes a redox reaction, when electrical energy is applied. Such a cell is used for the decomposition of chemical compounds, for instance, the decomposition of water into hydrogen and oxygen. The electrolytic cell also consists of two half-cells. In a galvanic cell, chemical energy is converted into electrical energy, and in an electrolytic cell, electrical energy is converted into chemical energy.

The overall potential of a galvanic cell (or an electrochemical cell including electrolytic cells) can be measured. Usually this is done against the SHE. However, there is no simple way to measure the potential between an electrode and the electrolyte, that is, the isolated potential of the half-cell. In this sense, the standard electrode potential is the electromotive force that can be measured in a galvanic cell that consists of one half-cell with the electrode under consideration and the other half-cell with the SHE.

The SHE was already mentioned in Section 3.2.2. As explained earlier, the standard electrode potentials are measured relative to the SHE, and for that matter, the SHE potential is defined as 0 V. For the sake of comparability, such measurements have to be carried out under standard conditions. The standard electrode potentials of metals and semimetals are arranged in the galvanic series.

## 3.5 Electrochemical energy storage

Electrochemical energy storage is directly related to galvanic cells. It is to underline that electrochemical energy storage should not be confused with electrochemical energy conversion. The latter one, that is, electrochemical energy conversion, is related to fuel cells. In addition, flow batteries or redox-flow batteries are related to electrochemical energy conversion – they are technically akin to a fuel cell.

To simplify matters, one can say that the chemical energy that is stored in a galvanic cell can be directly transformed into electrical energy by an electrochemical redox reaction. The colloquial designation of a galvanic cell is the battery, but one has to keep in mind that in general a battery means the combination of several galvanic cells. However, single galvanic cells are usually also called batteries. Moreover, one also has to keep in mind that battery is an umbrella term for various kinds of batteries, where essentially nonrechargeable (or primary batteries) and rechargeable batteries (or secondary batteries or accumulators) are distinguished. The details on batteries are extensively discussed in Chapter 4.

Chemical energy is the form of energy related to a chemical bond. The making or breaking of a chemical bond in a chemical compound involves energy that either can be absorbed or evolved from the chemical system. Very often, it is explained that chemical energy is released from the chemical system, if the bond is broken – however, this is a misconception and not a correct statement. Likewise, the statement that energy is required to break a chemical bond is a misconception, too. The chemical energy should also not be mixed up with the chemical binding energy. The chemical binding energy describes the stability of a certain chemical bond and specifies how much energy is necessary to break a chemical bond.

Hence, it is a better or a harmless statement that chemical energy is the potential of a substance to undergo a chemical reaction and to transform into another chemical substance. The energy that can be released or absorbed due to a chemical reaction between several substances is equal to the difference of the energy content of the educts (initial substances before the reaction) and the products (final substances after the reaction), if the initial and final temperatures have not changed. On a macroscopic level, chemical energy describes the type of energy related to the electromagnetic forces in atoms and molecules, and it affects chemical reactions. Here, the concept of a chemical energy also includes the kinetic and potential energies of electrons and electromagnetic interactions between electrons and the atomic nucleus. With this regard, chemical energy is an internal energy and chemical reactions that absorb or release energy can be characterized by a change of the internal energy $U$ in accordance with the thermodynamic equation (2.10).

One can state that in the scientific community – especially in chemical sciences – the term chemical energy is usually not used since only under well-defined conditions, it can be exactly defined and, in this case, specific notations are used. For instance, the chemical energy that is released due to a combustion reaction at constant pressure is called heat of combustion or combustion enthalpy.

Before we put the focus on a first example of a galvanic cell in Section 3.5.2, we briefly discuss a very simple model that illustrates electrochemical reactions in a very comprehensible way.

### 3.5.1 Electrochemical reactions

Chemical and electrochemical reactions can be described in a very simple and catchy way. In the following brief explanations, we first look at chemical reactions and subsequently extend our angle of view to the electrochemical reactions.

Chemical reactions can proceed in very different and complex ways. One can say that the morphologies of chemical reactions have many different characteristics. We start a very simple case with two substances A and B (i.e., the educts). We suppose that A and B are metallic materials. The educts are brought in contact to undergo a chemical reaction and form a new substance C (i.e., the product) that is also a metallic material. This simple chemical sample reaction can be described by the following formula:

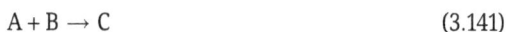

$$A + B \rightarrow C \tag{3.141}$$

The driving force of a chemical reaction is given by the change of the Gibbs free energy (per mole), that is, the difference between the change of Gibbs free energy during the formation of all products and of the change of the Gibbs free energy during the formation of all educts. Or in brief, the driving force of the chemical reaction is given by the difference between the Gibbs free energies of the involved parties after the chemical reaction and before the chemical reaction. With the Gibbs free energy $G^E_i$ of $i$ educts and the Gibbs free energy $G^P_j$ of $j$ products we obtain

$$\Delta G = \sum_j G^P_j - \sum_i G^E_i \tag{3.142}$$

This equation simplifies the simple reaction (3.141):

$$\Delta G = G^C - G^A - G^B \tag{3.143}$$

If the educts are simply elements, then $G^A = G^B = 0$; hence, for the general case according to eq. (3.142), we obtain

$$\Delta G = \sum_j G_j^P \tag{3.144}$$

Therefore, for our simple reaction (3.141), with the product C and the elemental educts A and B, eq. (3.144) is reduced to

$$\Delta G = G^C \tag{3.145}$$

If $\Delta G < 0$, then the reaction occurs spontaneously in the direction of the product.

The chemical sample reaction (3.141) might be called the one-dimensional case. Some very basic considerations must be considered. If the educts come in physical contact, the chemical reaction starts, and the product C is growing around the initial contact surface (Figure 3.29) – that is between the initial educts A and B. The premise that the chemical reaction can proceed is the ability of the A and B particles to diffuse through the C material. The A particles must reach the B–C interface, and the B particles must reach the A–C interface so that A and B can react, and thereby, C can grow. Hence, at the A–C and B–C interface regions, the chemical reactions occur, and the product C is formed and centrically grows between A and B and at the expense of the educts. The chemical reaction finally stops, when at least one of the educts is totally consumed by the reaction.

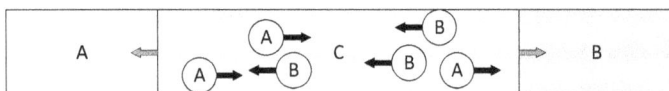

**Figure 3.29:** Chemical sample reaction between two initial educts A and B that form a product C.

This simple model can be easily extended to a schematic model for an electrochemical reaction (Figure 3.30). Taking two metallic educts A and B that now interact in an electrochemical reaction and form a metallic product C, the corresponding reaction can be described again by formula (3.141). However, in electrochemical reactions another player – the electrolyte E – is in the game. While the A, B and C are electrically conductive (with electrons as the carriers), the electrolyte operates as an electric filter that prevents electronic conductivity, but allows ionic conductivity. If we assume that $A^+$ is the predominant ion in the electrolyte, $A^+$ ions can migrate through the electrolyte, but the electrons cannot. The electrolyte is an electronic isolator. Finally, the electrochemical reaction – that is, the redox reaction – requires neutral Atoms A and B. Hence, the electrons must migrate via external wiring, that is, the electrochemical reaction that is schematically shown in Figure 3.30 occurs in a galvanic cell with electrically connected terminals (electrodes).

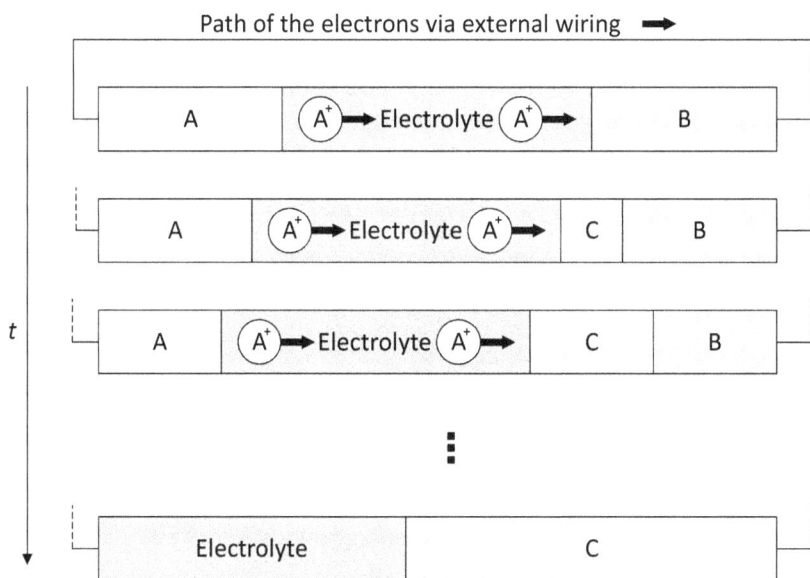

**Figure 3.30:** Time dependence of an electrochemical reaction between educts A and B that are separated by an electrolyte E. $A^+$ ions migrate through E; the product C starts to grow at the E–B interface.

At the interface between educt A and electrolyte E, $A^+$ ions enter the electrolyte leaving behind electrons in A, that is, oxidation occurs. In the very first beginning of the electrochemical reaction process, the $A^+$ ions migrate toward the B–E interface (diffusion process). At the B–E interface, the $A^+$ ions are neutralized by electrons that externally migrated to the B terminal. Hence, the neutral A atoms deposit onto the outer B layer at the B–E interface, and A and B react to form C. At the A–E interface, we observe the following oxidation reaction:

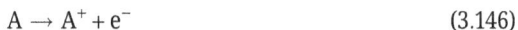

$$A \rightarrow A^+ + e^- \tag{3.146}$$

Simultaneously, at the B–E interface, the corresponding reduction reaction takes place:

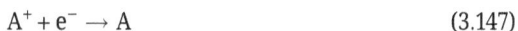

$$A^+ + e^- \rightarrow A \tag{3.147}$$

As the reaction proceeds, the product C centrically grows around the initial B–E interface. After the C layer has formed centrically to the initial B–E interface, the $A^+$ ions migrate through the electrolyte and are deposited onto the outer C layer at the C–E interface. Again, they are neutralized by the electrons that now migrated through the external wiring and the metallic B and C materials. Both A and B atoms diffuse through the product C; if they meet, they react and form C. Hence, C is growing at the expense of A and B (Figure 3.30). It is to underline that the reactions (3.146) and (3.147) occur simultaneously, both, if the A is deposited onto the surface of B in the beginning of the reaction and onto the surface of C during the proceeding reaction. Again, the chemical reaction finally stops, when one of the educts is totally consumed by the reaction.

We can say that during an electrochemical reaction in total a reaction between neutral species is occurring in accordance with the reaction formula (3.141). At the same time subprocesses occur. $A^+$ ions (or $B^+$ ions) diffuse through the electrolyte; and electrons migrate through the external wiring. This has several consequences. If the ion flow in the electrolyte is interrupted, the electrochemical reaction stops. Likewise, if the external electric current is interrupted, the electrochemical reaction stops as well. Hence, for open-circuit conditions, the forces on the ions in the electrolyte are balanced – that is, we face an equilibrium of forces since the chemical driving force is balanced by electrostatic driving forces according to eq. (3.134).

Finally, we can summarize that a chemical or electrochemical reaction between neutral atoms or species dictates the forces on charged particles – that is, the ions – in the electrolyte.

In the following section, we exemplarily discuss the copper–silver galvanic cell that spontaneously transforms electrochemical energy into electrical energy.

### 3.5.2 For instance, the copper–silver galvanic cell

For the discussion of the electrochemical properties and operation of the copper–silver galvanic cell (or copper–silver cell), we use the setup that is shown in Figure 3.31, where the two half-cells, that is, the anodic $Cu|Cu^{2+}$ and the cathodic $Ag^+|Ag$ half-cell (as written in cell notation) of this galvanic element are connected via a salt bridge.

However, before we discuss the copper–silver cell in detail, we briefly discuss the properties and mechanism of a salt bridge that is a significant part of galvanic cell with two separated half-cells.

The salt bridge connects both half-cells of a galvanic cell, as shown in Figure 3.31. It is an ionic conductor, but does not conduct electrons. In general, a salt bridge is a device for laboratory applications, and it does not appear in commercial batteries. However, for the discussion of the physical and chemical processes appearing in the galvanic cells, it is useful. In operation, when the anodic and cathodic electrodes are externally connected via electrical wiring and a load, the main function of the salt bridge is to prevent a liquid junction potential, that is, a potential difference arising between the two solutions in the half-cells, when they are in contact with each other. The salt bridge maintains the electrical neutrality of the solutions in the two half-cells, that is, within the internal electrical circuit of the galvanic cell. Moreover, the diffusion of the liquid solution from one half-cell to the other can be prevented by the salt bridge. If the electrodes are externally connected via electrical wiring and a load, the salt bridge completes the electrical circuit by connecting the electrolytes in the two spatially separated half-cells. As the electrochemical reactions proceed within the galvanic cell, without a salt bridge, the solution in the anodic half-cell would accumu-

$$\Delta E = 0{,}46 \text{ V}$$

Salt bridge

Ag

AgNO₃

Cu

Cu(NO₃)

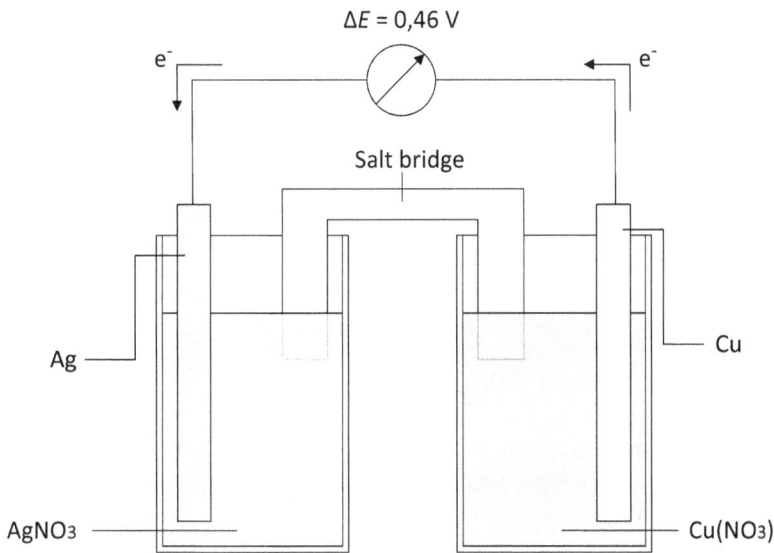

**Figure 3.31:** Copper–silver galvanic cell. The two half-cells are connected by a salt bridge that provides ion migration.

late positive charges, and the solution in the cathodic half-cell would accumulate negative charges. Therefore, further reactions would quickly be prevented, and hence, production of electricity would come to a stop rather soon.

Salt bridges can be made from glass tubes or simply from filter paper. The glass tube salt bridge consists of a U-shaped glass tube that is filled with a quite inert electrolyte. For instance, solutions with sodium chloride (NaCl), potassium chloride (KCl) or potassium nitrate (KNO$_3$) are used. In general, the electrolyte in the salt bridge should not react with the chemicals employed in the galvanic cell. Furthermore, the anions and cations of the electrolyte within the salt bridge should have a similar conductivity – so that they have nearly the same migration velocity. The conductivity of the salt bridge depends on the ion concentration and the diameter of the glass tube. If the concentration of electrolyte solution is not saturated, the ion concentration has a stronger impact on the conductivity. If the concentration of electrolyte solution is saturated, the diameter of the glass tube determines the conductivity. For a comfortable handling, usually the electrolyte in the salt bridge is gelatinized. The jelly-like substance agar-agar can be used for this purpose. Agar-agar is used in food production and can be derived from the polysaccharide agarose.

Filter paper that is soaked with quite inert electrolytes that do not react with the chemicals from the galvanic cell are the simple solution for a salt bridge. As in case of the glass tube bridge, solutions with sodium chloride, potassium chloride or potassium nitrate (KNO$_3$) might be used as electrolytes. Gelatinization is not necessary here since the filter paper provides a handy solid medium for the ion conduction. The conductivity of the salt bridge made of filter paper depends on the texture of the filter paper, its absorbing abilities and – of course – on the concentration of the electrolyte solution. Smooth textures and high absorbance provide enhanced ion conductivities.

Porous disks – for instance, ceramic material, providing ion conductivity between the two half-cells – might be also used instead of a salt bridge. Basically, they serve the same purpose.

Now we come back to the copper–silver galvanic cell. At the copper electrode, oxidation occurs since copper is the less noble metal in comparison to silver, and at the silver electrode, the reduction reaction occurs. According to the setup shown in Figure 3.31, the copper–silver cell can be described in cell notation as follows:

$$Cu_{(s)}|Cu(NO_3)_{2(aq)}||AgNO_{3(aq)}|Ag_{(s)} \tag{3.148}$$

The phase boundaries between the Cu or Ag electrodes and the corresponding copper(II) nitrate and silver nitrate electrolytic solutions in the half-cells, respectively, are marked by the vertical lines. The electrolytes in the two half-cells are separated by a salt bridge. Therefore, two vertical lines mark the separation in cell notation since they are not in direct contact. The cell notation (3.148) emphasizes the different aqueous solutions in the containers of the half-cells, that is, the copper(II) nitrate solution – $Cu(NO_3)_{2(aq)}$ – in the container of the anodic half-cell and a silver(I) nitrate solution – $AgNO_{2(aq)}$ – in the container of the cathodic half-cell. We can state that the copper–silver cell is characterized by the fact that two electrolytes – or conducting liquids – are employed.

It is to mention that the salt bridge can contain, for instance, an aqueous potassium nitrate solution ($KNO_{3(aq)}$). $KNO_3$ dissociates in water into positive potassium ($K^+$) ions and negative nitrate ions ($NO_3^-$).

The cell notation of the copper–silver cell can also be written (3.148) in another style that emphasizes the dissociated copper(II) nitrates and silver(I) nitrates in the aqueous solutions within the corresponding half-cell containers:

$$Cu_{(s)}|Cu^{2+}_{(aq)}, NO^-_{3(aq)}||Ag^+_{(aq)}, NO^-_{3(aq)}|Ag_{(s)} \tag{3.149}$$

The negative $NO_3^-$ ions can be ignored; they are only spectator ions since they do not participate in the redox reaction. Therefore, the cell notation (3.149) can be written in a reduced style that is the commonly used form:

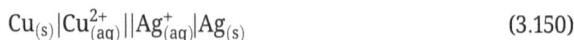

$$Cu_{(s)}|Cu^{2+}_{(aq)}||Ag^+_{(aq)}|Ag_{(s)} \tag{3.150}$$

The cell notation (3.150) exhibits the electrochemical players of the copper–silver cell.

Now we discuss the electrochemical mechanisms and the operation of the copper–silver cell with spatially separated half-cells that are connected via a salt bridge as shown in Figure 3.31. We start with the initial state, where the complete electric circuit of the galvanic cell is not yet closed – that is, the two half-cells are completely separated since the salt bridge is not in place.

The copper anode is immersed into an aqueous copper(II) nitrate solution. Analog, the silver cathode is dipped into an aqueous silver(I) nitrate solution. Both electrodes are electrically externally connected via a wire and a load. At the copper electrode, $Cu^{2+}$ ions are entering the solution according to the oxidation process:

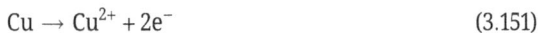

$$Cu \rightarrow Cu^{2+} + 2e^- \tag{3.151}$$

The amount of Cu ions in the anodic half-cell is larger than the amount of Ag ions in the aqueous silver(I) nitrate solution in the cathodic half-cell since Cu is the less noble element compared to Ag. Therefore, within the Cu electrode the negative charge is larger than within the Ag electrode. Hence, a potential appears pushing electrons via the external electrical wiring toward the Ag electrode.

In the anodic half-cell, that is, in the aqueous copper(II) nitrate solution a surplus of $Cu^{2+}$ ions arises. The copper(II) nitrate solution becomes more and more positively charged, until the solution of further copper ions from the anode comes to a stop since the electric field between the anode and the solution is balancing the solution pressure.

On the other hand, in the cathodic half-cell, that is, in the aqueous silver(I) nitrate solution a surplus of $NO_3^-$ ions arises. The excess electrons in the Ag electrode react with $Ag^+$ ions in the silver(I) nitrate solution, and neutral Ag atoms are deposited onto the surface of the silver cathode (also called plating). This is the reduction process according to the following formula:

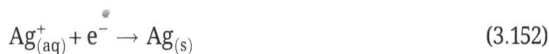

$$Ag^+_{(aq)} + e^- \rightarrow Ag_{(s)} \tag{3.152}$$

The silver(I) nitrate solution becomes more and more negatively charged, until the deposition of further silver ions onto the cathode comes to a stop since the electric field between the solution and the cathode keeps the electrons in the cathode away from the $Ag^+$ in the solution. Hence, we face a shortage of electrons at the silver electrode, and at the copper electrode a surplus of electrons occurs, that is, the Ag electrode is the positive terminal (cathode), and the Cu electrode is the negative terminal (anode) of the copper–silver cell.

The corresponding heterogeneous redox reaction that occurs in the copper–silver cell is given by

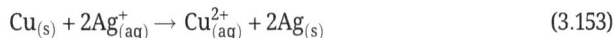

$$Cu_{(s)} + 2Ag^+_{(aq)} \rightarrow Cu^{2+}_{(aq)} + 2Ag_{(s)} \tag{3.153}$$

The Ag ions are singly positively charged, and the Cu ions exhibit a double positive charge. The oxidation number of silver (+1) on the reactants side (i.e., the left side) of formula (3.146) changes to zero on the right-hand products side. The oxidation number of copper changes from zero on the reactants side to +2 on the products side; this shows that copper is oxidized since the oxidation number increases during the reaction.

More precisely, taking into account the copper(II) nitrate solution in the anodic half-cell and the silver(I) nitrate solution in the cathodic half-cell, the redox reaction of the copper–silver cell can be written as follows:

$$Cu_{(s)} + (2Ag^+, NO_3^-)_{(aq)} \rightarrow (Cu^{2+}, NO_3^-)_{(aq)} + 2Ag_{(s)} \tag{3.154}$$

However, the negative $NO_3^-$ ions can be ignored; they are only spectator ions that play no role for the redox reaction.

At this state, the electrodes of the copper–silver cell act as a terminal for the electrons. Although the electrodes are externally connected via wiring and a load, the cell is not working since the salt bridge is not yet installed, that is, the half-cells are still separated, and the electric circuit is not closed. If the electrodes are only connected via external wiring, a potential or voltage appears between the two terminals, but no current can flow.

The cell voltage of the copper–silver cell is given by the difference of the half-cell potentials:

$$E = \Delta\Phi = \Phi_{Ag} - \Phi_{Cu} \tag{3.155}$$

For the copper–silver cell, the half-cell reactions are

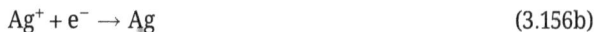

$$Cu \rightarrow Cu^{2+} + 2e^- \tag{3.156a}$$

$$Ag^+ + e^- \rightarrow Ag \tag{3.156b}$$

The corresponding standard potentials (measured in relation to the SHE) are $\Phi_{Cu} = 0.35$ V for the oxidation part and $\Phi_{Ag} = 0.81$ V for the reduction part, respectively. Here, $\Phi_{Cu}$ and $\Phi_{Ag}$ are absolute values, that is, positive numbers. Equation (3.155) can also be written as a sum rather than a difference. In this case, one has to take the sum of the oxidation and reduction potentials that have different algebraic signs ($\Phi_{red} = -\Phi_{ox}$ or for our example $\Phi_{Ag} = -\Phi_{Cu}$, respectively). Both approaches are equivalent. One only has to take care about the consistency.

The salt bridge is not yet installed, that is, the electrical circuit is not closed, and the open-circuit voltage for the copper–silver cell is given by

$$E_{oc} = \Delta\Phi \, (I = 0) \tag{3.157}$$

For the copper–silver cell, the open-circuit voltage has a value of

$$E_{oc,298K} = 0.46V \tag{3.158}$$

This value holds for a temperature $T = 298$ K, a pressure $p = 1$ bar (100 kPa) and an activity of $a = 1$ or a concentration of 1 mol/L. For strongly diluted solutions, the activity is approximately equal to the concentration, and the standard state of a substance in a solution is 1 mol/L.

If the salt bridge is installed between the two half-cells, a path for ions to move between the two half-cells is provided, and the solutions within the half-cells are kept electrically neutral. The electrons also can start to flow, and a current appears that might be used in a load. Copper is oxidized, releasing electrons that flow through the wire to the silver electrode. At the surface of the silver electrode, the electrons meet the $Ag^+$ ions of the electrolyte in the cathodic half-cell (i.e., in the aqueous silver(I) nitrate solution). Hence, neutral silver atoms are deposited onto the electrodes surface. In total, silver ions from the silver(I) nitrate solution are deposited onto the silver electrode, while the copper electrode is being consumed, that is, more and more copper is dissolved in the solution of the anodic half-cell (i.e., in the aqueous copper(II) nitrate solution).

To keep the system running, the cations in the salt bridge migrate to the cathodic half-cell and the anions to the anodic half-cell. In the solution of the cathodic half-cell, the cations replace the silver ions that were deposited onto the silver electrode. The concentration of $Ag^+$ ions decreases in the aqueous solution of the cathodic half-cell, and the concentration of $NO_3^-$ ions increases. Therefore, to maintain the electrical neutrality, the salt bridge provides positive ions. In case of the copper–silver cell, the salt bridge contains potassium nitrate ($KNO_3$). Hence, positive $K^+$ cations moved toward the cathodic half-cell, where they keep the solution electrically neutral. On the other hand, negative $NO_3^-$ anions in the salt bridge move toward the anodic half-cell. In the anodic half-cell that contains more and more newly formed $Cu^{2+}$ cations originating from the Cu anode, the $NO_3^-$ ions from the salt bridge keep the solution electrically neutral.

The electrical work that is supplied by the copper–silver cell is given by

$$\delta W_{el} = -E \cdot dQ \tag{3.159}$$

If the redox reactions are completely executed, we get

$$W_{el} = -|z| \cdot F \cdot E \tag{3.160}$$

Here $z$ mol electrons are transported to the galvanic cell. $F$ is the Faraday constant. If the temperature and pressure are constant, the change of the molar Gibbs free energy $G_m$ is given by the electrical work $W_{el}$, that is,

$$\Delta G_m = W_{el} = -|z| \cdot F \cdot E \tag{3.161}$$

If $E > 0$ – this is the case for the copper–silver cell – then $\Delta G_m < 0$ and hence, the electrochemical reactions occur spontaneously, if the circuit of the cell is closed, that is, if the anode and the cathode are connected externally via wiring and the half-cells internally via the salt bridge.

The cell voltage can be related to the thermodynamic potentials. For the molar entropy $S_m$ we get the following equation:

$$\Delta S_m = -\left(\frac{\partial \Delta G_m}{\partial T}\right)_p = |z| \cdot F \cdot \left(\frac{\partial E}{\partial T}\right)_p \tag{3.162}$$

The molar enthalpy $H_m$ is given by

$$\Delta H_m = \Delta G_m + T \cdot \Delta S_m = -|z| \cdot F \cdot \left( E - T \cdot \left( \frac{\partial E}{\partial T} \right)_p \right) \tag{3.163}$$

Moreover, the molar volume $V_m$ is described by the following equation:

$$\Delta V_m = \left( \frac{\partial \Delta G_m}{\partial p} \right)_T = -|z| \cdot F \cdot \left( \frac{\partial E}{\partial p} \right)_T \tag{3.164}$$

The following brief discussion holds for the copper–silver cell as well as for any galvanic cell. If the electrical circuit is not closed via external wiring and/or a load or if the salt bridge is not installed yet, the open-circuit voltage $E_{oc}$ appears at the electrodes of the galvanic cell – see eqs. (3.157) and (3.158). If the electrical circuit is closed via external wiring and the salt bridge, a current can flow through the system, that is, electrons through the external wiring and ions through the salt bridge and the electrolytes. Then chemical energy – that was stored in the galvanic cell – directly converts into electrical energy.

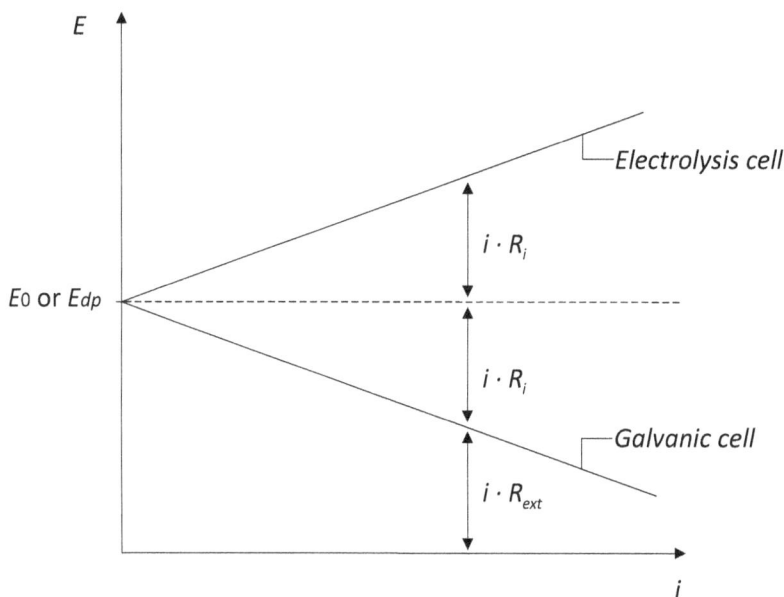

**Figure 3.32:** Voltages $E$ of galvanic and electrolysis cells in dependence on the electric current $i$.

The voltage at the electrodes of the galvanic cell depends on the current flow and the internal resistance $R_i$ of the cell (Figure 3.32):

$$E = E_{oc} - i \cdot R_i \tag{3.165}$$

The power that can be delivered by the galvanic cell is given by

$$P = E \cdot i = E_{oc} \cdot i - i^2 \cdot R_i \tag{3.166}$$

If we set $i \cdot R_i = x \cdot E$, eqs. (3.165) and (3.166) can be written as follows:

$$E = E_{oc} - x \cdot E \tag{3.167}$$

$$P = \frac{E_0^2}{R_i} \cdot x - \frac{E_0^2}{R_i} \cdot x^2 \tag{3.168}$$

Hence, the maximum power that can be delivered by the galvanic cell is given for $x = 0.5$, or $i \cdot R_i = E_0/2$. This is just the case, if the internal resistance $R_i$ and the external resistance $R_{ext}$ of the wiring/load are equal (see Figure 3.32), that is,

$$R_i = R_{ext} \Rightarrow P = P_{max} \tag{3.169}$$

Finally, it is mentioned that the voltage at the electrodes of an electrolysis cell similarly depends on the current flow and the internal resistance $R_i$ of the cell – only with opposite algebraic sign (Figure 3.32):

$$E = E_{dp} + i \cdot R_i \tag{3.170}$$

In eq. (3.170), $E_{dp}$ is called the decomposition potential. Electrolysis uses a direct electric current (DC) to drive an electrochemical reaction that does not occur spontaneously, that is, $\Delta G_m > 0$.

# 4 Batteries

Initially, the term "battery" originates from the technical terminology of the military sector. In the military, a battery comprises a clustered unit of artillery, either in a temporary and shiftable position on a battlefield or at a fixed position for defense, for instance, a fortress. Following this military definition, the term "battery" means the connection of several galvanic cells to one device for electrochemical energy storage. However, devices for electrochemical energy storage that are made from a single galvanic cell are also called a battery. This chapter is devoted to the various types of batteries for electrochemical energy storage.

If we speak about batteries, we have to keep in mind that we have to distinguish two families of batteries. These are, on the one hand, the so-called primary cells or primary batteries, and on the other hand, the so-called secondary cells or secondary batteries. The first ones are not rechargeable, while the latter ones can be recharged. Colloquially speaking, the primary cells are often simply called batteries, whereas the secondary cells are named rechargeable batteries or accumulators. The electrochemical reactions in primary batteries cannot be reversed. During operation – that is, when the electric circuit of the galvanic cells is closed – the chemical reactions in primary cells produce reaction products that are stable and no longer electrochemically active. Therefore, they cannot be transformed back to the educts, if an electrical current is applied in the reverse direction. In reality, this means that electrodes – usually the anodes – are destroyed or dissolved during battery operation. Hence, in primary batteries, electrochemical energy irreversibly converts into electrical energy. Secondary batteries, can be recharged, that is, if a reversed current is employed to the accumulator, it can be recharged again. Hence, electrical energy can be converted back into chemical energy. The electrochemical reactions can be reversed, since the products appearing during discharge of the accumulator are still electrochemically active, if electrical energy is introduced into the system. In secondary batteries, electrochemical energy can reversibly – or nearly reversible – be converted into electrical energy and vice versa.

There are also tertiary batteries, which are better known under the name fuel cells. To be precise, fuel cells are not electrochemical devices for energy storage but rather electrochemical energy converters.

Before we concentrate on the various types of batteries – that is, primary and secondary batteries – in Section 4.1, we summarize the most important physical, chemical and technical parameters that describe the properties of batteries.

https://doi.org/10.1515/9783111618531-004

## 4.1 Battery parameters

Most of the relevant physical, chemical and technical parameters that specify or describe the properties of batteries are nominal parameters. These nominal parameters are defined for standard conditions or nominal operation conditions, respectively. The standard conditions are well-defined sets of conditions for experimental setups and measurements that have to be established to allow for the comparison between different sets of data. The standard parameters can deviate more or less strongly from the actual parameters under normal real-life measurements and/or real-life operating conditions. Anyway, for comparison, the nominal parameters are extremely important and useful.

The standard conditions are defined, for instance, by the "International Union of Pure and Applied Chemistry (IUPAC)," the "National Institute of Standards and Technology (NIST)" in the USA, or "Deutsches Institut für Normung (DIN)" in Germany. The standard conditions are not generally accepted standards; that is, various organizations have established different alternative definitions for their standard reference conditions.

For instance, according to the IUPAC regulations, the standard temperature and pressure (STP) is defined as a temperature of 273.15 K (or 0 °C or 32 °F) and an absolute pressure of exactly $10^5$ Pa (i.e., 100 kPa or 1 bar). These values have been valid since 1982. Before 1982, the STP was defined as a temperature of 273.15 K (or 0 °C or 32 °F) and an absolute pressure of exactly 1 atm (i.e., $1.01324 \times 10^5$ Pa).

The NIST defines its standard by the normal temperature and pressure (NTP). The NTP is given by a temperature of 20 °C (i.e., 293.15 K or 68 °F) and an absolute pressure of 1 atm (i.e., $1.01324 \times 10^5$ Pa or 14.696 psi).

In Germany, the physics community uses the DIN 1343 standard, that is, a temperature of 273.15 K (i.e., 0 °C) and an absolute pressure of $1.01324 \times 10^5$ Pa (or 1 atm), while the chemistry community uses the IUPAC STP.

Therefore, we are dealing with several definitions of STPs. Often, these official STPs are significantly different from actual conditions within laboratories. For instance, the room temperature (RT) is usually assumed to be 298 K or 25 °C. Therefore, every once in a while, we also speak of standard laboratory conditions and mean, for instance, a temperature of 25 °C and a normal pressure of 100 kPa or 1 bar. However, this is not a real standard, since these values are inevitably related to a geographic location, as different locations all over the world vary in climate and especially in altitude, which has a significant impact on the pressure. Hence, what RT and pressure mean might vary. Therefore, it is always necessary to specify the temperature and pressure precisely when speaking of a certain physical or chemical parameter.

Concerning electrochemistry, the standard redox potential is usually given for an activity of $\alpha = 1$. For strongly diluted solutions, the activity is approximately equal to the concentration, and the standard state of a substance in a solution is 1 mol/L.

In the following sections, we list and briefly discuss the relevant battery parameters of interest.

### 4.1.1 Nominal current, discharging, and charging currents

Batteries are designed for a certain current called nominal current, $I_N$, which is measured in amps (A). The average discharging currents – and charging currents in the case of secondary cells or accumulators – are frequently related to the nominal current. Likewise, the permissible maximal discharging and charging currents are related to the nominal current.

If the discharging current is larger than the nominal current ($I_{discharge} > I_N$), then the electric charge that can be taken from a battery is smaller than the extractable electric charge under nominal operation conditions.

If the charging current is larger than the nominal current, the lifetime of a secondary battery can be reduced. Usually, the charging current, measured in amps, should not exceed 10% of the battery capacity, measured in amp-hours (Ah). For example, if a secondary cell has a capacity of $C = 30$ Ah, its charging current should not be larger than $I_{charge} = 3$ A. More explanations about the capacity of a galvanic cell are given in Section 4.1.3.

If rapid charging is necessary, the value of the charging current, measured in amps, should not exceed one-third of the capacity value, measured in amp-hours; for example, for a battery with a capacity of 30 Ah, the charging current for rapid charging should not exceed 10 A. Frequent rapid charging comes at the expense of the accumulator's lifetime.

### 4.1.2 Nominal discharging time

The nominal discharging time, $t_N$, is the mean discharge time under nominal operation conditions for what a battery is designed for – $t_N$ is given in hours.

If the discharging time is shorter than the nominal discharging time ($t_{discharge} < t_N$), then the electric charge that can be taken from a battery is smaller than the extractable electric charge under nominal operation conditions. Hence, rapid discharging is at the expense of the electric charge that can be taken from a battery.

### 4.1.3 Capacity and nominal capacity

The capacity of a battery – that is, primary or secondary cells – is the amount of electrical charge that can be stored within the battery, and it is measured in amp-hours (Ah), although the corresponding SI unit is amp-seconds (As). It should not be confused with the electrical capacity of a capacitor. The electrical capacity of a capacitor is given in amp-seconds per volt (As/V) or coulomb per volt, that is, in farad (1 F = 1 C/V = 1 As/V). By the way, batteries also have an electrical capacity.

The nominal capacity, $C_N$, is the capacity of the battery under nominal operation conditions, that is, for a certain temperature and discharge current, a specified discharge time, and so on. The product of the nominal current, $I_N$, and the nominal discharging time, $t_N$, gives $C_N$:

$$C_N = I_N \cdot t_N \tag{4.1}$$

At lower temperature ($T < T_N$, with the nominal temperature $T_N$), high discharging currents ($I_{discharge} > I_N$) and shorter discharging times ($t_{discharge} < t_N$), the capacity of any battery will be reduced compared to the nominal capacity $C_N$.

In general, the capacity of a battery significantly depends on the rate of discharge. The higher the discharge, the lower the capacity that can be delivered by the battery. Since different types of discharging are employed (e.g., the discharge of a battery at a constant discharge current or at constant power, and so forth), it is necessary to specify exactly the discharge rate, the discharging current, the final discharge voltage, and others to get a realistic information about the capacity and the nominal capacity of a battery.

The capacity of the battery is reduced with increasing discharge current due to an increasing voltage drop at the internal resistance of the battery. This implies a reduction in the output voltage of the battery, and the final discharge voltage is reached earlier than under nominal conditions. Furthermore, the limitation of the electrochemical reaction rates also causes a reduction in the battery capacity at excess discharging currents ($I_{discharge} > I_N$).

Moreover, operation temperatures have a significant impact on the capacity of batteries. At higher temperatures, the mobility of electrons and ions in the electrodes and the electrolyte is enhanced, respectively. This should be favorable; however, at enhanced temperatures, irreversible chemical side reactions at the electrodes and/or within the electrolyte can also occur, reducing the lifetime and capacity of the batteries.

Eventually, batteries are subject to deterioration, which downgrades the capacity of the battery. In real life, the charging and discharging of secondary cells are not completely reversible, since electrochemical side reactions that occur impede complete charging or discharging. For instance, in the case of lead–acid batteries, sulfating leads to the formation of lead(II) sulfate ($PbSO_4$) crystals – the so-called whiskers – at the surface of the electrodes. Whisker formation causes, on the one hand, a reduction of the active surface area of the electrodes, leading to a reduction in capacity, and on the other hand, they can even cause short circuits that finally destroy the lead–acid battery.

Moreover, it has to be mentioned that the interconnection of batteries also has an impact on the capacity. In a serial connection of several batteries, the voltages are summed up, while parallel connections result in a summation of the capacity.

### 4.1.4 C-rate

The C-rate or C-factor is a measure of the rate at which a battery can be discharged or charged. It is defined by the maximally allowed discharging or charging currents – that are given for certain boundary conditions – normalized to the nominal capacity $C_N$ of the battery:

$$C = \frac{I_{max}}{C_N} \tag{4.2}$$

Usually, the C-rate is measured in 1/hours ($h^{-1}$), although the corresponding SI unit is 1/second ($s^{-1}$). It is always positive; hence, only from the context one can state whether it describes a charge or discharge process. One has to be careful not to confuse the C-rate with the capacity $C$ of a battery.

The C-rate gives the reciprocal value of the time for that a battery can be discharged (or charged) with the maximal discharging current (or charging current). For example, a certain battery with a capacity of 1 Ah and a maximal discharge current of 4 A yields a C-rate of 4 $h^{-1}$. Hence, the battery can be discharged in 1/4 h or 15 min with a current of 4 A. In other words, with a discharging current of 4 A, four batteries of the mentioned type can be discharged within 1 h. On the other hand, for the same battery, a charging current of 500 mA results in a C-rate of 0.5 $h^{-1}$. This means that in 1 h a charging current of 500 mA increases the charge state of the (secondary) battery by 50%.

If the C-rates are high ($C \geq 1\,h^{-1}$), one has to carefully charge the batteries to avoid overcharging, overheating and damage. Batteries can be even set on fire or explode, if the C-rates are too high. Therefore, to prevent overcharging and damage, the charging process must be monitored carefully with regard to temperature or terminal voltage.

### 4.1.5 Nominal voltage

The nominal voltage $U_N$ (V) – also identified by $E_N$ (V) – is the mean voltage that appears during the discharge of a galvanic cell under nominal operation conditions. $U_N$ is an approximation value. For a battery – that is, the series connection of several galvanic cells – the nominal voltage $U_N$ is given by the total voltage of the galvanic cells connected in series and measured under nominal operation conditions.

The nominal voltage $U_N$ – or $E_N$ – of a battery or galvanic cell is always smaller than the corresponding open-circuit voltage $U_{oc}$ (or $E_{oc}$).

$$U_{oc} > U_N \ (\text{or } E_{oc} > E_N) \tag{4.3}$$

The open-circuit voltage is dependent on the type, or rather the materials, of the electrodes and the galvanic series. It can be calculated using the Nernst equations (3.119) or (3.136). The Nernst equation also depends on the materials of the electrodes, or rather the galvanic series. Moreover, it depends on the concentrations of the aqueous solutions in the particular half-cells and on the temperature. Therefore, calculations of the Nernst equation have to be carried out for standard or nominal operation conditions (e.g., at $T = 298.15$ K, $p = 1.01324 \times 10^5$ Pa).

### 4.1.6 Nominal energy density and nominal specific or volumetric energy densities

The nominal energy density $W_N$ (Wh) is the amount of energy that is stored in a battery under nominal operation conditions. It is given by the product of the nominal capacity $C_N$ and the nominal voltage $U_N$, that is,

$$W_N = C_N \cdot U_N \tag{4.4}$$

Usually, the amount of energy that is stored in a battery is related to the mass $m$ or to the volume $V$ of the battery. In the former case, we speak about the specific or gravimetric energy density $w_{NM}$ of a battery, where the physical unit of $w_{NM}$ is given in Wh/kg:

$$w_{NM} = \frac{W_N}{m} = \frac{C_N \cdot U_N}{m} \tag{4.5}$$

If the nominal amount of energy that is stored in a battery is related to its volume, we speak about the volumetric energy density $w_{NV}$ of a battery, that is:

$$w_{NV} = \frac{W_N}{V} = \frac{C_N \cdot U_N}{V} \tag{4.6}$$

The physical unit of $w_{NV}$ is given in Wh/L or Wh/dm$^3$ (where dm$^3$ stands for cubic decimeter); that is, the volume is given in L or dm$^3$.

### 4.1.7 Nominal power and nominal specific or volumetric power densities

The nominal power $P_N$ (W) is given by the product of the nominal voltage $U_N$ and the nominal current $I_N$; that is:

$$P_N = U_N \cdot I_N = \frac{U_N \cdot C_N}{t_N} \tag{4.7}$$

According to eq. (4.1), the nominal current is given by $I_N = C_N/t_N$, where $t_N$ is the nominal discharging time.

Usually, the power of a battery is related to the mass $m$ or to the volume $V$ of the battery. If power is related to the battery's mass, then we speak about the specific or gravimetric power density $p_{NM}$ of a battery, and $p_{NM}$ is given in W/kg:

$$p_{NM} = \frac{P_N}{m} = \frac{U_N \cdot C_N}{t_N \cdot m} \tag{4.8}$$

If power is related to the volume of the battery, then we speak about the volumetric power density $p_{NV}$ of a battery, and $p_{NM}$ is given in W/L or W/dm$^3$:

$$p_{NV} = \frac{P_N}{V} = \frac{U_N \cdot C_N}{t_N \cdot V} \tag{4.9}$$

### 4.1.8 Ragone diagram

The Ragone diagram – or Ragone chart – is named after the US American metallurgist David V. Ragone. It is used to compare the performance of various batteries – and other energy storage devices – with regard to their power density and energy density characteristics. Usually, for this comparison, the gravimetric power and energy densities are used. As shown in Figure 4.1, the Ragone diagram is logarithmically scaled on both axes; the gravimetric power densities are plotted on the vertical axis, and the gravimetric energy densities on the horizontal axis (and sometimes vice versa).

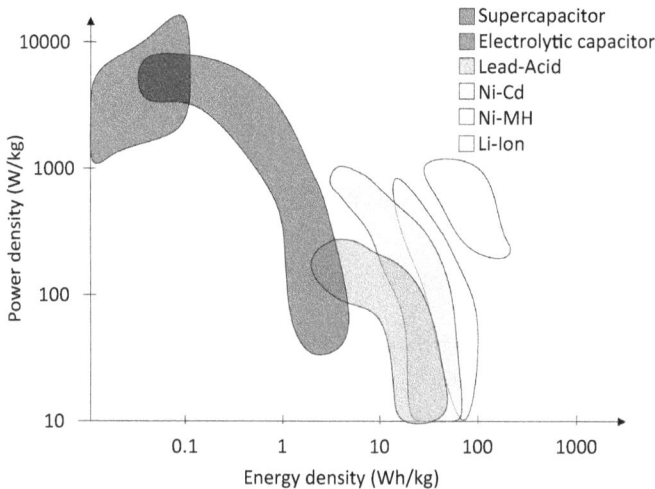

**Figure 4.1:** Ragone diagram with gravimetric power and energy densities.

The horizontal axis describes how much energy is available for a certain storage device or battery, and the vertical axis contains the information about how quickly that energy can be delivered. A certain point in a Ragone diagram represents the period during which the specific gravimetric energy of a considered device can be delivered at a specific gravimetric power. This period (in hours) – that is, the nominal discharging time $t_N$ – can be calculated from the ratio of the (gravimetric or volumetric) energy and power densities:

$$t_N = \frac{E_N}{P_N} = \frac{E_{NM}}{P_{NM}} = \frac{E_{NV}}{P_{NV}} \tag{4.10}$$

According to eq. (4.10), the isocurves, that is, the curves of constant values in the Ragone diagram are straight lines with unity slopes. For instance, if a battery should power a not-too-bright flashlight over a longer period, we need high energy densities but not so much power (or power densities). On the other hand, fast electronic switches do not need too much energy (or energy densities) but require high power densities to provide very short switching times. The appropriate battery or storage devices for the first application typically would be located in the lower right area of the Ragone diagram, whereas the battery or storage devices for the second application should be located in the upper left area of the Ragone diagram.

## 4.1.9 Discharging and charging

During the discharging process of an electrochemical energy storage system (e.g., a battery or a supercapacitor), chemical energy is widely transformed into electrical energy. Here, chemical energy is the form of energy that is stored in the system. Electrical energy can be extracted at the terminals of the energy storage system or battery.

The transformation of chemical to electrical energy is – of course – a lossy process. During the discharge process, thermal and electrical losses can occur, as well as chemical losses. Thermal losses can be attributed, for instance, to heat dissipation during the electrochemical reactions. Chemical losses can be related to chemical side reactions that do not contribute to the transformation of chemical energy to electrical energy. Electrical losses can be ascribed, for instance, to ohmic losses, contact resistances or leakage currents.

During the charging process of an electrochemical energy storage system (e.g., a secondary battery or a supercapacitor), electrical energy is widely transformed into chemical energy – now we just face the opposite reaction compared to the discharging process. Here, electrical energy is introduced to the electrochemical storage device at the terminals of the system – that is, a current is flowing through the system – and provokes chemical or electrochemical reactions that come to a stop when the current flow is interrupted. At this point, the former electrical energy is stored as chemical energy within the system.

Of course, the charging process is also a lossy process. Again, thermal and electrical losses can occur, as well as chemical losses, in a similar manner to the discharging processes.

The peculiarities of discharging and charging processes – for example, overcharging and overdischarging – can have a strong impact on the lifetime and/or recovery of the electrochemical energy storage system.

### 4.1.10 Efficiencies

The efficiency of batteries is defined in two ways; that is, we distinguish an ampere-hour efficiency ($\eta_{Ah}$) and a watt-hour efficiency ($\eta_{Wh}$). The ampere-hour and watt-hour efficiencies are defined by the following equations:

$$\eta_{Ah} = \frac{\displaystyle\int_0^{t_{discharge}} i_{discharge}(t)dt}{\displaystyle\int_0^{t_{charge}} i_{charge}(t)dt} \tag{4.11}$$

$$\eta_{Wh} = \frac{\displaystyle\int_0^{t_{discharge}} u_{discharge}(t) \cdot i_{discharge}(t)dt}{\displaystyle\int_0^{t_{charge}} u_{charge}(t) \cdot i_{charge}(t)dt} \tag{4.12}$$

In eqs. (4.11) and (4.12), $i_{charge}(t)$ and $i_{discharge}(t)$ are the time-dependent charging and discharging currents; $u_{charge}(t)$ and $u_{discharge}(t)$ are the time-dependent charging and discharging voltages, and $t_{charge}$ and $t_{discharge}$ are the charging and discharging times, respectively. With mean charging and discharging currents $I_{charge}$ and $I_{discharge}$ and the mean charging and discharging voltages $U_{charge}$ and $U_{discharge}$, we can reduce eqs. (4.11) to the following approximated equation for the ampere-hour efficiency $\eta_{Ah}$:

$$\eta_{Ah} \approx \frac{I_{discharge} \cdot t_{discharge}}{I_{charge} \cdot t_{charge}} \tag{4.13}$$

For the approximated watt-hour efficiency $\eta_{Wh}$, we obtain:

$$\eta_{Wh} \approx \frac{U_{discharge} \cdot I_{discharge} \cdot t_{discharge}}{U_{charge} \cdot I_{charge} \cdot t_{charge}} \tag{4.14}$$

Since the charging voltage is larger than the discharging voltage, that is, $U_{charge} > U_{discharge}$, due to an internal voltage drop in the battery, we obtain:

$$\eta_{Ah} > \eta_{Wh} \qquad (4.15)$$

The reciprocal ampere-hour efficiency, $1/\eta_{Ah}$, is called the charge factor.

## 4.1.11 Lifetime

The lifetime of a battery is mainly determined by the deterioration of the electrodes and the electrolyte. As per definition, the end of life of such an electrochemical energy storage device is accomplished, if the capacity of the battery becomes smaller than 80% of the nominal capacity, that is,

$$C < 0.8 \cdot C_N \qquad (4.16)$$

In the case of a nonrechargeable primary battery, the lifetime of the battery refers to the duration the load can run on a fully charged battery – where fully charged means that the capacity of the battery is greater than 80% of the nominal capacity (i.e., $C > C_N$). The lifetime might be also called calendrical lifetime. However, in the case of a rechargeable secondary battery, the lifetime of the battery can have two meanings. Besides the calendrical lifetime, we can also refer to the total number of charging/discharging cycles the secondary battery can withstand until it fails – or better, until the capacity of the secondary battery becomes less than 80% of the nominal capacity. With regard to this latter meaning, the lifetime of a (secondary) battery might be called the cycle life or cycle lifetime of the battery. In the case of a primary battery, the calendrical lifetime equals the cycle lifetime, since the battery lasts for just one cycle. The cycle life of a secondary battery depends primarily on the depth of discharge during the single charging/discharging cycles. In general, fast charging/discharging processes can increase unwanted chemical side reactions, and, hence, very often result in a reduced lifetime of the battery.

## 4.1.12 Self-discharge

Internal chemical reactions within batteries can reduce the stored charge, that is, the capacity of the batteries, even under open-circuit conditions, when the electrodes are not externally connected via wiring and load. Self-discharge is a phenomenon that reduces the so-called shelf life of a battery. Shelf life describes the duration of how long a battery will keep its desired performance after the production, that is, how long a battery can be stored without significant deterioration until it is used. Self-discharge directly starts after the end of production of a battery and results in a less-than-full charge when we make use of it and the battery comes into operation. However, the extent of self-discharge can vary considerably depending on the type of battery. Of course, the development of batteries always includes the optimization of the battery with regard to self-discharge that is as low as possible. However, primary batteries usually have lower self-discharge tendencies than secondary batteries, since they are intended for a nonrecurring discharge cycle without a chance of recharging.

Self-discharge can have various reasons, for example, thermal or electrical reasons, and various reasons in combination. The battery type, the ambient temperature, the state of charge, and other external and internal factors can have a strong impact on the self-discharge characteristics of a battery. Self-discharge is always related to a chemical reaction (or to various chemical reactions); therefore, the temperature – that is, the general energy insertion – plays an important role. In general, one can say that higher temperatures assist a faster discharge of the battery.

## 4.2 Historically important batteries

Before we focus on the description and properties of technically important batteries, in the following sections, we first take a brief look at some historic milestones that mark the development of battery technology.

We start with the voltaic pile and the Daniell cell. The voltaic pile is the ancestor of our modern batteries. It was invented at the beginning of the nineteenth century and initiated a large variety of discoveries and developments in the context of electrochemistry – and especially for the development of battery technologies. Until the development of electrical generators in the last quarter of the nineteenth century, the electrical power supply in the field of research as well as in the electric industry was dependent on the voltaic pile and its improvements, for instance, the Daniell cell. Hence, one can say that nearly the entire nineteenth century was electrically powered by the voltaic pile and its successors. The Daniell cell was invented in 1836. Like the voltaic pile, it consists of copper and zinc electrodes, but in comparison to the voltaic pile, its performance was much better. All the early batteries were primary cells, that is, not rechargeable.

Another important milestone concerning the development of battery technology was the invention of the Leclanché cell in 1866, since in 1886, it was further developed into the first dry-cell battery. It is also a primary cell. The negative terminal of the Leclanché cell, that is, the anode, was also made from zinc, similar to the anodes in the voltaic pile or the Daniell cell mentioned earlier. However, the positive terminal, that is, the cathode, was made from carbon (graphite) and manganese dioxide. The manganese dioxide works as a depolarizer, changing its oxidation state in a charge-transfer step of electrochemical reactions. Since the optimized Leclanché cell was the first dry cell, it was much more useful than earlier battery types; hence, it was very successful. Enhancements of the Leclanché cell – that is, the zinc–carbon cells – were well important into the twentieth century, and even play a minor role today.

The first rechargeable battery, or secondary cell, was the Planté cell, better known under the name lead–acid battery. It was invented already in 1859. The lead–acid battery could have been a technological sensation; however, it was invented too early, since the generators for the charge process were not yet invented at that time. Hence, the industrial utilization of the lead–acid cell started only 20 years after its invention. Anyway, until now modern types of the lead–acid battery are very important (e.g., as a starter battery in the automotive industry).

### 4.2.1 Voltaic pile and Daniell cell

The voltaic pile can be regarded as the first electrical battery, and it was invented in 1799/1800 by the Italian physicist and chemist Alessandro Volta, who did pioneering research, especially in the field of electricity. Volta's pile was the first device that delivers significant electrical currents over a longer period. In fact, the voltaic pile was really a groundbreaking device, promoting a variety of discoveries and developments in the nineteenth century – for instance, electrolysis, that is, the electrical decomposition of water into hydrogen and oxygen. Nearly during the entire nineteenth century, the early electrical industry was powered by the voltaic pile, and especially by its technological enhancements (e.g., the Daniell cell was the first useful battery).

Eventually, the voltaic pile was a result of a scientific debate that was stimulated by the discoveries of Luigi Galvani, who was an Italian physician, physicist, biologist and philosopher. In 1780, Galvani dissected a frog. During the dissection, the frog was affixed to a brass hook. When Galvani touched the leg of the frog with an iron scalpel, the leg vellicated. Galvani's interpretation of this experimental observation was that the energy that drove such contractions originated from the leg itself, and hence, he called it the "animal electricity". However, his friend and fellow scientist Alessandro Volta disagreed. Through experimental analysis, Volta showed that this phenomenon was caused by the two different types of metals that came in contact via the moist intermediary tissue of the frog's leg, that is, in modern words the two different metals were in contact via an electrolyte. His interpretation of Galvani's observations was published in 1791. Finally, as a result of the disagreement over the galvanic response suggested by Galvani, in 1799/1800, Volta invented the voltaic pile, and with this invention, Volta proved that electricity can be generated chemically – or, in other words, chemical energy can be transformed into electrical energy.

The voltaic pile consists of copper and zinc disks separated by a sheet of cardboard or cloth soaked in a salt solution (or brine), that is, a galvanic cell is established with copper and zinc electrodes and an electrolyte in between. These galvanic cells are stacked on top of each other and connected in series to increase the voltage of the voltaic pile. The zinc electrode of a single galvanic cell is in direct contact with the copper electrode of the neighboring cell (see Figure 4.2); hence, the overall voltage of the voltaic pile is the sum of the voltages of each single cell. Theoretically, a voltage of about 1.1 V is expectable in case of the copper–zinc electrode pair due to the standard potentials (voltages) of these metals. Stacking $x$ galvanic cells, the voltage of the voltaic pile is $E^0_{pile} = x \cdot E^0_{cell}$. Volta believed that the electromotive force – that is, the voltage developed by a source of electrical energy such as a battery or a dynamo – occurred at the contact between the two metals. Therefore, Volta's original piles had an extra copper disk in contact with the upper zinc electrode and an extra zinc disk in contact with the bottom copper disk (Figure 4.2). Volta did not see that the electrolyte – that is, the brine – was a significant player in these phenomena. However, it was rather soon realized that water in the electrolyte (brine) was involved in the chemical reactions of the voltaic pile.

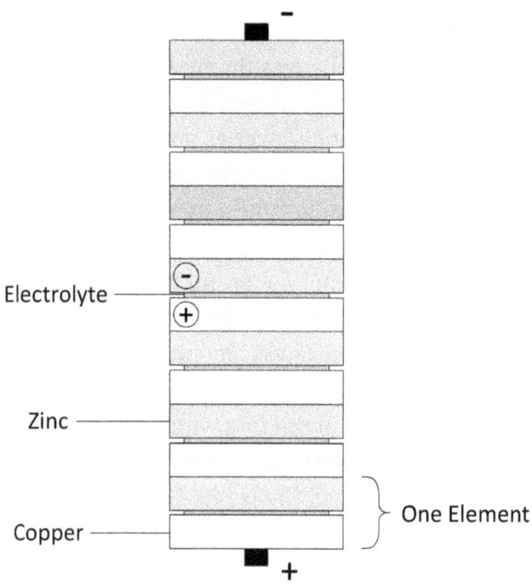

**Figure 4.2:** Voltaic pile: galvanic copper–zinc elements were stacked in series. Between the copper and zinc plates, a cardboard or cloth soaked in a salt solution provides the electrolyte.

In detail, the following chemical reactions occur in the voltaic pile with zinc and copper electrodes. Both the Zn and Cu electrodes are electrically connected by an external wire. Zinc is the less noble metal; hence, at the zinc electrode – the anode, that is, the negative pole – oxidation occurs:

$$Zn \rightarrow Zn^{2+} + 2e^- \tag{4.17}$$

Metallic zinc at the surface of the zinc electrode dissolves in the form of doubly positively charged $Zn^{2+}$ ions into the electrolyte, leaving two negatively charged electrons ($e^-$) behind in the metal (therefore, the Zn anode is negatively charged) that can migrate through the external wire to the cathode.

The nobler Cu electrode is the positive cathode, where the reduction occurs. However, various reactions might occur at the cathode depending on the conditions. If the Cu electrode is not highly polished, you can always find oxidation products at the cathode surface, that is, copper oxides originated from the contact of the Cu electrode with air. First, at the cathode surface, Cu ions that are located can be reduced by the electrons that migrated from the anode to the cathode, that is:

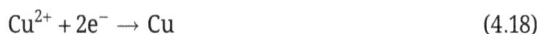

$$Cu^{2+} + 2e^- \rightarrow Cu \tag{4.18}$$

However, in the voltaic pile, $Cu^{2+}$ ions are exhausted quite soon, and another reaction becomes dominant at the Cu electrode. At the surface of the cathode, two positively charged hydrogen ions ($H^+$) from the electrolyte are neutralized by two electrons ($e^-$) delivered via the external electric wire, and eventually, a neutral, uncharged hydrogen molecule ($H_2$) is formed:

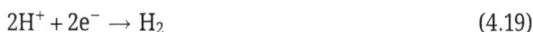

$$2H^+ + 2e^- \rightarrow H_2 \tag{4.19}$$

The $H_2$ molecules formed at the Cu surface of the cathode by the reduction reaction bubbles away as hydrogen gas. Finally, the overall redox reaction of the dominant chemical reaction in the voltaic pile is given by

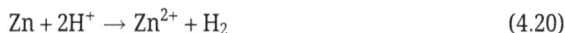

$$Zn + 2H^+ \rightarrow Zn^{2+} + H_2 \tag{4.20}$$

that is, an electronic transfer occurs from metallic Zn in the anode to the hydrogen ions in the electrolyte.

Furthermore, to a minor degree, a reaction with oxygen from the air can also occur at the Cu electrode:

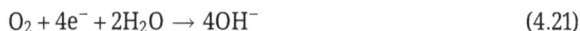

$$O_2 + 4e^- + 2H_2O \rightarrow 4OH^- \tag{4.21}$$

This reaction occurs mainly for a fresh setup of the pile, since in the beginning, the electrolyte is saturated with oxygen from the air. When the pile is installed and the external electric current is flowing, this oxygen gets rapidly exhausted. New oxygen from the air diffuses only very weakly from the sides into the interior of the sheet of cardboard or cloth soaked with electrolyte; hence, this reaction is not very significant for the operation of the battery. Although oxygen can enhance the effectiveness of the voltaic pile (at least at the beginning of its operation), it also has some drawbacks, as it can consume Zn from the anode due to a corrosive reaction:

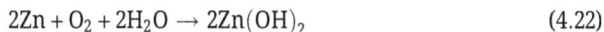

$$2Zn + O_2 + 2H_2O \rightarrow 2Zn(OH)_2 \tag{4.22}$$

Zinc hydroxide ($Zn(OH)_2$) is a powder that does not easily dissolve in water, and it segregates in an aqueous solution or the electrolyte. Hence, the Zn atoms that are consumed according to the corrosion reaction (4.22) no longer contribute to the delivery of electrical current by the voltaic pile; rather, the Zn anode can be destroyed by the corrosive effect of oxygen and the segregation of zinc hydroxide.

Anyway, the voltaic pile continuously produces electricity, and it delivers a quite stable current during the discharge process – at least from the nineteenth-century point of view. Moreover, its loss of charge over time when not in use is quite low. However, the maximal voltages that are achievable are limited due to the internal resistance of the device. Each galvanic cell of the voltaic pile, consisting of a zinc|electrolyte|copper sequential arrangement, generates about 0.76 V with a brine electrolyte. The theoretical voltage of 1.10 V that should occur for a Zn–Cu electrode pair is, in reality, not achieved, since the reduction reaction is dominated by eqs. (4.19) or (4.21), respectively, and not by eq. (4.18). Moreover, the voltaic pile was not able to deliver currents for a longer time duration.

If we are dealing with the oxidation and reduction reactions (3.141) and (3.142) that occur at the beginning of the operation and current delivery of the voltaic pile, its cell notation is given as follows:

$$Zn_{(s)} \mid Zn^{2+}_{(aq)} \mid Cu^{2+}_{(aq)} \mid Cu_{(s)} \tag{4.23}$$

The phase boundaries are marked by vertical lines; that is, the boundaries between the Zn or Cu electrodes and the electrolytes in between. The cell notation of the voltaic pile reflects a galvanic cell, where the two half-cells are not spatially separated from each other.

Significant improvements could be achieved with the so-called Daniell cell (or Daniell element) that overcame the limitations of the voltaic pile. It was named after the English chemist and physicist John Frederic Daniell. The Daniell cell faces similar chemical reactions to the voltaic pile; however, the reactions in the Daniell cell last much longer, since cuprous salt was added to the cell. Therefore, the copper ion supply in the liquid solution is much larger than in the case of the voltaic pile, where the amount of copper ions was exhausted quite fast – as mentioned earlier. As a consequence of this improvement, the Daniell cell is dominated by the redox reaction, where Zn is oxidized and Cu is reduced. Hence, according to the galvanic series and the standard potentials of Zn and Cu, the Daniell cell can deliver a voltage of 1.1 V.

So, in 1836, the English chemist and physicist John Frederic Daniell developed an electrochemical cell that had some similarities with the voltaic pile, since zinc and copper electrodes were used. However, both electrodes were located in compartments or half-cells, which were separated by a salt bridge or a porous membrane that allows ionic transport and prevents the mixture of the two different electrolytes. Electronic transportation – that is, electrical currents – does, of course, not internally occur within the cell.

One of the early original designs of the Daniell cell is sketched in Figure 4.3. The following description of this cell is given in the own words of John Frederic Daniell (*Daniell's Introduction to Chemical Philosophy*, 2nd Edition, London, John W. Parker, West Strand, 1843, page 505):

One of the cells of the constant battery is here represented. a b c d is a copper cylinder, in which is placed a smaller cylinder of porous earthenware. Upon the upper part of the copper cylinder rests a perforated colander, i k, through which the earthenware cylinder passes. l m is a cast rod of amalgamated zinc, resting upon the top of the interior cylinder by a cross piece of wood, and forming the axis of the arrangement. The cell is charged by pouring into the earthenware cylinder water acidulated with one-eighth part of its bulk of oil of vitriol, the space between the earthenware tube and the copper being filled with the same acidulated water saturated with sulfate of copper; and a solid sulfate of copper being placed the colander.

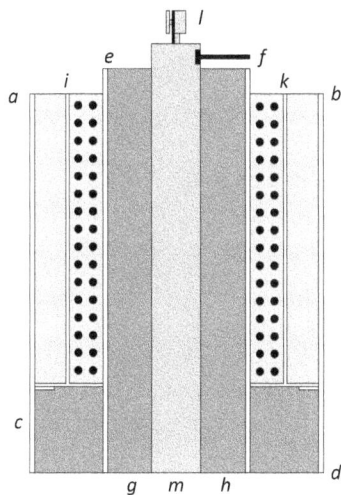

**Figure 4.3:** Original design of the Daniell cell, following Daniell's *Introduction to Chemical Philosophy*, 2nd Edition, London, John W. Parker, West Strand, 1843, page 505 (description in the text).

A more practical design of the Daniell cell is shown in Figure 4.4. Sometimes, this type of Daniell cell is also called a porous pot cell. Here, the Daniell cell consists of a zinc anode that is submerged in an earthenware pot filled with an aqueous zinc(II) sulfate ($ZnSO_4$) solution. The earthenware pot itself is immersed in a larger copper can that is filled with an aqueous copper(II) sulfate ($CuSO_4$) solution. The earthenware pot acts as a porous membrane, which allows ions to migrate through and prevents the mixture of the two different electrolytes in the anodic and cathodic half-cells that contain $Zn^{2+}/Zn$ and $Cu^{2+}/Cu$ redox couples, respectively. In operation, the electrons can migrate from the Zn anode to the Cu cathode via an external electrical connection.

The chemistry and operation of the Daniell cell will now be discussed in detail. For this purpose, we use a setup of the Daniell cell that is sketched in Figure 4.5, where the two half-cells – represented in cell notation by $Zn|Zn^{2+}$ and $Cu^{2+}|Cu$ – are connected via a salt bridge. Alternatively, the Daniell cell can be assembled with a porous membrane, as sketched in Figure 4.6.

According to such constructions, the Daniell cell can be described in cell notation as follows (three types of cell notations are listed):

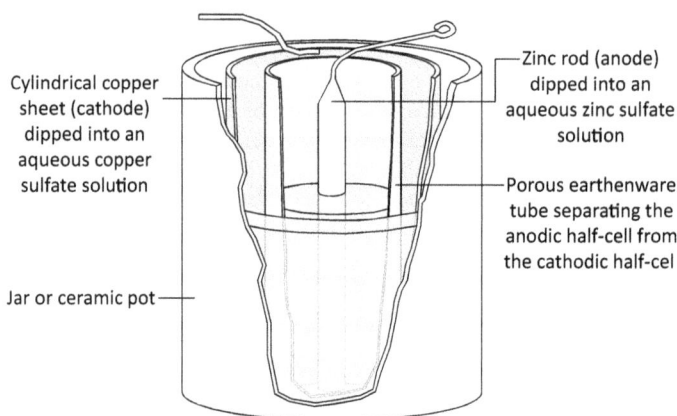

**Figure 4.4:** A more practical design of the Daniell cell – also called the porous pot cell.

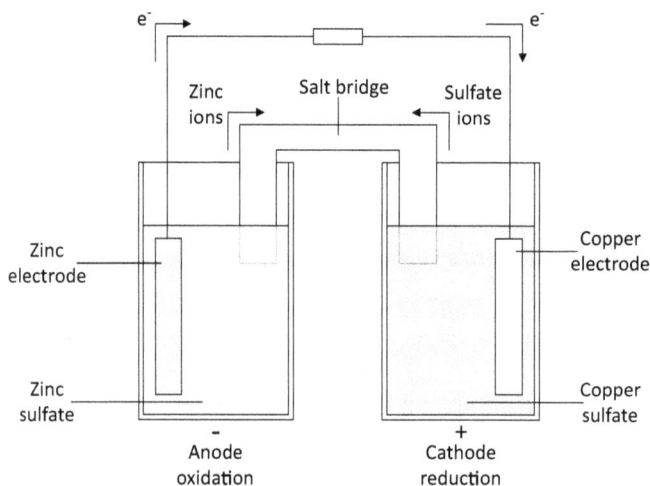

**Figure 4.5:** Sketch of the Daniell cell that illustrates the mechanisms of the discharge process in this galvanic cell. The separated half-cells are connected by a salt bridge.

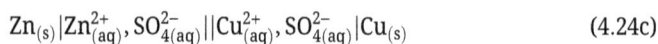

$$Zn_{(s)}|Zn_{(aq)}^{2+}||Cu_{(aq)}^{2+}|Cu_{(s)} \tag{4.24a}$$

$$Zn_{(s)}|ZnSO_{4(aq)}||CuSO_{4(aq)}|Cu_{(s)} \tag{4.24b}$$

$$Zn_{(s)}|Zn_{(aq)}^{2+}, SO_{4(aq)}^{2-}||Cu_{(aq)}^{2+}, SO_{4(aq)}^{2-}|Cu_{(s)} \tag{4.24c}$$

Again, phase boundaries are marked by the vertical lines. These are the boundaries between the Zn or Cu electrodes and the corresponding electrolytes in the half-cells, respectively. Since the electrolytes in the two half-cells are separated by a salt bridge (or alternatively by a porous membrane), that is, since they are not in direct contact,

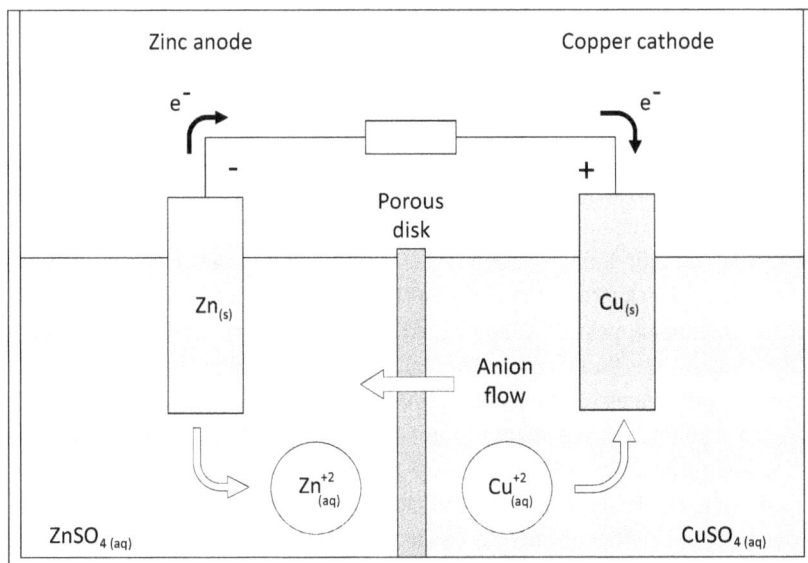

**Figure 4.6:** Sketch of the Daniell cell with a porous membrane that separates the half-cells.

two vertical lines mark their separation in cell notation. The cell notation (4.24a) exhibits the electrochemical players in the Daniell cell. The second cell notation (4.24b) emphasizes the different aqueous solutions in the containers of the half-cells, that is, zinc(II) sulfate solution in the container of the anodic half-cell and copper(II) sulfate solution in the container of the cathodic half-cell. Finally, cell notation (4.24c) emphasizes the dissociated zinc(II) and copper(II) sulfates in aqueous solutions within the corresponding half-cell containers. However, $SO_4^{2-}{}_{(aq)}$ ions in aqueous solutions do not play a role in the electrochemical redox reactions. Earlier, we called such ions spectator ions.

We should add that the Daniell cell was one of the first electrochemical or galvanic cells that used amalgamated zinc as an electrode, that is, an alloy of mercury and zinc, to reduce the corrosion of the zinc anode, when the battery was not in use – as we could see already from the earlier citation. Anyway, Zn is the main player in the redox reactions of the two half-cells consisting of copper and zinc electrodes, respectively. From the electrochemical point of view, mercury does not play a role. Hence, for the ongoing discussions, we simply speak about Zn electrodes.

So, the Daniell cell is characterized by the fact that two electrolytes – or conducting liquids – were used. A zinc metal strip is dipped in a solution of zinc(II) sulfate ($ZnSO_4$) in one container, and a copper metal strip is dipped in a solution of copper(II) sulfate ($CuSO_4$) in another container.

Now we discuss the electrochemical mechanisms and the operation of the Daniell cell with spatially separated half-cells connected via a salt bridge. We start with the initial state, where the complete electric circuit of the galvanic cell is not closed; that is, the salt bridge is not in place; hence, the two half-cells are completely separated.

The Zn anode is dipped into an aqueous zinc(II) sulfate solution, and the Cu cathode is dipped into an aqueous copper(II) sulfate solution. Both electrodes are electrically connected via a load by a wire. At the zinc electrode, more and more $Zn^{2+}$ ions enter the solution. This is the oxidation process according to formula (4.17). The amount of Zn ions in the electrolyte of the anodic half-cell is larger than the amount of Cu ions in the electrolyte of the cathodic half-cell, since Cu is the nobler element compared to Zn. Therefore, within the Zn electrode, the negative charge is larger than within the Cu electrode. Hence, a potential appears, pushing electrons via the electrical wiring toward the Cu electrode. Then, at the Cu electrode, the solution of Cu ions comes to a stop. The excess electrons in the Cu electrode react with $Cu^{2+}$ ions in the aqueous copper(II) sulfate solution, and neutral Cu is plated on the electrode surface. This is the reduction process according to formula (4.18). Hence, we face a shortage of electrons at the Cu electrode, and at the Zn electrode, a surplus of electrons occurs, that is, the Cu electrode is the positive terminal (cathode), and the Zn electrode is the negative terminal (anode) of the Daniell cell.

Within the anodic half-cell, the oxidation occurs, that is,

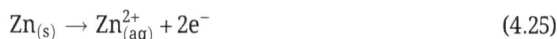

$$Zn_{(s)} \rightarrow Zn^{2+}_{(aq)} + 2e^- \tag{4.25}$$

And within the cathodic half-cell, the corresponding reduction occurs, that is,

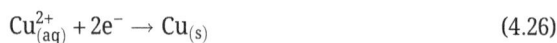

$$Cu^{2+}_{(aq)} + 2e^- \rightarrow Cu_{(s)} \tag{4.26}$$

Hence, the corresponding heterogeneous redox reaction that occurs in the Daniell cell is given by

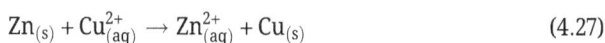

$$Zn_{(s)} + Cu^{2+}_{(aq)} \rightarrow Zn^{2+}_{(aq)} + Cu_{(s)} \tag{4.27}$$

More precisely, taking into account the zinc(II) sulfate solution in the anodic half-cell and the copper(II) sulfate solution in the cathodic half-cell, the heterogeneous redox reaction of the Daniell cell can be written as follows, keeping in mind that the $SO_4^{2-}{}_{(aq)}$ ions are spectator ions:

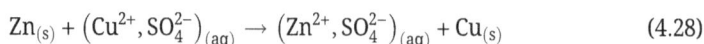

$$Zn_{(s)} + \left(Cu^{2+}, SO_4^{2-}\right)_{(aq)} \rightarrow \left(Zn^{2+}, SO_4^{2-}\right)_{(aq)} + Cu_{(s)} \tag{4.28}$$

At this stage, the electrodes act as a terminal for the electrons. Although the wire connects the electrodes via a load, nothing happens since the salt bridge is not yet installed. Both half-cells are spatially separated; if the electrodes are only electrically connected via wiring, a potential or voltage appears between the two terminals, but no current can flow.

In the anodic half-cell, that is, in the zinc(II) sulfate solution, a surplus of $Zn^{2+}$ ions arises. The zinc(II) sulfate solution becomes more and more positively charged until the dissolution of further zinc ions from the anode comes to a stop, as the electric field between the anode and the solution is balancing the solution pressure. On the other hand, in the cathodic half-cell, that is, in the copper(II) sulfate solution, a surplus of $SO_4^{2-}$ ions arises. Here, the copper(II) sulfate solution becomes more and more negatively charged until the deposition of further copper ions onto the cathode (also called plating) comes to a stop, as the electric field between the solution and the cathode keeps the electrons in the cathode away from $Cu^{2+}$ in the solution.

The cell voltage of the Daniell cell is given by the difference in the half-cell potentials:

$$E = \Delta\Phi = \Phi_{cathode} - \Phi_{anode} \tag{4.29}$$

And if the electric circuit is not closed (i.e., if the salt bridge is not installed yet), the open-circuit voltage is given by

$$E_{oc} = \Delta\Phi(I = 0) \tag{4.30}$$

For the Daniell cell, the open-circuit voltage at 298 K has the value of

$$E_{oc, 298K} = 1.0183V \tag{4.31}$$

Now we install the salt bridge between the two half-cells that serves as an ionic conductor. Hence, we get a closed circuit. The salt bridge provides a path for ions to move between the two half-cells and keeps the solutions within them electrically neutral. If the salt bridge connects the two half-cells, the electrons can also start to flow, and a current appears that might be used in a load. Zinc is oxidized, releasing electrons that flow through the wire to the copper electrode. At the surface of the copper electrode, the electrons meet the $Cu^{2+}$ ions of the electrolyte, that is, the copper(II) sulfate solution in the cathodic half-cell; hence, neutral copper atoms are deposited onto the electrode surface. In total, copper ions from the copper(II) sulfate solution are deposited onto the copper electrode, while the zinc electrode is being consumed, that is, more and more zinc is dissolved in the solution of the anodic half-cell.

To keep the system running, the cations in the salt bridge migrate to the cathodic half-cell. In the solution of the cathodic half-cell, the cations replace the copper ions that were deposited onto the Cu electrode. The concentration of $Cu^{2+}$ ions decreases in the solution of the cathodic half-cell, and the concentration of $SO_4^{2-}$ sulfate ions increases. Therefore, to maintain electrical neutrality, the salt bridge provides positive ions. For instance, these could be positive $Na^+$ or $K^+$ ions if the salt bridge contains solutions with sodium chloride (NaCl), potassium chloride (KCl), or potassium nitrate (KNO₃), as mentioned earlier. On the other side, anions in the salt bridge – for instance, negative $Cl^-$ or $NO_3^-$ ions – move toward the anodic half-cell, where they keep the solution that contains more and more newly formed $Zn^{2+}$ cations (originating from the Zn anode) that are electrically neutral.

Similarly, as for the copper–silver cell discussed in Section 3.5.2, the electrical work that is supplied by the Daniell cell is given by

$$\delta W_{el} = -E \cdot dQ \tag{4.32}$$

And if the redox reactions are completely executed, we obtain:

$$W_{el} = -|z| \cdot F \cdot E \tag{4.33}$$

where $z$ mol electrons are transported in the Daniell cell, and $F$ is the Faraday constant. If the temperature and pressure are constant, the change in the molar Gibbs free energy, $G_m$, is given by the electrical work, $W_{el}$, that is,

$$\Delta G_m = W_{el} = -|z| \cdot F \cdot E \tag{4.34}$$

If $E > 0$ – this is the case for the Daniell cell – then $\Delta G_m < 0$; hence, the electrochemical reactions occur spontaneously.

## 4.2.2 Leclanché cell

The Leclanché cell is a primary battery that was invented and patented in 1866 by the French chemist and engineer Georges Leclanché. The Leclanché cell featured a zinc anode and a graphite cathode with a surrounding manganese dioxide ($MnO_2$) depolarizer. A depolarizer is a substance that changes its oxidation state; that is, simply expressed, a depolarizer is involved in the formation or breaking of chemical bonds in a charge-transfer step of electrochemical reactions. The substance working as a depolarizer is located at the cathode – that is, it surrounds the graphite electrode – and picks up electrons during the discharge process. Like all galvanic cells and batteries, the initial version of the Leclanché cell was a wet cell that used a liquid electrolyte, that is, an aqueous solution of ammonium chloride ($NH_4Cl$).

Ammonium chloride is the ammonium salt of hydrochloric acid (HCl). In general, such chloride salts are often highly soluble in water – this applies to ammonium chloride, too. In water, $NH_4Cl$ dissociates into a positive ammonium ion ($NH_4^+$) and a negative chloride ion ($Cl^-$). This solution is a mild acid. Ammonium chloride dissociates according to the following dissociation reaction:

$$NH_4Cl_{(s)} \longrightarrow NH_{4(aq)}^+ + Cl_{(aq)}^- \tag{4.35}$$

The Leclanché cell was an enhancement of earlier developments in battery technology. The zinc electrode was adopted from the voltaic pile and the Daniell cell.

The graphite electrode with a depolarizer was adopted from the Poggendorff cell and the Fuller cell – the latter was a further development of the Poggendorff cell. The chemistry of both galvanic cells was, in principle, the same. Both cells use a zinc anode and a carbon (graphite) cathode. In both cases, the electrolyte was diluted sulfuric acid ($H_2SO_4$), and the depolarizer was chromic acid ($H_2CrO_4$). However, in the Poggendorff cell, the half-cells of the battery are not separated, while in case of the

Fuller cell both half-cells are separated (Figures 4.7 and 4.8). In addition, the Fuller cell has an improved amalgamated zinc electrode.

The Poggendorff cell was set up in a glass bottle with a long neck. A zinc plate (the anode) was placed between two carbon plates (the cathodes); a sketch is shown in Figure 4.7. In the Poggendorff cell, the electrolyte and depolarizer were mixed. When the Poggendorff cell was not in use, the zinc electrode was pulled out of the liquid mixture of the electrolyte and the depolarizer, and it was stored in the neck of the bottle. This was necessary since the chromic acid in the mixture is very aggressive and a strong, corrosive oxidizing agent that would dissolve the zinc electrode even when the primary battery is not in use. The Fuller cell avoids the problems with the chromic acid solution since it uses amalgamated zinc electrodes that are more resistant. It was assembled in a glass or glazed earthenware container containing the chromic acid solution and the graphite plate (cathode), as well as a smaller porous pot that was filled with diluted sulfuric acid and a zinc rod (anode). The setup of the Fuller cell with two separated half-cells is shown in Figure 4.8. The inner porous pot also contained a small amount of mercury (Hg) that forms an amalgam with the zinc; hence, the unwanted dissolution of the zinc electrode is reduced, and the anode gets more resistant.

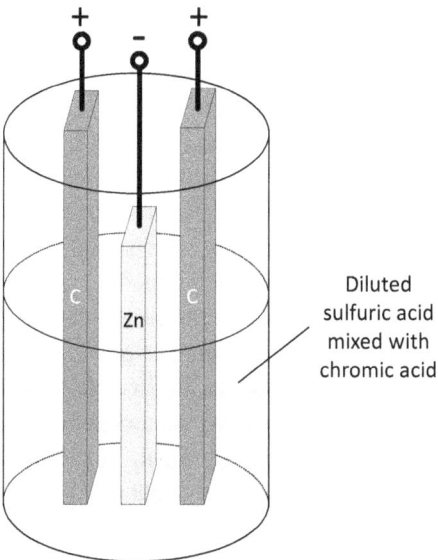

Diluted
sulfuric acid
mixed with
chromic acid

**Figure 4.7:** Sketch of the setup of the Poggendorff cell. A zinc plate (anode) is placed between two carbon plates (cathodes); the electrolyte – diluted sulfuric acid ($H_2SO_4$) – is mixed with the depolarizer – chromic acid ($H_2CrO_4$).

Both the Poggendorff cell and the Fuller cells are called chromic acid cells. Sometimes they are also named dichromate cells, since chromic acid can be made by acidifying a solution of sodium dichromate ($Na_2Cr_2O_7$) or potassium dichromate ($K_2Cr_2O_7$) with sulfuric acid. However, in principle, the Poggendorff and the Fuller cells contained no dichromate.

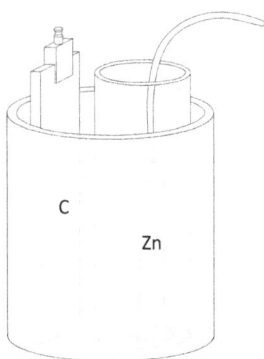

**Figure 4.8:** Sketch of the setup of the Fuller cell. The graphite plate (cathode) is placed in the outer glass or glazed earthenware pot. A zinc rod (anode) is placed in the inner porous pot. Both half-cells are separated from each other. The outer half-cell contains diluted sulfuric acid ($H_2SO_4$) mixed with the depolarizer (chromic acid ($H_2CrO_4$)). In the inner half-cell with the zinc rod, chromic acid is not added to the electrolyte.

Now we come back to the Leclanché cell, that uses manganese dioxide ($MnO_2$) as the depolarizer. The initial construction of the Leclanché cell is shown in Figure 4.9. A glass jar was used as the exterior container of the setup. The glass jar was filled with an aqueous solution of ammonium chloride ($NH_4Cl$). A smaller porous earthenware pot was also placed into the glass jar. The porous earthenware pot was densely filled with a mixture of powdered manganese dioxide ($MnO_2$) – the depolarizer – and some granulated carbon, that increases the conductivity. Furthermore, a carbon (graphite) rod was inserted into the porous earthenware pot. The graphite rod works as an electrode, that is, it is the positive terminal – that is, the cathodic side – of the Leclanché cell. However, the reduction reaction in the cathodic half-cell that is installed within the porous earthenware pot, occurs with the depolarizer – not at the graphite electrode, as will be discussed later. A zinc rod was used as the anode. It was dipped into the aqueous solution of ammonium chloride in the glass jar – along with the porous earthenware pot. The porous earthenware pot functions as a separator between the two half-cells of the Leclanché cell. However, the liquid solution – that is, the electrolyte – permeates through the pot; hence, the ionic contact to the graphite cathode could be established.

The internal resistance of the Leclanché cell is rather high, as the porous earthenware pot has a relatively large resistance to ion conductivity within the galvanic cell. Several modifications were developed to reduce the internal resistance of the Leclanché cell – for instance, the so-called agglomerate block cell and the sack cell.

In 1871, Leclanché replaced the porous earthenware pot with a pair of so-called agglomerate blocks that were attached to a carbon plate (the electrode) by rubber bands. The agglomerate blocks consist of a mixture of the depolarizer $MnO_2$ and binding agents. This mixture was pressed into ingot molds to form the agglomerate blocks. In the sack cell, the porous earthenware pot was replaced by a wrapping of canvas or sacking. Moreover, the surface of the anode was enhanced by replacing the zinc rod by a zinc cylinder. The internal resistance of the Leclanché cell could be reduced by these constructive improvements.

However, a real technological breakthrough was achieved several years later, when the Leclanché cell was significantly improved in 1886 by the German scientist

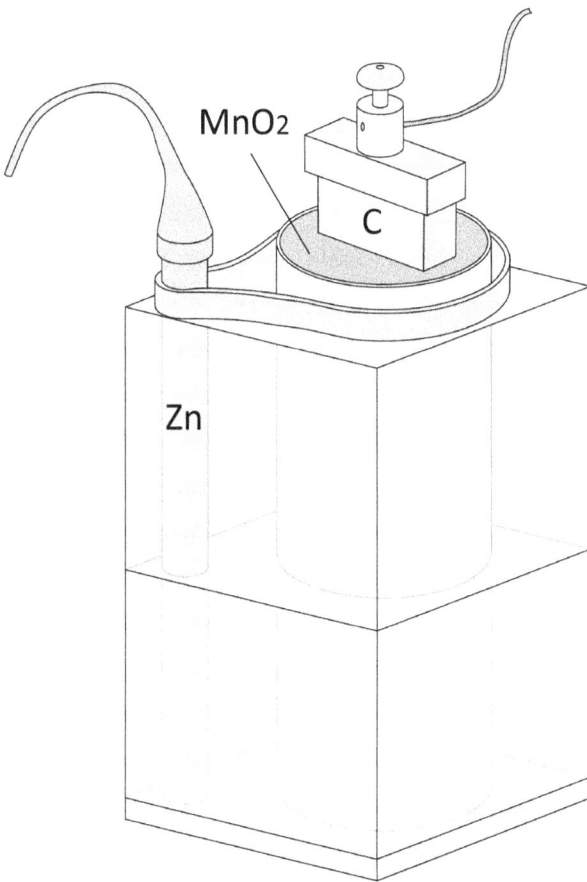

**Figure 4.9:** Sketch of the setup of the Leclanché cell. A zinc rod (anode) is placed in the outer glass jar. The glass jar also contains an aqueous solution of ammonium chloride ($NH_4Cl$). A small earthenware pot is placed in the glass jar as well. The small earthenware pot is densely filled with a mixture of manganese dioxide ($MnO_2$) – the depolarizer – and granulated carbon. A graphite rod (C) is also placed in the inner earthenware pot and provides the positive terminal of the Leclanché cell.

Carl Gassner. He could manage to realize an electrolyte paste – for instance, by gelatinization; hence, the first dry cell was born. Gassner's invention received a German patent in 1886 and a corresponding US patent in 1887. According to his improvement of the initial wet cell, the Leclanché cell is considered as the parent of modern dry cells. The big advantage of dry cells using a paste electrolyte is the much better portability as compared to early batteries with liquid electrolytes, since a dry cell can operate in any orientation without any leakage. The early wet cells were rather fragile and difficult to handle, since they were usually made from glass containers; and the electrodes were hanging in the form of rods or sheets from the open top of the container. Hence, the dry cell was a tremendous technological progress in battery technology.

The first dry cell of the Leclanché cell – that is, the Gassner cell – used a zinc container that contained all other components of the battery. Of course, the zinc container also works as the anode, that is, the negative electrode of the cell. The positive electrode was a carbon (graphite) rod immersed in a mixture of manganese dioxide (depolarizer) and carbon black mixture. Carbon black can be produced by an incomplete combustion of heavy petroleum products (e.g., coal tar). It has a high surface area-to-volume ratio and reduces the electrical resistance of the cathode. The electrolyte – an aqueous solution of ammonium chloride ($NH_4Cl$) – was absorbed in the porous material of the cell ($MnO_2$, carbon black). The depolarizer and the graphite electrode – that is, the cathodic half-cell – were separated from the zinc container – that is, the anodic half-cell – by a folded paper sack that was also soaked in the electrolyte. The Gassner cell was sealed across the top.

Now, we will discuss the electrochemical reactions within the Leclanché cell – that is, the redox reactions that occur when the electric circuit is closed by the external electrical connection between the electrodes. Side reactions will be briefly discussed; reactions that cause fatigue of the Leclanché cell will be presented as well. The structure of the Leclanché cell can be described by the following cell notation:

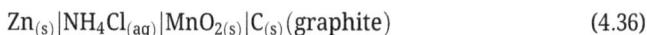

$$Zn_{(s)}|NH_4Cl_{(aq)}|MnO_{2(s)}|C_{(s)}(\text{graphite}) \tag{4.36}$$

At the negative zinc anode of the Leclanché cell, the same oxidation reaction occurs which we already know from the Daniell cell, that is,

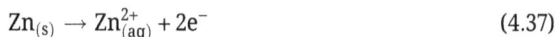

$$Zn_{(s)} \rightarrow Zn^{2+}_{(aq)} + 2e^- \tag{4.37}$$

At the positive cathode of the Leclanché cell, the reduction of $MnO_2$ occurs; that is, the depolarizer changes its oxidation state. The manganese in $MnO_2$ is in the tetravalent state; therefore, here we also speak about Mn(IV) oxide. In detail, in the cathodic half-cell of the Leclanché cell, the tetravalent Mn(IV) in the manganese dioxide is reduced to manganese oxide-hydroxide with trivalent Mn(III). This reduction can be described by the following reaction formula:

$$2MnO_{2(s)} + 2H_2O_{(l)} + 2e^- \rightarrow 2MnO(OH)_{(s)} + 2OH^-_{(aq)} \tag{4.38}$$

The reduction reaction (4.38) works since the water molecule undergoes a so-called self-ionization (also called self-dissociation or molecular autoionization). This is an ionization reaction in which a water molecule ($H_2O$) deprotonates, meaning that the water molecule loses a proton ($H^+$) leaving a negatively charged hydroxide ion ($OH^-$) behind, that is,

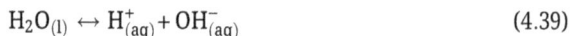

$$H_2O_{(l)} \leftrightarrow H^+_{(aq)} + OH^-_{(aq)} \tag{4.39}$$

The self-ionization process can run in both directions, and it can be found in pure water and aqueous solutions, respectively. Usually, the proton directly protonates another hydrogen molecule, resulting in a positively charged hydronium ion ($H_3O^+$).

Hence, in this case, the self-ionization reaction can be described by the reaction formula $2H_2O \rightarrow H_3O^+ + OH^-$. Anyway, instead of protonating another hydrogen molecule, the proton can also be used for the formation of manganese oxide-hydroxide in the reduction reaction (4.38), that is, $MnO_2 + H^+ + e^- \rightarrow MnO(OH)$, which is sometimes called a one-electron reduction.

If we keep in mind the self-ionization of water molecules in the aqueous solution (i.e., in the electrolyte), the oxidation reaction (4.37) and the reduction reaction (4.38) of the Leclanché cell can be merged to the overall redox reaction:

$$Zn_{(s)} + 2MnO_{2(s)} + 2H_2O_{(1)} \rightarrow 2MnO(OH)_{(s)} + Zn(OH)_{2(aq)} \qquad (4.40)$$

Water molecules that are consumed in the reduction reaction (4.38), or rather in the redox reaction (4.40), can be delivered at a later stage, for instance, by the following reaction that occurs in the aqueous solution (electrolyte):

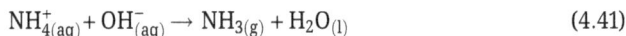

$$NH_{4(aq)}^+ + OH_{(aq)}^- \rightarrow NH_{3(g)} + H_2O_{(1)} \qquad (4.41)$$

In this reaction, the positive ammonium ion ($NH_4^+$) originates from the electrolyte, that is, from the aqueous solution of ammonium chloride ($NH_4Cl$), where $NH_4Cl$ molecules dissociate into positive ammonium ions ($NH_4^+$) and negative chloride ions ($Cl^-$) – see reaction (4.35). The negatively charged hydroxide ion ($OH^-$) is supplied, for instance, by the reactions (4.38) or (4.39).

The zinc hydroxide ($Zn(OH)_2$) in the aqueous solution (electrolyte) is dissociated according to the following dissociation reaction:

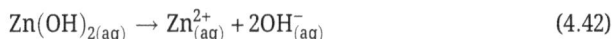

$$Zn(OH)_{2(aq)} \rightarrow Zn_{(aq)}^{2+} + 2OH_{(aq)}^- \qquad (4.42)$$

These dissociated zinc hydroxide ions ($Zn^{2+}$, $OH^-$) react with the dissociated ammonium chloride ions ($NH_4^+$, $Cl^-$, see reaction (4.35)) and form diamine zinc chloride – [$Zn(NH_3)_2$]$Cl_2$ – according to the following formula:

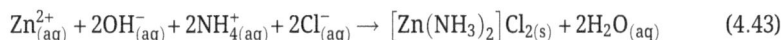

$$Zn_{(aq)}^{2+} + 2OH_{(aq)}^- + 2NH_{4(aq)}^+ + 2Cl_{(aq)}^- \rightarrow [Zn(NH_3)_2]Cl_{2(s)} + 2H_2O_{(aq)} \qquad (4.43)$$

The diamine zinc chloride is a complex salt with very low solubility. Therefore, it precipitates onto the electrodes; hence, the internal cell resistance increases. Therefore, the performance of the Leclanché cell is reduced, since energy dissipation is growing due to the enhanced cell resistance, and it degrades.

The reduction in the cathodic half-cell of the Leclanché cell can also occur in accordance with the following reaction formula:

$$2MnO_{2(s)} + 2NH_{4(aq)}^+ + 2e^- \rightarrow 2MnO(OH)_{(s)} + 2NH_{3(g)} \qquad (4.44)$$

The positive ammonium ions ($NH_4^+$) for this reaction originate from the electrolyte, as mentioned earlier.

The ammonia molecules ($NH_3$) that are formed in the cathodic half-cell during the reduction reaction (4.44) are bound to the doubly positively charged $Zn^{2+}$ ions in the electrolyte that originate from the oxidation reaction (4.37), and form a doubly positively charged diamine zinc complex:

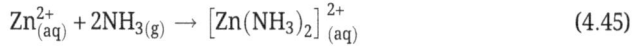

$$Zn^{2+}_{(aq)} + 2NH_{3(g)} \rightarrow \left[Zn(NH_3)_2\right]^{2+}_{(aq)} \tag{4.45}$$

Subsequently, the $[Zn(NH_3)_2]^{2+}$ complexes in the aqueous solution can react with the negatively charged chloride ions ($Cl^-$), and diamine zinc chloride is formed:

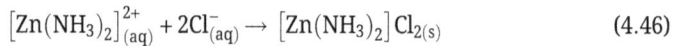

$$\left[Zn(NH_3)_2\right]^{2+}_{(aq)} + 2Cl^-_{(aq)} \rightarrow \left[Zn(NH_3)_2\right]Cl_{2(s)} \tag{4.46}$$

The overall redox reaction that corresponds to the reduction reaction (4.44) can be expressed as follows:

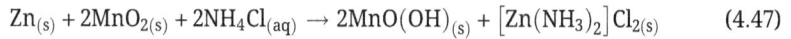

$$Zn_{(s)} + 2MnO_{2(s)} + 2NH_4Cl_{(aq)} \rightarrow 2MnO(OH)_{(s)} + \left[Zn(NH_3)_2\right]Cl_{2(s)} \tag{4.47}$$

During this reaction in the Leclanché cell, diamine zinc chloride finally drops out resulting in an increase of the internal cell resistance of the Leclanché cell, as mentioned earlier.

Reaction eq. (4.44) is a little simplified and a reduced version of the reduction reaction that is somewhat more complex, as will be briefly shown now. The more detailed description of the reduction process is given by

$$2MnO_{2(s)} + 2NH_4Cl_{(aq)} + 2e^- \rightarrow Mn_2O_{3(s)} + 2NH_{3(g)} + H_2O_{(l)} + 2Cl^-_{(aq)} \tag{4.48}$$

Concerning the reduction reaction (4.48), one has to keep in mind that ammonium chloride in the aqueous solution – $NH_4Cl_{(aq)}$ – dissociates in accordance with reaction (4.35), as mentioned earlier. Then, the respective formulation of the reduction reaction (4.48) can be written as follows:

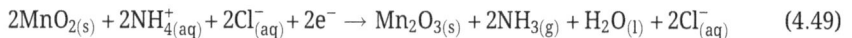

$$2MnO_{2(s)} + 2NH^+_{4(aq)} + 2Cl^-_{(aq)} + 2e^- \rightarrow Mn_2O_{3(s)} + 2NH_{3(g)} + H_2O_{(l)} + 2Cl^-_{(aq)} \tag{4.49}$$

or rather:

$$2MnO_{2(s)} + 2NH^+_{4(aq)} + 2Cl^-_{(aq)} + 2e^- \rightarrow 2MnO(OH)_{(s)} + 2NH_{3(g)} + 2Cl^-_{(aq)} \tag{4.50}$$

Keeping in mind reaction (4.37) for the oxidation in the anodic half-cell of the Leclanché element, we now derive the following equation for the redox reaction:

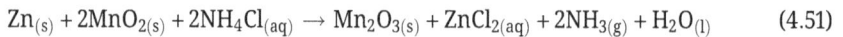

$$Zn_{(s)} + 2MnO_{2(s)} + 2NH_4Cl_{(aq)} \rightarrow Mn_2O_{3(s)} + ZnCl_{2(aq)} + 2NH_{3(g)} + H_2O_{(l)} \tag{4.51}$$

or rather:

$$Zn_{(s)} + 2MnO_{2(s)} + 2NH_4Cl_{(aq)} \rightarrow 2MnO(OH)_{(s)} + ZnCl_{2(aq)} + 2NH_{3(g)} \tag{4.52}$$

In the aqueous solution, zinc chloride ($ZnCl_2$) dissociates according to the following dissociation reaction:

$$ZnCl_{2(aq)} \rightarrow Zn^{2+}_{(aq)} + 2Cl^{-}_{(aq)} \tag{4.53}$$

The zinc and chloride ions merge with the ammonia molecules and form diamine zinc chloride again, according to the reactions (4.45) and (4.46) that were shown earlier. Hence, formula (4.52) can again be converted into the form (4.47), describing the redox reaction of the Leclanché cell, where finally diamine zinc chloride, that is, [Zn $(NH_3)_2]Cl_2$, drops out.

As we have seen, the chemistry of the Leclanché cell can be rather complex. A variety of side reactions – also concerning the reduction in the cathodic half-cell of the Leclanché cell – are possible, and we have only discussed some of them. However, the underlying electrochemical reactions in the Leclanché cell that are responsible for the operation of this primary battery can be represented by the following set of reaction formulas:

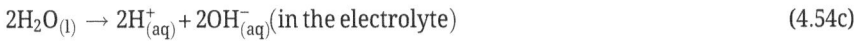

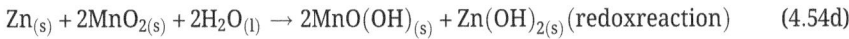

$$Zn \rightarrow Zn^{2+}_{(aq)} + 2e^{-} (\text{oxidation at the anode}) \tag{4.54a}$$

$$2MnO_{2(s)} + H^{+}_{(aq)} + 2e^{-} \rightarrow 2MnO(OH)_{(s)} (\text{reduction at the cathode}) \tag{4.54b}$$

$$2H_2O_{(l)} \rightarrow 2H^{+}_{(aq)} + 2OH^{-}_{(aq)} (\text{in the electrolyte}) \tag{4.54c}$$

$$Zn_{(s)} + 2MnO_{2(s)} + 2H_2O_{(l)} \rightarrow 2MnO(OH)_{(s)} + Zn(OH)_{2(s)} (\text{redox reaction}) \tag{4.54d}$$

In addition, we have side reactions that finally degrade the Leclanché cell:

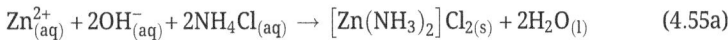

$$Zn^{2+}_{(aq)} + 2OH^{-}_{(aq)} + 2NH_4Cl_{(aq)} \rightarrow [Zn(NH_3)_2]Cl_{2(s)} + 2H_2O_{(l)} \tag{4.55a}$$

$$Zn^{2+}_{(aq)} + 2OH^{-}_{(aq)} + 2NH_4Cl_{(aq)} + 2NH_{3(g)} \rightarrow [Zn(NH_3)_4]Cl_{2(s)} + 2H_2O_{(l)} \tag{4.55b}$$

In the case of a slow or moderate discharge of the Leclanché cell, the diamine zinc chloride also precipitates according to reaction (4.55a). The latter side reaction (4.55b) only occurs in the case of rapid discharging, when the temperature in the Leclanché cell increases due to higher ionic currents and energy dissipation. At these elevated temperatures, ammonium chloride ($NH_4Cl$) transforms according to the following reaction:

$$NH_4Cl_{(aq)} \rightarrow NH_{3(g)} + HCl_{(g)} \tag{4.56}$$

Summarizing reactions (4.54a)–(4.54d) and (4.55a), we finally receive the overall redox reaction (4.47) of the Leclanché cell that was already presented. Finally, we can add that, if all nitrogen, or more precisely ammonia, has been consumed, alkaline zinc salts drop out, and the current flow stops.

The main electrochemical reactions occurring in the Leclanché cell are summarized in Figure 4.10.

The cell voltage of the Leclanché cell is given by the difference of the half-cell potentials. If the electric circuit is not closed ($I = 0$), the open-circuit voltage $E_{oc}$ of the Leclanché cell at RT has a value of

$$E_{oc, 298K} = \Delta\Phi(I = 0) \approx 1.4\text{V} \qquad (4.57)$$

where $\Delta\Phi = \Phi_{cathode} - \Phi_{anode}$ at zero current.

In general, the Leclanché cell could not provide a stable, ongoing current for a long time. Chemical side reactions within the cell that we discussed earlier, increase the internal resistance and, hence, lower the voltage of the battery. The Leclanché cell was mainly used in the late nineteenth century and the early twentieth century for telegraphy, signaling, and electric bells. Since, during operation, the internal electrical resistance steadily increases, for instance, during lengthy conversations the battery fatigues, that is, the provided voltage goes down and makes a conversation inaudible. Since the chemical side reactions can be reversed, at least partly reversed, when the battery is left idle, the Leclanché cell can recover to some extent. Thus, the Leclanché cell could be used better for a shorter intermittent than for a lengthy steady operation.

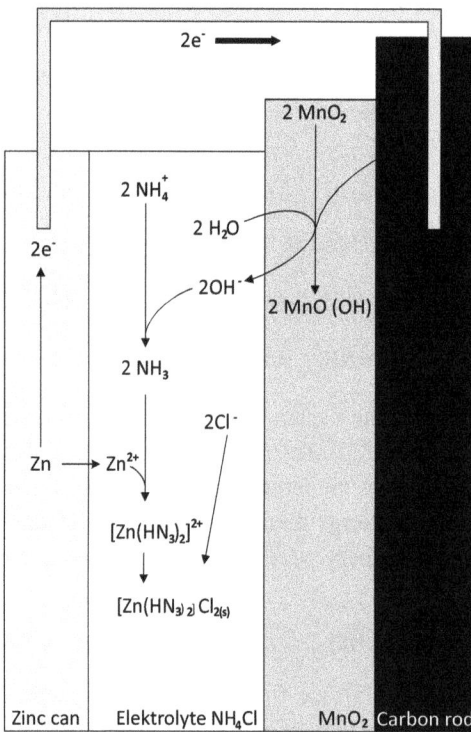

**Figure 4.10:** Visualization of the main electrochemical reactions that occur in the Leclanché cell.

### 4.2.3 Planté cell

In 1859, the French physicist Gaston Planté developed the first rechargeable battery, that is, the first accumulator or secondary battery. It is named after the inventor, that is, the Planté cell; and its successors are known today as the lead–acid battery or lead–acid accumulator. Recharging of the Planté cell can be achieved when a reverse current passes through the cell. The cell voltage of the charged Planté cell under open-circuit conditions is about 2 V at RT.

The lead–acid battery, in its original form as presented by Planté, consists of two sheets of lead. These two sheets were separated by rubber strips – to avoid short-circuit faults – and stacked. Then, the stacked and separated Pb sheets are rolled into a spiral and immersed into a sulfuric acid solution ($H_2SO_4$), as shown in Figure 4.11. The Planté cell was a rather voluminous and heavy device. Its specific and volumetric energies are rather low. However, at this time – in the nineteenth century – the Planté cell could deliver quite large currents, especially in surges. Therefore, its specific power was not so bad. Moreover, the internal resistance of the Planté cell was rather low, which is useful for various applications.

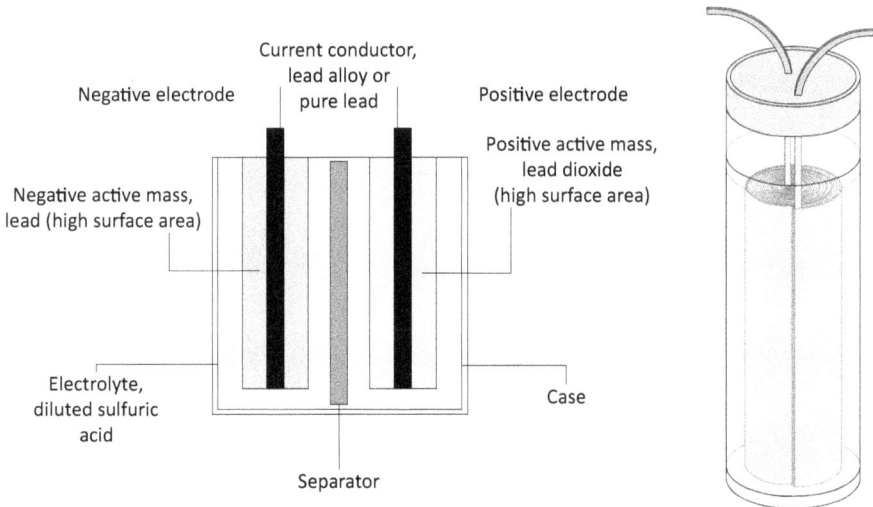

**Figure 4.11:** Visualization of the principal setup of the Planté cell (left side). Practical setup (right side): The stacked and separated Pb sheets are rolled into a spiral form and immersed into a sulfuric acid ($H_2SO_4$) solution.

In 1881, the French chemical engineer Camille Alphonse Faure introduced a significantly improved version of the Planté cell. Faure could manage to coat metallic lead with a paste of lead oxides. He patented this method; and it finally led to the improved Planté cell (Figure 4.12) that is characterized by a grid lattice made of metallic lead, into which a paste of lead oxide was pressed and cured by gentle warming in an ambient humid atmosphere (steam). During the curing process, lead sulfates ($PbSO_4$, i.e., divalent Pb(II) sulfate) were formed that adhered to the metallic lead. Hence, the cured paste could be converted into an electrochemically active material called "active mass." From this $Pb-PbO_2$ structure, a plate was formed, operating as the positive electrode of the lead–acid battery, whereas a metallic Pb plate acts as a negative electrode. Several alternating plates of Pb and $Pb-PbO_2$, that is, negative and positive electrodes, respectively, can be stacked to increase the cell capacity of this battery type. A big advantage was that this design matches mass production and led to the industrial fabrication of lead–acid batteries. Moreover, the Faure cell was smaller and, therefore, lighter than earlier lead–acid cells based on the design of Gaston Planté. Such improved accumulators could be used, for instance, as automotive batteries, where the main application was to start internal combustion engines of automobiles, which was extremely useful and simplified the start of an automotive engine. Therefore, one might say that the Planté cell – or, better, the Faure cell – leveraged the automobile industry, as they were rather reliable and efficient; in particular, they delivered sufficient high currents. Moreover, early electric cars were successfully powered by the Planté/Faure cell.

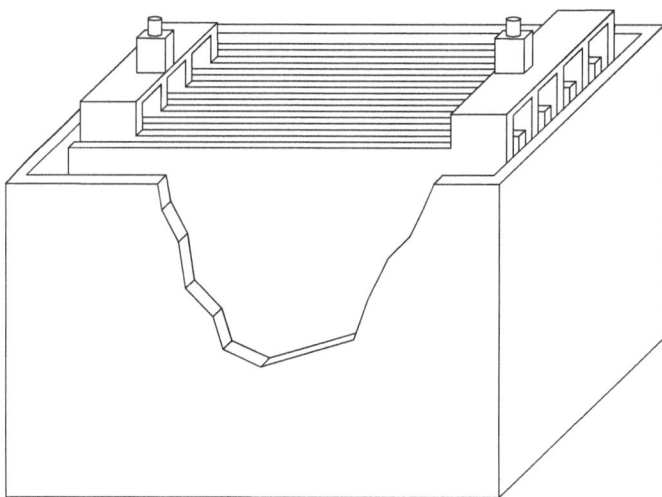

**Figure 4.12:** Faure improved the Planté cell; several alternating plates of Pb and $Pb-PbO_2$, that is, negative and positive electrodes, respectively, were stacked to increase the cell capacity of the lead–acid battery.

Now, we will discuss the electrochemical reactions within the Planté cell. These reactions are basically found also in modern lead–acid batteries. The cell notation of the Planté cell is given as follows:

$$Pb_{(s)}|H_2SO_{4(aq)}|PbO_{2(s)}|Pb_{(s)} \qquad (4.58)$$

Both electrodes of the Planté cell are immersed in an aqueous solution of sulfuric acid ($H_2SO_4$) and react with the dilute acid, that is, the negative Pb electrode at the $Pb_{(s)}|$ $H_2SO_{4(aq)}$ interface on the left side of the cell notation (4.58), and the positive Pb-$PbO_2$ electrode at the $H_2SO_{4(aq)}|PbO_{2(s)}$ interface on the right side. In case of the discharge process, oxidation occurs on the left side of the cell notation (4.58), where electrons are released; the reduction occurs on the right side, where electrons are consumed.

Thus, considering the discharging process, the negative electrode of the Planté cell is made from metallic lead (Pb), where oxidation occurs, and reduction occurs at the positive electrode made from lead dioxide ($PbO_2$). The lead in lead dioxide is tetravalent; hence, we also speak about Pb(IV) oxide. In fact, the positive plate also consists of metallic Pb, but here the outer layers of the electrode that are in contact with the electrolyte consist of the oxidized form – $PbO_2$. The outer $PbO_2$ layers of the positive electrode are formed during the charging process, as will be elucidated later. Hence, both terminals of the Planté cell are made from the same material – Pb (see Figure 4.13).

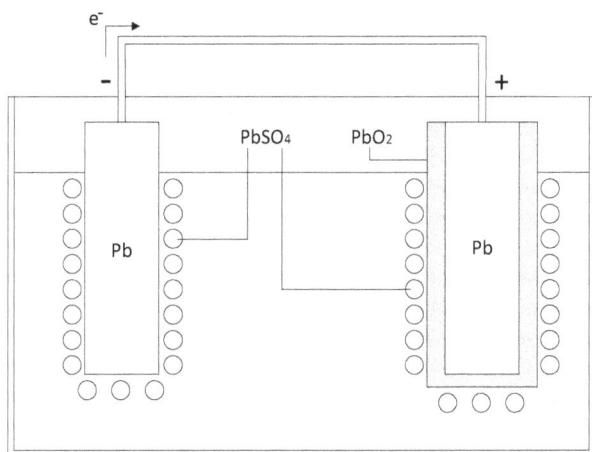

**Figure 4.13:** Principal setup and composition of the lead–acid cell (Planté cell).

In the case of the charging process, where the electrochemical reactions are reversed, a forced reduction occurs on the left side of the cell notation (4.58), that is, at the negative Pb electrode; and a forced oxidation takes place on the right side of the cell notation (4.58) at the positive Pb–$PbO_2$ electrode. Thus, in the case of the charging process, the anode is the positive Pb–$PbO_2$ electrode, and the cathode is the negative Pb electrode. To avoid confusion when dealing with secondary batteries in the following discussions, we prefer to use the designations negative and positive electrodes or terminals, and we do not speak about anodes and cathodes anymore.

The electrochemical reactions in a Planté cell can be summarized by the relatively simple reaction formula:

$$Pb_{(s)} + PbO_{2(s)} + 2H_2SO_{4(aq)} \leftrightarrow 2PbSO_{4(s)} + 2H_2O_{(l)} \tag{4.59}$$

Reaction (4.59) characterizes a redox reaction that can run in both directions. The left side of the reaction indicates the charged state of the Planté cell, and the right side represents the corresponding discharged state. The reaction from the left side to the right side of the reaction stands for the discharge process that spontaneously occurs if the electric circuit over the galvanic cell is closed; and the reaction from the right side to the left side represents the forced redox reaction that stands for the charging process.

If the Planté cell is completely charged, the negative plate consists of metallic lead, and the positive plate consists of lead dioxide, where in between the electrolyte exhibits an aqueous solution of maximally concentrated sulfuric acid (Figure 4.14). Lead dioxide is a very good oxidizing agent (also called an oxidant or oxidizer). In general, an oxidizing agent is a chemical substance (or an atom or molecule) that undergoes a chemical reaction that removes one or more electrons from another substance (or atom or molecule). In other words, the oxidizing agent is a substance that has the ability to cause other substances to lose electrons, that is, to oxidize the other substances. Oxygen is a common oxidizing agent; in fact, it stimulated the term "oxidizing agent" (or oxidant and oxidizer). In a redox reaction, the oxidizing agent is the substance that is reduced – that is, it is the electron acceptor. On the other hand, a reducing agent is the electron donor, that is, the substance that is oxidized during the redox reaction and loses one or more electrons.

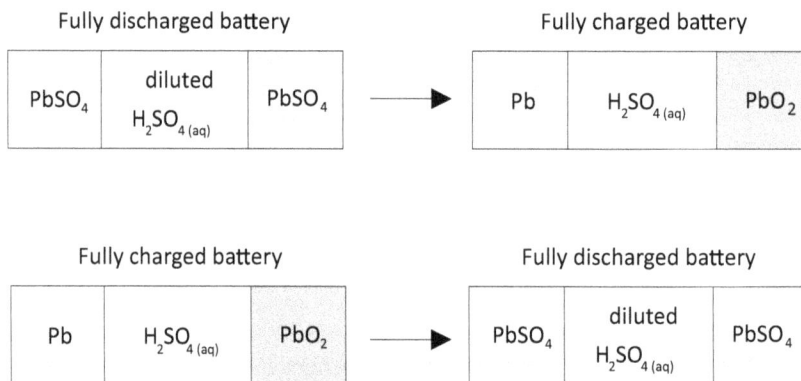

Fully discharged battery                                    Fully charged battery

| $PbSO_4$ | diluted $H_2SO_{4\,(aq)}$ | $PbSO_4$ | ⟶ | Pb | $H_2SO_{4\,(aq)}$ | $PbO_2$ |

Fully charged battery                                        Fully discharged battery

| Pb | $H_2SO_{4\,(aq)}$ | $PbO_2$ | ⟶ | $PbSO_4$ | diluted $H_2SO_{4\,(aq)}$ | $PbSO_4$ |

Figure 4.14: Completely charged Planté cell: the negative plate consists of metallic lead (Pb), and the positive plate consists of lead dioxide ($PbO_2$); in between, the aqueous solution exhibits a maximal concentration of sulfuric acid ($H_2SO_4$). Completely discharged Planté cell: the negative plate and the positive plates are consisting of lead(II) sulfate ($PbSO_4$); in between, the aqueous solution exhibits a dilute concentrated sulfuric acid. The arrows mark the charging (above) and the discharging (below) process.

If the external electrical circuit is closed, during the discharge process at the negative plate of the Planté cell, the following oxidation reaction occurs, leading to the formation of lead(II) sulfate ($PbSO_4$):

$$Pb_{(s)} + SO_{4(aq)}^{2-} \rightarrow PbSO_{4(s)} + 2e^- \qquad (4.60)$$

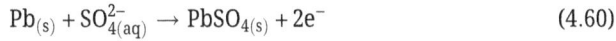

The two electrons that are released in this reaction remain in the metallic Pb electrode and migrate to the positive electrode if the electric circuit over the whole cell is closed. We see that, in the case of the Planté cell, where the negative plate is made of metallic lead, the situation is not as simple as, for instance, in the case of the zinc anodes of the Daniell cell or the Leclanché cell that are directly oxidized according to the oxidation equations (4.25) or (4.37), respectively.

It is to add that $H_2SO_4$ in aqueous solution dissociates according to the following dissociation reaction:

$$H_2SO_{4(aq)} \rightarrow SO_{4(aq)}^{2-} + 2H_{(aq)}^+ \qquad (4.61a)$$

or rather:

$$2H_2SO_{4(aq)} + 4H_2O_{(l)} \rightarrow 2SO_{4(aq)}^{2-} + 4H_3O_{(aq)}^+ \qquad (4.61b)$$

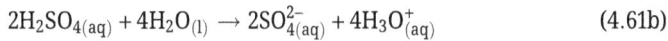

The hydrogen ions ($H^+$) that are produced according to the dissociation reaction (4.61a) of sulfuric acid enter the aqueous solution of the Planté cell, directly attach to water molecules, and form hydronium ions ($H_3O^+$). Hence, the dissociation of sulfuric acid can be better described by eq. (4.61b).

At the positive plate, the $H_3O^+$ ions are subsequently consumed according to the following reduction reaction:

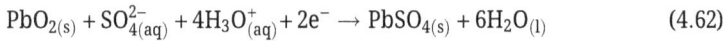

$$PbO_{2(s)} + SO_{4(aq)}^{2-} + 4H_3O_{(aq)}^+ + 2e^- \rightarrow PbSO_{4(s)} + 6H_2O_{(l)} \qquad (4.62)$$

Sulfuric acid ($H_2SO_4$), or rather the $SO_4^{2-}$ ions in the aqueous solution, are consumed at both the negative and the positive plates, respectively. Therefore, $PbSO_4$ is formed at both electrodes – the negative and the positive ones.

In summary, the discharging process of the Planté cell can be described by the following redox reaction:

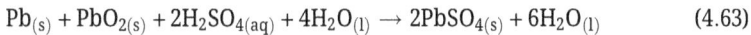

$$Pb_{(s)} + PbO_{2(s)} + 2H_2SO_{4(aq)} + 4H_2O_{(l)} \rightarrow 2PbSO_{4(s)} + 6H_2O_{(l)} \qquad (4.63)$$

Of course, this reaction is implied in the above-mentioned reaction formula (4.59) that describes the discharging and charging of the Planté cell.

During the discharging process – besides lead(II) sulfate ($PbSO_4$) – water is generated, as can be seen in redox reactions (4.59) and (4.63). In the discharged state, both the positive and the negative plates are converted into lead(II) sulfate (Figure 4.14) – at least the significant outer parts of them that are in contact with the aqueous solution. The electrolyte has lost much of its dissolved sulfuric acid due to the formation of $PbSO_4$ at the electrodes; therefore, the aqueous solution exhibits a minimal sulfuric acid concentration.

During the charging process, at the negative lead electrode, a forced reduction of $Pb^+$, and therefore, its deposition as metallic lead onto lead electrode occurs:

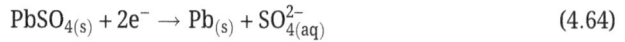

$$PbSO_{4(s)} + 2e^- \rightarrow Pb_{(s)} + SO_{4(aq)}^{2-} \tag{4.64}$$

On the other hand, at the positive electrode, a forced oxidation of divalent Pb(II) to tetravalent Pb(IV) occurs, resulting in a deposition of $PbO_2$ onto the positive electrode:

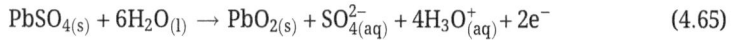

$$PbSO_{4(s)} + 6H_2O_{(l)} \rightarrow PbO_{2(s)} + SO_{4(aq)}^{2-} + 4H_3O_{(aq)}^+ + 2e^- \tag{4.65}$$

In summary, for the Planté cell, the charging process can be described by the following redox reaction:

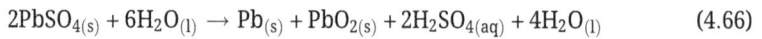

$$2PbSO_{4(s)} + 6H_2O_{(l)} \rightarrow Pb_{(s)} + PbO_{2(s)} + 2H_2SO_{4(aq)} + 4H_2O_{(l)} \tag{4.66}$$

Again, this is implied in the reaction formula (4.59) of the Planté cell mentioned earlier.

The thermodynamic equilibria of the electrode reactions in the Planté cell are directed parallel to the discharge direction; that is, they point from the left to the right side in reaction (4.59), since, from the thermodynamic point of view, the discharged state is more stable. The charged Planté cell tends to self-discharge, as lead and lead dioxide are thermodynamically unstable in the sulfuric acid solution. Hence, under open-circuit conditions, when no external load is connected to the Planté cell, both lead and lead dioxide react with the electrolyte. Thus, self-discharge occurs at both electrodes according to the following reactions:

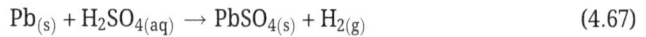

$$Pb_{(s)} + H_2SO_{4(aq)} \rightarrow PbSO_{4(s)} + H_{2(g)} \tag{4.67}$$

and

$$2PbO_{2(s)} + H_2SO_{4(aq)} \rightarrow 2PbSO_{4(s)} + 2H_2O_{(l)} + O_{2(g)} \tag{4.68}$$

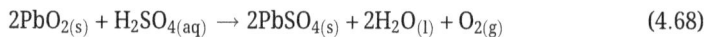

These reactions are similar to the oxidation reaction (4.60) and the reduction reaction (4.62) during the discharge process. However, under self-discharge conditions, the external electrical circuit is not closed; hence, electron migration through the external wiring cannot occur. Therefore, at the negative electrode, the $H^+$ ions and electrons – appearing in the oxidation reaction (4.60) during the discharging process – combine and form $H_2$ gas molecules during the self-discharge process (reaction (4.67)). At the positive electrode, lead dioxide reacts with the sulfuric acid to form lead sulfate, where the remaining constituents form water and oxygen gas ($O_2$).

The successors of the Planté cell, that is, modern lead–acid accumulators, will be discussed in Section 4.4.1. Now, we stop our reflections on historically important batteries, and we will continue with technically important primary batteries.

## 4.3 Technically important primary batteries

A primary battery or primary cell is a nonrechargeable galvanic cell, since, in general, the electrochemical reactions occurring in a primary cell cannot be reversed. The electrochemical reactions in a primary battery consume the substances that are responsible for the electrochemical energy storage. If those substances are completely exhausted, the electrochemical reactions stop (i.e., the final thermodynamic equilibrium is reached), and the primary battery has finished its operation as a voltage source. Although primary batteries are losing their technical importance more and more in favor of secondary batteries, and their market share steadily declines, they are still useful for a variety of applications. Actually, for specific applications, they still have some advantages.

Primary batteries can be designed in such a way that their self-discharge rate is very low – much lower than that of secondary batteries. Thus, primary batteries can be stored for a long period. If the self-discharge is low, their cell capacity can be nearly completely used for the desired purpose, even after a long storage period. Moreover, hearing aid devices require only very small electrical currents over a long period. The same holds true, for instance, for smoke detectors. For such applications, primary batteries can be employed – in particular, since, in general, their nominal energy densities are significantly higher than those of secondary batteries are. Thus, primary batteries can be the better choice for long-term usage.

A disadvantage of primary batteries is that they are environmentally unfriendly. Batteries contain significant amounts of harmful substances, that is, toxic heavy metals or rather strong acids. Hence, after the nonrecurring usage of primary batteries, they have to be carefully disposed – unfortunately, this is done not always in a responsible way. Moreover, the fabrication of batteries – in particular the fabrication of primary batteries – needs a big amount of energy. This amount of energy is much higher than the energy that is stored for a meaningful usage inside the battery. Hence, their ecological footprint is rather bad. With this regard, rechargeable secondary batteries have a somewhat better reputation, although they also can contain a variety of problematic substances.

In the following sections, we will describe those primary batteries that were important in the past and those that are still useful now. We will start with three primary cells that are based on the historic Leclanché cell. Later on, we will discuss other solutions.

### 4.3.1 Zinc–carbon cell

The zinc–carbon cell (or zinc–carbon battery) is the modern form of the historic Leclanché cell. In fact, it is a Leclanché cell with a modified and improved setup. The basic electrochemical reactions in the zinc–carbon cell are the same as in the Leclanché cell; that is, reactions (4.54a)–(4.55b) are valid also for the zinc–carbon cell. Here, we just repeat the redox equation that describes the zinc–carbon cell as well as the historic Leclanché cell:

$$Zn_{(s)} + 2MnO_{2(s)} + 2H_2O_{(l)} \rightarrow 2MnO(OH)_{(s)} + Zn(OH)_{2(s)} \qquad (4.69)$$

Both the oxidation at the Zn anode and the reduction of tetravalent manganese(IV) oxide ($MnO_2$) to trivalent manganese(III) oxide-hydroxide (MnO(OH)) occur in the same manner as they do in the historic Leclanché cell (Section 4.2.2). The main side reaction that results in the formation of the hardly soluble diamine zinc chloride – [$Zn(NH_3)_2$]$Cl_2$ – and that finally degrades the zinc–carbon cell is given as follows:

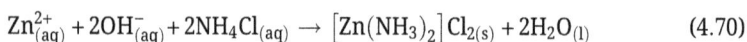

$$Zn^{2+}_{(aq)} + 2OH^-_{(aq)} + 2NH_4Cl_{(aq)} \rightarrow \left[Zn(NH_3)_2\right]Cl_{2(s)} + 2H_2O_{(l)} \qquad (4.70)$$

As well, the side reaction (4.70) is the same as it appears in the historic Leclanché cell (reaction (4.55a)). The details and peculiarities were already described in Section 4.2.2 and will not be repeated here.

A zinc–carbon cell is typically constructed as shown in Figure 4.15. The cell is assembled in a container shaped like a can that is made of zinc and already functions as the anode. The zinc container (or zinc can) is attached to a metal bottom that is operating the negative terminal of the battery. For electrical isolation, the outer surface of the zinc can is coated with a nonconductive paint. In the middle of the zinc can, a carbon rod is located with a metallic cap on top that acts as the positive terminal. A mixture of manganese(IV) oxide ($MnO_2$) and powdered graphite is packed around the carbon rod, operating as the cathode of the battery. Paper soaked with an aqueous paste of ammonium chloride ($NH_4Cl$) works as the electrolyte between the two electrodes. The top of the can is sealed with a cap in such a way that the zinc can and the positive terminal of the battery are not electrically connected.

The zinc–carbon battery exhibits some problematic properties. As we already mentioned, when we discussed the historic Leclanché cell in Section 4.2.2, the formation of the hardly soluble diamine zinc chloride precipitates at the electrodes of the zinc–carbon cell enhancing the internal ionic resistance of the cell. Hence, energy dissipation is increased, and the power of the battery is reduced.

Moreover, the zinc–carbon cell is not leak proof, since the zinc anode – that is, the container of the zinc–carbon cell – erodes during operation, and leakage can occur after longer operation. If the zinc–carbon cell leaks and electrolyte runs out of the battery case, then electrical contacts erode, too, and even the circuitry of the electric device that is powered by the zinc–carbon cell can be damaged. Hence, zinc–carbon batteries should not be used in expensive devices. Likewise, the zinc–carbon cell exhibits a rather large self-discharge, too. Therefore, it should also not be used in devices, where it ought to remain for a longer period, for instance, smoke detectors.

If the electrolyte in the Leclanché cell or the zinc–carbon cell is changed, that is, if zinc chloride ($ZnCl_2$) substitutes the ammonium chloride ($NH_4Cl$) in the aqueous solu-

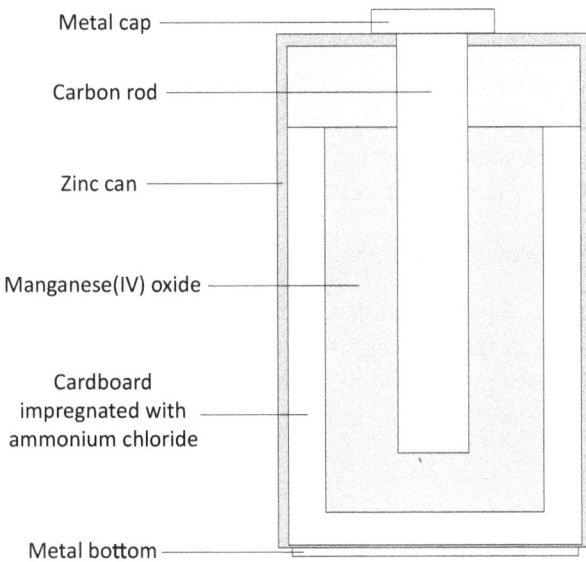

**Figure 4.15:** Construction of a zinc–carbon cell.

tion, some improvements can be achieved. In this case, we do not speak of a zinc–carbon cell anymore but about a zinc-chloride cell. This successor of the Leclanché cell will be presented in the next section.

### 4.3.2 Zinc-chloride cell

The zinc-chloride cell is an improved successor of the Leclanché cell that exhibits less problematic properties than the zinc–carbon cell. It has also a zinc anode and a manganese(IV) oxide cathode. The production of the zinc-chloride cells is done with purer chemicals than those used for its antecessors. In particular, in zinc-chloride cells, high-grade manganese dioxide that is electrolytically prepared is used. Moreover, a greater amount of carbon black is mixed with the manganese dioxide that reduces the electrical resistance and promotes the electrochemical reactions at the cathode. Concerning the construction of zinc-chloride cells, the seal design is improved, too, since it must be more carefully protected from drying out than the zinc–carbon cell, which will be mentioned later. Zinc-chloride cells exhibit a longer lifetime and a higher cell capacity than zinc–carbon cells, and they deliver a steadier voltage, too. Compared to the zinc–carbon cell, the better performance of the zinc-chloride cell compensates for its higher production costs.

The major difference between the zinc-chloride cell and its antecessor the zinc–carbon cell is the modified electrolyte; that is, the aqueous ammonium chloride solution that is used in the historic Leclanché cell and in the zinc–carbon cell is replaced by a zinc chloride ($ZnCl_2$) solution. This somewhat simplifies the electrochemical reactions for the zinc-chloride cell. Moreover, the formation of the hardly soluble diamine zinc chloride – $[Zn(NH_3)_2]Cl_2$ – is avoided, which caused an enhancement of the internal cell resistance of the zinc–carbon cell and finally degrades its performance.

$ZnCl_2$ is hygroscopic, attracting water molecules, and in an aqueous solution, it dissociates according to the following dissociation reaction, as mentioned already:

$$ZnCl_{2(aq)} \rightarrow Zn^{2+}_{(aq)} + 2Cl^-_{(aq)} \tag{4.71}$$

In the zinc-chloride cell, the oxidation reaction at the Zn anode is the same as for the historic Daniell and Leclanché cells or the zinc–carbon cell, that is, $Zn \rightarrow Zn^{2+}_{(aq)} + 2e^-$. Besides this standard oxidation reaction, the following oxidation reaction might also occur in parallel:

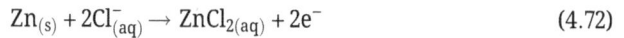

$$Zn_{(s)} + 2Cl^-_{(aq)} \rightarrow ZnCl_{2(aq)} + 2e^- \tag{4.72}$$

Here, at the surface of the anode, $Cl^-$ ions in the electrolyte can react with metallic zinc, and $ZnCl_2$ is formed that enters the solution, where it dissociates according to the dissociation reaction (4.71). During this reaction, two electrons again remain in the electrode.

At the cathode, reduction occurs according to the following reaction formula:

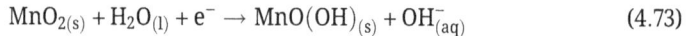

$$MnO_{2(s)} + H_2O_{(1)} + e^- \rightarrow MnO(OH)_{(s)} + OH^-_{(aq)} \tag{4.73}$$

Thus, the reduction of tetravalent manganese(IV) oxide to trivalent manganese(III) oxide-hydroxide occurs in the same manner as repeatedly mentioned earlier.

In the electrolyte, a complex formation – that is, the formation of zinc oxychloride – occurs:

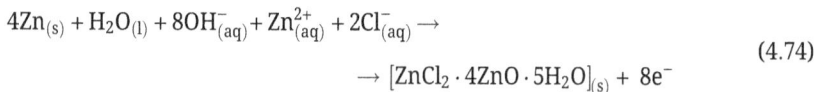

$$4Zn_{(s)} + H_2O_{(1)} + 8OH^-_{(aq)} + Zn^{2+}_{(aq)} + 2Cl^-_{(aq)} \rightarrow$$
$$\rightarrow [ZnCl_2 \cdot 4ZnO \cdot 5H_2O]_{(s)} + 8e^- \tag{4.74}$$

Zinc oxychloride binds water. Hence, during the discharge process, the water is consumed. On the one hand, this is very useful, since the zinc-chloride cells dry out during operation, and therefore, the complex formation according to (4.74) prevents leakage of the cell. On the other hand, the housings of the zinc-chloride cell must be done carefully to prevent dehydration of the cell, since the redox reaction needs water, and it is not useful to lose extra water due to bad housings.

Finally, the overall redox reaction of the zinc-chloride cell can be written as follows:

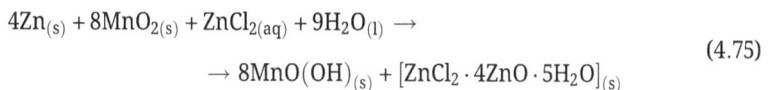

$$4Zn_{(s)} + 8MnO_{2(s)} + ZnCl_{2(aq)} + 9H_2O_{(1)} \rightarrow$$
$$\rightarrow 8MnO(OH)_{(s)} + [ZnCl_2 \cdot 4ZnO \cdot 5H_2O]_{(s)} \tag{4.75}$$

The zinc oxychloride drops out and deposits onto the electrodes. Therefore, during the discharge process, the cell resistance increases, and thus, the delivered voltage decreases. However, the degradation of the zinc-chloride cell due to the dropout of zinc oxychloride is less distinct than the degradation of the zinc–carbon cell, where diamine zinc chloride drops out.

At this point, it has to be mentioned that, very often, both the zinc–carbon cell and the zinc-chloride cell are summarized under the generic term zinc–carbon cell. However, for a better discrimination of both successors of the Leclanché cell, we prefer, in this textbook, the usage of a distinct nomenclature with the two different designations.

Compared to the primary zinc–carbon and zinc-chloride cells, alkaline batteries exhibit much better properties, particularly concerning the lifetime of the battery. This primary cell is discussed in the next section.

### 4.3.3 Alkaline batteries

The alkaline battery (or alkaline cell) with a nominal voltage of 1.5 V is a primary cell that is based on the electrochemical redox reactions between metallic zinc and manganese(IV) oxide ($MnO_2$) – similar to the zinc–carbon and zinc-chloride cell. Therefore, it can also be regarded as a successor of the historical Leclanché cell. However, in the case of the alkaline battery, the alkaline electrolyte potassium hydroxide (KOH) is used instead of the acidic ammonium chloride or zinc chloride used for the zinc–carbon and zinc-chloride cells, respectively. The alkaline electrolyte is also responsible for the naming of the alkaline battery. Other galvanic cells or batteries use alkaline electrolytes too, but they employ electrodes with other active materials, that is, not Zn or $MnO_2$. In general, alkaline batteries exhibit better cell performance – especially with regard to lifetime and energy density – than its antecessors the zinc–carbon and zinc-chloride cells. In addition, alkaline cells can be stored much longer than zinc–carbon and zinc-chloride cells, that is, the so-called shelf life is longer, since the self-discharge is less pronounced.

Although the outer dimensions and the locations of the positive and negative terminals of alkaline batteries are the same as those for zinc–carbon or zinc-chloride cells, the general construction is significantly modified, as shown in Figure 4.16. Alkaline batteries exhibit a cylindrical housing, where the negative terminal is located at the bottom of the device and the positive terminal at the top. However, the zinc anode of the alkaline battery is located at the center of the cylindrical housing and the manganese dioxide cathode at the outer side of the container (compare Figures 4.15 and 4.16).

The anode is composed of a dispersion of zinc powder in a gel that also contains the KOH electrolyte. Since zinc powder is used, the surface area at the zinc–electrolyte interface is strongly increased, as compared to the zinc–carbon and zinc-chloride cells with the metal zinc can as the anode (Figure 4.15); therefore, the internal resistance of the alkaline battery is much lower. A metallic pin – the collector – serves as the electri-

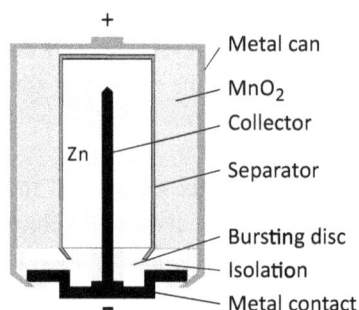

**Figure 4.16:** Construction of an alkaline cell. Zn and $MnO_2$ powders are used to increase the contact surfaces of the anode and cathode materials, respectively. The powders are soaked in the KOH electrolyte solution. The separator is an ionic conductor and an electrical insulator.

cal contact between the zinc powder and the negative terminal at the bottom of the cell. Since the zinc anode that erodes during operation, is located in the center of the cell container, the leakage tendency of the alkaline battery is reduced. The outer metallic container of the alkaline battery, that is in close contact with the manganese(IV) oxide, acts as the electrical contact to the positive terminal on top of the cell.

A separator electrically isolates the negative and the positive electrodes (Figure 4.16); that is, it prevents electric short-circuit faults within the alkaline battery. However, of course, the separator has to be a well ionic conductor to provide good cell properties with low internal ionic resistance. Furthermore, it has to be chemically resistant against the strongly alkaline electrolyte. The separator is made of a synthetic polymer or a cellulose layer.

Quite often, a bursting disk is introduced at the bottom of the cell construction (Figure 4.16). It works as a protection against overpressure within the cell that can arise, if the discharge current is too high; therefore, the temperature of the cell is strongly increased.

Alkaline batteries use the alkaline potassium hydroxide (KOH) electrolyte. KOH is a strong base; and it dissociates according to the following dissociation reaction:

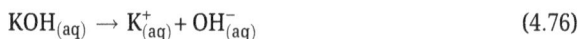

$$KOH_{(aq)} \rightarrow K^+_{(aq)} + OH^-_{(aq)} \tag{4.76}$$

At the anode of the alkaline battery, oxidation occurs. As mentioned already, metallic zinc atoms are oxidized, leaving two electrons per atom in the anode. Hence, the oxidation number of zinc increases from 0 to +2. However, the particular product of the oxidation reaction depends on the oxidation conditions, and various reactions occur. In the beginning of the discharging process, the concentration of $OH^-$ ions in the aqueous solution (electrolyte) is rather high. Under these early discharge conditions, the following oxidation reaction occurs:

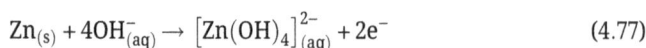

$$Zn_{(s)} + 4OH^-_{(aq)} \rightarrow \left[Zn(OH)_4\right]^{2-}_{(aq)} + 2e^- \tag{4.77}$$

At this stage of the oxidation reaction, where one Zn atom reacts with four $OH^-$ ions, a zincate ion is formed, that is, the doubly charged tetrahydroxozincate ion [Zn

$(OH)_4]^{2-}$. If the electrolyte is oversaturated with tetrahydroxozincate ions, zinc oxide (ZnO) drops out, that is:

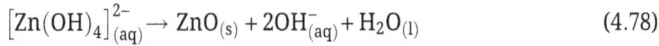

$$\left[Zn(OH)_4\right]^{2-}_{(aq)} \rightarrow ZnO_{(s)} + 2OH^-_{(aq)} + H_2O_{(l)} \tag{4.78}$$

If the discharging process in the alkaline battery continues at lower $OH^-$ concentrations, zinc hydroxide, $Zn(OH)_2$, is formed according to the following oxidation reaction:

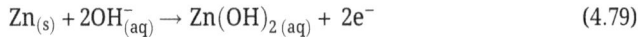

$$Zn_{(s)} + 2OH^-_{(aq)} \rightarrow Zn(OH)_{2\,(aq)} + 2e^- \tag{4.79}$$

Now, at lower $OH^-$ concentrations, one Zn atom reacts with only two $OH^-$ ions. Afterward, the zinc hydroxide slowly converts into water and zinc oxide, whereupon the latter one drops out, that is:

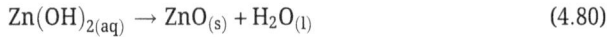

$$Zn(OH)_{2(aq)} \rightarrow ZnO_{(s)} + H_2O_{(l)} \tag{4.80}$$

The overall oxidation reaction can be summarized as follows:

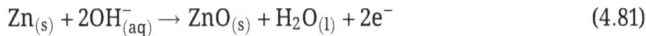

$$Zn_{(s)} + 2OH^-_{(aq)} \rightarrow ZnO_{(s)} + H_2O_{(l)} + 2e^- \tag{4.81}$$

Beside the formation of zinc hydroxide according to the above-mentioned reactions, the doubly charged tetrahydroxozincate ion $[Zn(OH)_4]^{2-}$ can combine with two potassium ($K^+$) ions in the electrolyte and form potassium tetrahydroxozincate (or, in brief, potassium zincate) that contributes to the battery degradation:

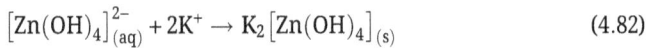

$$\left[Zn(OH)_4\right]^{2-}_{(aq)} + 2K^+ \rightarrow K_2\left[Zn(OH)_4\right]_{(s)} \tag{4.82}$$

Alternatively, potassium tetrahydroxozincate can also be formed by the reaction of zinc hydroxide and potassium hydroxide:

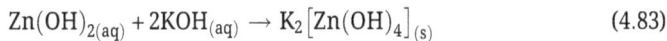

$$Zn(OH)_{2(aq)} + 2KOH_{(aq)} \rightarrow K_2\left[Zn(OH)_4\right]_{(s)} \tag{4.83}$$

At the cathode of the alkaline battery, tetravalent manganese(IV) oxide ($MnO_2$) is reduced to trivalent manganese(III) oxide-hydroxide ($MnO(OH)$). This reaction is equal to the reduction reactions (4.38) and (4.73) in the Leclanché cell (zinc–carbon cell) and the zinc-chloride cell shown earlier:

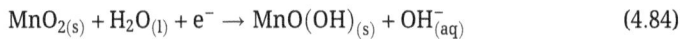

$$MnO_{2(s)} + H_2O_{(l)} + e^- \rightarrow MnO(OH)_{(s)} + OH^-_{(aq)} \tag{4.84}$$

Again, the oxidation number of manganese reduces from +4 to +3, and an additional $H^+$ ion is built into the crystal lattice of $MnO_2$ and forms $MnO(OH)$. It is interesting that both manganese(IV) oxide and manganese(III) oxide-hydroxide exhibit the same crystal structure.

In summary, the redox reaction of the alkaline battery cell can be written as follows:

$$Zn_{(s)} + 2MnO_{2(s)} + 2H_2O_{(l)} + 2OH^-_{(aq)} \rightarrow \left[Zn(OH)_4\right]^{2-}_{(aq)} + 2MnO(OH)_{(s)} \qquad (4.85)$$

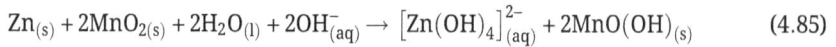

As mentioned in the redox reaction (4.85) and reaction (4.78), the alkaline electrolyte of potassium hydroxide (KOH) plays only an intermediate role in the electrochemical reactions of the alkaline battery. In particular, we have seen in eq. (4.77) that during the oxidation reaction, one Zn atom reacts with four $OH^-$ ions, and the doubly charged tetrahydroxozincate ion $[Zn(OH)_4]^{2-}$ is formed. As mentioned earlier, if, after a while, the electrolyte is oversaturated with tetrahydroxozincate ions, zinc oxide (ZnO) drops out. Therefore, if we integrate reaction (4.78) into the redox reaction (4.85), we obtain:

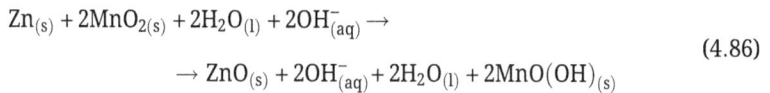

$$Zn_{(s)} + 2MnO_{2(s)} + 2H_2O_{(l)} + 2OH^-_{(aq)} \rightarrow$$
$$\rightarrow ZnO_{(s)} + 2OH^-_{(aq)} + 2H_2O_{(l)} + 2MnO(OH)_{(s)} \qquad (4.86)$$

We can see that the overall amount of KOH – or better, the concentration of the $OH^-$ ions in the electrolyte – remains constant in the alkaline battery. In particular, during the reactions, equal amounts of $OH^-$ ions are consumed and produced.

Here, we should add that during a very mild (or slow) discharging process, trivalent manganese(III) oxide-hydroxide can be reduced further by a rather slow reduction reaction to divalent manganese(II) hydroxide. This second step of the reduction reaction can be described by the following reaction formula:

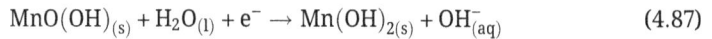

$$MnO(OH)_{(s)} + H_2O_{(l)} + e^- \rightarrow Mn(OH)_{2(s)} + OH^-_{(aq)} \qquad (4.87)$$

During this reaction, the divalent manganese(II) hydroxide – $Mn(OH)_2$ – drops out. It has a poor solubility in water, and hence, also in the aqueous potassium hydroxide solution (electrolyte).

Reaction (4.87) works since the water molecule undergoes a self-ionization reaction, as we have explained in Section 4.2.2, that is, $H_2O$ deprotonates, meaning that it loses a proton ($H^+$) and leaves a negatively charged hydroxide ion ($OH^-$) behind (Figure 4.10), that is:

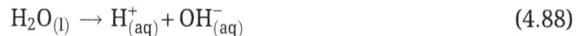

$$H_2O_{(l)} \rightarrow H^+_{(aq)} + OH^-_{(aq)} \qquad (4.88)$$

Instead of protonating another hydrogen molecule according to the well-known formula $2H_2O \rightarrow H_3O^+ + OH^-$, the proton is added to the trivalent manganese(III) oxide-hydroxide (MnO(OH)); hence, divalent manganese(II) hydroxide (Mn(OH)$_2$) is formed that drops out.

Finally, yet importantly, it has to be mentioned that zinc is not stable in a strongly alkaline solution. Therefore, the following redox side reaction occurs during the storage of charged or partly discharged alkaline batteries:

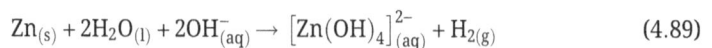

$$Zn_{(s)} + 2H_2O_{(l)} + 2OH^-_{(aq)} \rightarrow \left[Zn(OH)_4\right]^{2-}_{(aq)} + H_{2(g)} \qquad (4.89)$$

Here, the zinc anode oxidizes, and water reduces to gaseous hydrogen ($H_2$). If the anode is made from very pure zinc, this reaction runs quite slowly and does not play a significant role. However, even small contaminations, for instance, with heavy metals like iron, copper, or nickel, significantly increase the production of $H_2$ gas. During mass production, such contaminations are hardly avoidable – especially for the low-cost production of cheap batteries. Therefore, the bursting disk at the bottom of the cell construction (Figure 4.16) is introduced also for the reason that overpressure within the cell occurs due to the side reaction (4.89), where $H_2$ gas is formed.

It has been mentioned that the zinc anodes of dry cells like the zinc–carbon or the alkaline cells were amalgamated with mercury (Hg). This widely reduces side reactions of the metallic zinc with the electrolyte. Those side reactions reduce the lifetime and, especially, the storage time of the battery. Zinc amalgam, denoted as Zn (Hg), is an alloy of zinc and mercury. As mentioned earlier, in the past, the zinc anodes were amalgamated with mercury to prevent damage to the electrode during storage. Zinc amalgam protects zinc electrodes from corrosion; in particular, zinc amalgam does not release hydrogen or hydrogen gas that can appear by side reactions in galvanic cells, for instance, according to reaction (4.89). The mercury itself does not participate in the electrochemical redox reactions responsible for the operation of batteries with zinc anodes. Nowadays, amalgamation of the zinc anodes is not necessary anymore, since much purer zinc is employed for the anodes that significantly reduce unwanted side reactions by itself. Due to its harmfulness, nowadays mercury is widely eliminated in batteries.

### 4.3.4 Mercury batteries

A mercury battery is a primary battery that delivers a nearly constant voltage of 1.35 V until close to the end of its life. The voltage change remains in the order of 1% for several years under light load and over a wide temperature range. These properties make the mercury cell a very useful device. However, the mercury within the cell is rather problematic, and nowadays batteries that contain mercury are widely banned.

The electrochemical cell reaction is based on the redox reaction between a zinc anode and a mercuric oxide (HgO) cathode. Mercury is divalent in the mercuric oxide, that is, we speak about mercury(II) oxide. HgO is hardly soluble in water. Potassium hydroxide (KOH) is used as an electrolyte, but sodium hydroxide (NaOH) might also be used. Mercury batteries using a sodium hydroxide electrolyte provide a nearly constant voltage at low discharge currents. They were useful, for instance, for the application in hearing aids, watches or calculators. On the other hand, mercury batteries using potassium hydroxide as the electrolyte provide a constant voltage at higher currents that was useful for the application in photographic cameras with a flash. They also exhibit better performances at lower temperatures than those mercury batteries using a NaOH electrolyte. Upon dissociation, both electrolytes provide $OH^-$ ions – as already mentioned – according to the dissociation reactions: $NaOH \rightarrow Na^+ + OH^-$ and $KOH \rightarrow K^+ + OH^-$, respectively.

The cross-sectional layout of a button-type mercury battery is shown in Figure 4.17. The layout is rather simple. The mercuric oxide cathode is located in the lower part, and the zinc anode in the upper part of the primary cell. Hence, the positive terminal, that is, the cathode is located at the bottom and the negative terminal at the top of the mercury battery. A separator that is soaked with the potassium or sodium hydroxide electrolyte, separates the anode and the cathode. It is an ion conductor and is made of a synthetic polymer or a nonwoven cellulose layer. The separator has to be stable in the alkaline electrolyte solution.

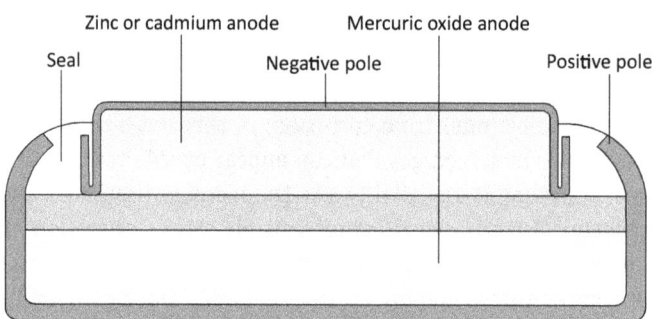

**Figure 4.17:** Cross-sectional layout of a bottom-type mercury battery.

The oxidation at the zinc anode of the mercury battery occurs in the same way as was already shown for the alkaline battery, that is, reactions (4.77)–(4.81) can be applied again and finally result in the same oxidation reaction as presented in the last section.

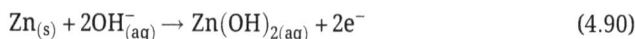

$$Zn_{(s)} + 2OH^-_{(aq)} \rightarrow Zn(OH)_{2(aq)} + 2e^- \tag{4.90}$$

Taking into account that zinc hydroxide – $Zn(OH)_2$ – slowly converts into water ($H_2O$) and zinc oxide (ZnO) that drops out, we get for the final oxidation reaction:

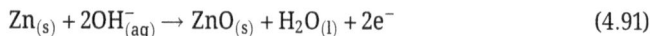

$$Zn_{(s)} + 2OH^-_{(aq)} \rightarrow ZnO_{(s)} + H_2O_{(l)} + 2e^- \tag{4.91}$$

Both oxidation reactions (4.90) and (4.91) were already presented in the last section for the alkaline battery (reactions (4.79) and (4.81)).

At the cathode, divalent mercury(II) oxide (HgO) is reduced to metallic mercury (oxidation number 0) according to the following reduction equation:

$$HgO_{(s)} + H_2O_{(l)} + 2e^- \rightarrow Hg_{(s)} + 2OH^-_{(aq)} \tag{4.92}$$

The overall redox reaction summarizing the anodic and cathodic half-cell reactions can be written as

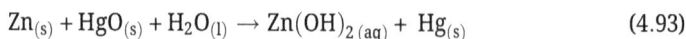

$$Zn_{(s)} + HgO_{(s)} + H_2O_{(l)} \rightarrow Zn(OH)_{2\,(aq)} + Hg_{(s)} \tag{4.93}$$

Taking into account again that zinc hydroxide slowly converts into water and zinc oxide, we get the following final redox reaction:

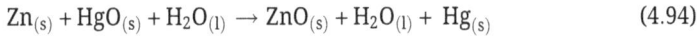

$$Zn_{(s)} + HgO_{(s)} + H_2O_{(l)} \rightarrow ZnO_{(s)} + H_2O_{(l)} + Hg_{(s)} \tag{4.94}$$

or rather:

$$Zn_{(s)} + HgO_{(s)} \rightarrow ZnO_{(s)} + Hg_{(s)} \tag{4.95}$$

During the discharging process in the anodic half-cell of the mercury battery, metallic zinc is oxidized, losing two electrons per zinc atom. The $Zn^{2+}$ ion enters the aqueous solution (electrolyte). On the other hand, in the cathodic half-cell of the mercury battery, mercuric oxide (HgO) is reduced, gaining two electrons and neutral elemental mercury forms, whereupon the redundant oxygen ion ($O^{2-}$) enters the aqueous solution (electrolyte), joins the $Zn^{2+}$ ion and forms ZnO that drops out.

The redox reaction of the mercury battery can be abbreviated as follows:

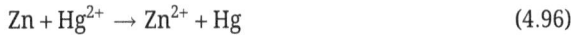

$$Zn + Hg^{2+} \rightarrow Zn^{2+} + Hg \tag{4.96}$$

Hence, during the discharging process of mercury batteries, an electron transfer from zinc to mercury occurs.

Similar to the situation in the alkaline battery, the evolution of hydrogen gas can occur in mercury batteries at the end of their life as well. As already mentioned, zinc is not stable in a strongly alkaline solution; therefore, the side reaction (4.89) occurs, where zinc oxidizes and water reduces to gaseous hydrogen ($H_2$). This unwanted reaction can be prevented by adding a small surplus of mercuric oxide to the mercury battery, that amalgamates the zinc electrode and suppresses the unwanted $H_2$ formation. However, better results without the usage of mercury can also be obtained when purer zinc anodes are employed, as mentioned in the previous section.

### 4.3.5 Silver oxide batteries

The silver oxide battery is a primary cell that exhibits an energy density of about 130 Wh/kg. The nominal voltage is 1.55 V. The cross-sectional layout of a button-type silver oxide battery is similar to the setup of the mercury battery shown in Figure 4.17. The anode is made, again, from zinc powder that is oxidized during the discharging process. But the cathode is now made from monovalent silver(I) oxide ($Ag_2O$) that is reduced to metallic silver during the discharging process. An ion-conducting separator (e.g., polymeric cloth) separates the anodic and cathodic half-cells. In silver oxide batteries, an alkaline electrolyte is used. The electrolyte can be an aqueous solution of potassium hydroxide (KOH); sodium hydroxide (NaOH) can likewise be employed.

Again, the oxidation reaction is given by

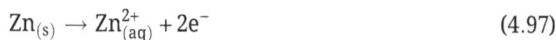

$$Zn_{(s)} \rightarrow Zn^{2+}_{(aq)} + 2e^- \tag{4.97}$$

Taking into account that the doubly charged $Zn^{2+}$ ions enter the aqueous solution and reacts with the hydroxide ions, we get for the oxidation reaction the following equation, where finally zinc oxide drops out:

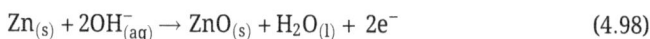

$$Zn_{(s)} + 2OH^-_{(aq)} \rightarrow ZnO_{(s)} + H_2O_{(l)} + 2e^- \tag{4.98}$$

These oxidation processes and the involved reactions were already presented in detail in the previous sections.

The reduction at the $Ag_2O$ cathode of the silver oxide battery occurs according to the following reaction:

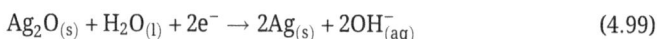

$$Ag_2O_{(s)} + H_2O_{(l)} + 2e^- \rightarrow 2Ag_{(s)} + 2OH^-_{(aq)} \tag{4.99}$$

In brief, the electron transfer from the metallic zinc atom to the silver atom defines the redox reaction of the silver oxide battery:

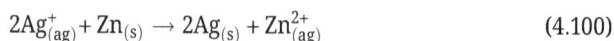

$$2Ag^+_{(ag)} + Zn_{(s)} \rightarrow 2Ag_{(s)} + Zn^{2+}_{(ag)} \tag{4.100}$$

The overall situation can be summarized by the following redox reaction:

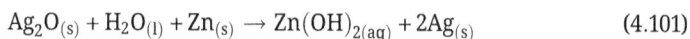

$$Ag_2O_{(s)} + H_2O_{(l)} + Zn_{(s)} \rightarrow Zn(OH)_{2(aq)} + 2Ag_{(s)} \tag{4.101}$$

Moreover, taking into account again that zinc hydroxide slowly converts into water and zinc oxide, we get the following final redox reaction, where the zinc oxide drops out:

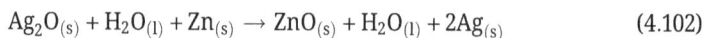

$$Ag_2O_{(s)} + H_2O_{(l)} + Zn_{(s)} \rightarrow ZnO_{(s)} + H_2O_{(l)} + 2Ag_{(s)} \tag{4.102}$$

or rather:

$$Ag_2O_{(s)} + Zn_{(s)} \rightarrow ZnO_{(s)} + 2Ag_{(s)} \tag{4.103}$$

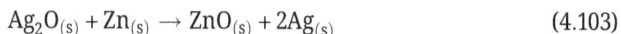

The side and intermediate reactions are the same as those explained in the previous sections. Side reactions like the formation of potassium tetrahydroxozincate according to reactions (4.82) and (4.83), can occur in silver oxide batteries too. In addition, if a sodium hydroxide (NaOH) electrolyte is used, sodium tetrahydroxozincate – or, in brief, sodium zincate – can be formed according to the following reaction:

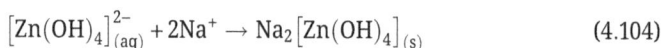

$$[Zn(OH)_4]^{2-}_{(aq)} + 2Na^+ \rightarrow Na_2[Zn(OH)_4]_{(s)} \tag{4.104}$$

Alternatively, sodium tetrahydroxozincate can also be formed by the reaction of zinc hydroxide and sodium hydroxide:

$$Zn(OH)_{2(aq)} + 2NaOH_{(aq)} \rightarrow Na_2[Zn(OH)_4]_{(s)} \tag{4.105}$$

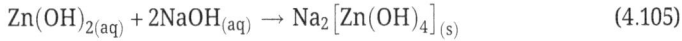

Potassium and sodium tetrahydroxozincates do not participate in the electrochemical reactions for battery operation anymore; therefore, their formation contributes to the degradation of silver oxide cells.

It is to add that in the past, amalgamated zinc anodes were used for the silver oxide battery to inhibit corrosion of the anode in the alkaline environment. Hence, due to the mercury content in the cell, the batteries become hazardous, when leakage occurs. Nowadays, silver oxide batteries without added mercury are fabricated using purer zinc anodes, as was previously mentioned.

### 4.3.6 Zinc–air batteries

Zinc–air batteries are primary cells that provide, in real applications, voltages around 1.2 V, although the theoretical nominal voltage should be higher (i.e., 1.6 V). The energy density of a zinc–air battery is about 1.3 kWh/kg. Battery operation is based on a redox reaction between zinc and oxygen. Since, in general, the oxygen reduction at the cathode is somewhat retarded, the theoretical voltage of 1.65 V cannot be achieved in practical applications. The zinc–air cells provide high energy densities, and their production is not very expensive. These properties make them quite useful primary cells – in particular, since the zinc–air battery is the type of battery that could replace the mercury battery that is very problematic due to the harmful mercury inside. The construction of the zinc–air battery can be explained with the help of Figure 4.18, where a cross-sectional layout of a button-type zinc–air battery is shown. However, the construction and sizes of zinc–air batteries are rather variable and can range from small button cells up to larger batteries.

As shown in Figure 4.18, the construction of the zinc–air battery is rather simple. A paste of zinc powder that is mixed with an aqueous solution of potassium hydroxide – the KOH electrolyte – forms the anode. A carbon lattice, where activated carbon absorbs and binds oxygen from the air, constitutes the cathode. An ion-conducting separator (e.g., polymeric cloth) separates the anodic and cathodic half-cells of the zinc–air battery. Air can enter the setup through a vent and delivers oxygen to the system. During storage, the vent has to be closed to prevent self-discharge – this is simply done with tape. During operation, the vent of the zinc–air battery has to be open.

During discharge, the zinc particles in the paste that is formed by the mixture of zinc powder and the KOH electrolyte are oxidized according to the following oxidation reaction:

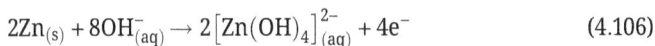

$$2Zn_{(s)} + 8OH^-_{(aq)} \rightarrow 2[Zn(OH)_4]^{2-}_{(aq)} + 4e^- \tag{4.106}$$

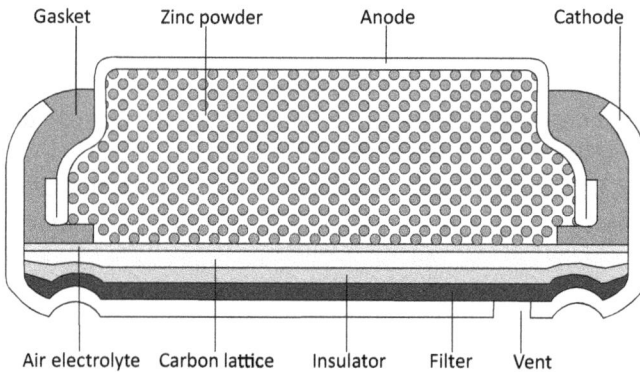

**Figure 4.18:** Cross-sectional layout of a bottom-type zinc–air battery.

Via the negative terminal of the zinc–air battery and the external load, the electrons are wandering to the cathode.

At the cathode of the zinc–air battery, oxygen from the ambient air reacts according to the following reduction reaction:

$$O_{2(g)} + 2H_2O_{(l)} + 4e^- \rightarrow 4OH^-_{(aq)} \tag{4.107}$$

During this reaction, hydroxide ions are formed that migrate through the zinc/electrolyte paste.

In the electrolyte, the doubly charged tetrahydroxozincate $[Zn(OH)_4]^{2-}$ ion decomposes to zinc oxide, water and two hydroxide ions:

$$2[Zn(OH)_4]^{2-}_{(aq)} \rightarrow 2ZnO_{(s)} + 2H_2O_{(l)} + 4OH^-_{(aq)} \tag{4.108}$$

In summary, the redox reaction can be written as follows:

$$2Zn_{(s)} + O_{2(g)} + 2H_2O_{(l)} \rightarrow 2Zn(OH)_{2(aq)} \tag{4.109}$$

Zinc hydroxide decomposes into zinc oxide and water, that is,

$$2Zn(OH)_{2(aq)} \rightarrow 2ZnO_{(s)} + 2H_2O_{(l)} \tag{4.110}$$

Therefore, the redox reaction (4.109) can be extended to

$$2Zn_{(s)} + O_{2(g)} + 2H_2O_{(l)} \rightarrow 2ZnO_{(s)} + 2H_2O_{(l)} \tag{4.111}$$

or rather

$$2Zn_{(s)} + O_{2(g)} \rightarrow 2ZnO_{(s)} \tag{4.112}$$

Since, for this redox reaction, air or rather oxygen has to enter the galvanic cell, zinc–air batteries cannot be used in a sealed battery holder or in a vacuum environment. The equivalent of 1,000 cm$^3$ (1 L) of air is required for every ampere-hour of the capacity that is provided by the zinc–air battery.

Zinc–air batteries have some properties that are comparable to the properties of fuel cells. In this regard, zinc can be considered the fuel. The reaction rate might be controlled by the airflow that can vary. Finally, the paste of a mixture of oxidized zinc powder and electrolyte might be replaced with a fresh paste of zinc powder and electrolyte. This is called mechanical recharge. Such mechanically rechargeable batteries can be used for electric vehicles. Vehicles can be recharged via exchanging the paste of exhausted electrolyte and (oxidized) zinc for fresh reactants at a service station. For that matter, small zinc granules serve as the reactant.

Zinc–air batteries can be assigned to the family of metal–air batteries. Another family member will be discussed in the following section.

### 4.3.7 Aluminum–air batteries

The aluminum–air battery is a primary cell that has one of the highest energy densities of all batteries. According to theory, the specific energy density has a value of about 8 kWh/kg; however, practical values reach up to about 1.3 kWh/kg. The theoretical cell voltage is 2.7 V; however, the practical operation voltages delivered by aluminum–air batteries are 1.2–1.6 V. Therefore, the nominal voltage is set to 1.2 V if a potassium hydroxide (KOH) electrolyte is used. Aluminum–air batteries using an aqueous NaCl solution as an electrolyte provide lower voltages; here, the nominal voltage is 0.7 V. In the present case, for the description of aluminum–air batteries, we only consider the KOH electrolyte.

The aluminum–air battery can be constructed similar to the zinc–air battery that was shown in Figure 4.18. Here, in the case of the aluminum–air battery, the zinc powder has to be replaced by aluminum powder, as shown in Figure 4.19. The aluminum powder is mixed with the KOH electrolyte; hence, a paste of aluminum powder and an aqueous solution of potassium hydroxide form the anode. The cathode is constructed in a similar way as for the zinc–air battery; that is, it consists of a carbon lattice, where activated carbon absorbs and binds oxygen from the air that can enter the galvanic cell through a vent. Then, the battery operation is based on the reaction of aluminum with the oxygen in the air. A separator separates both half-cells of the aluminum–air battery.

In the anodic half-cell of the aluminum–air battery, metallic aluminum is oxidized according to the following formula:

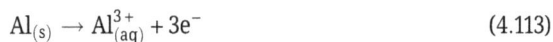

$$Al_{(s)} \rightarrow Al^{3+}_{(aq)} + 3e^-$$

(4.113)

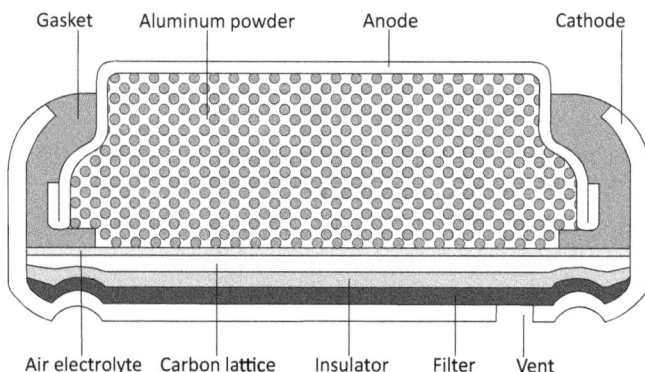

**Figure 4.19:** Cross-sectional layout of a bottom-type aluminum–air battery.

The triply charged aluminum ion enters the aqueous solution, reacts with the $OH^-$ ions from the dissociated KOH molecules, and forms aluminum hydroxide ($Al(OH)_3$). Hence, the anodic half-cell reaction (oxidation) can be written as

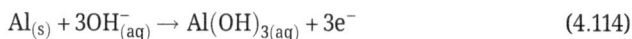

$$Al_{(s)} + 3OH^-_{(aq)} \rightarrow Al(OH)_{3(aq)} + 3e^- \tag{4.114}$$

At the cathode of the aluminum–air battery, exactly the same reaction occurs as was presented for the zinc–air battery (compare eq. (4.107)); that is, the reduction reaction in the cathodic half-cell of the aluminum–air battery is given by

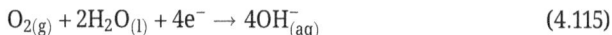

$$O_{2(g)} + 2H_2O_{(l)} + 4e^- \rightarrow 4OH^-_{(aq)} \tag{4.115}$$

The hydroxide ions that are formed during the reduction reaction migrate through the aluminum/electrolyte paste.

In summary, the redox reaction for the aluminum–air battery can be written as follows:

$$4Al_{(s)} + 3O_{2(g)} + 6H_2O_{(l)} \rightarrow 4Al(OH)_{3(aq)} \tag{4.116}$$

In aluminum–air batteries, the following side reaction can occur:

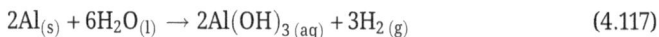

$$2Al_{(s)} + 6H_2O_{(l)} \rightarrow 2Al(OH)_{3\,(aq)} + 3H_{2\,(g)} \tag{4.117}$$

This side reaction is problematic since it causes corrosion and releases hydrogen gas.

Another problem with aluminum–air batteries is the thermodynamically favorable formation of protective aluminum oxide films ($Al_2O_3$) that form spontaneously on aluminum surfaces when they are exposed to air or an aqueous solution. $Al_2O_3$ formation at the surface of the aluminum particles of the anode suppresses the oxidation process and degrades the negative electrode of the aluminum–air battery. The aluminum in $Al_2O_3$ is trivalent; therefore, we speak about aluminum(III) oxide.

Moreover, in a base, that is, in the aqueous solution (the KOH electrolyte) of the aluminum–air battery, aluminum hydroxide that is formed in the redox reaction (4.116) and in the side reaction (4.117), acts as a Lewis acid, taking an electron pair from the hydroxide ion:

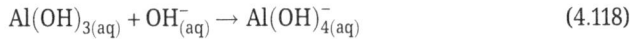

$$Al(OH)_{3(aq)} + OH^-_{(aq)} \rightarrow Al(OH)^-_{4(aq)} \tag{4.118}$$

Afterward, two tetrahydroxoaluminate ($Al(OH)_4^-$) ions can decompose and form aluminum(III) oxide, water and hydroxide ions, that is:

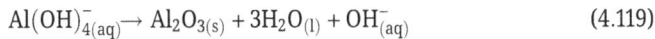

$$Al(OH)^-_{4(aq)} \rightarrow Al_2O_{3(s)} + 3H_2O_{(l)} + OH^-_{(aq)} \tag{4.119}$$

The aluminum(III) oxide is insoluble. It is deposited on the surface of the aluminum particles of the anode; hence, it degrades the electrode.

If the aluminum anode is consumed, that is, if according to the above-mentioned redox and side reactions in the aluminum–air cell, aluminum was completely transformed to aluminum hydroxide or aluminum(III) oxide, and the battery does not produce electricity anymore.

It has been mentioned that it is possible to mechanically recharge the aluminum–air battery with new anodes (i.e., a paste of aluminum and KOH electrolyte). Hereby, aluminum can be employed that was recycled from the aluminum hydroxide produced within the galvanic cell during the discharging process.

Aluminum–air batteries exhibit promising properties (especially the high energy density), but up to now, they are not widely used because of the mentioned problems. Moreover, the fabrication of the anodes is rather expensive. Mechanical recharging that in principle is possible, is quite expensive too.

## 4.3.8 Lithium batteries

Lithium batteries are primary cells that should not be confused with the rechargeable lithium-ion batteries (LIBs) that will be discussed later in detail (Section 4.4.4). The anodes of lithium batteries are always made from metallic lithium (Li). Lithium exhibits one of the highest negative values of the standard potential, that is, $E_{Li}^0 = -3.05$ V against the standard hydrogen electrode. Therefore, lithium is an excellent material for use as an anode in galvanic cells. A variety of cathode materials and electrolytes are used for the construction of lithium batteries; hence, a quite large number of different versions of lithium primary cells can be found, and the family of lithium batteries is quite large. Cell voltages between about 1.5 and 3.8 V are possible depending on the materials and chemical compounds that are employed for the lithium batteries – especially for the cathode materials. Despite rather large cell voltages, the capacities and gravimetric and/or volumetric energy densities of lithium batteries are also quite high – this is beneficial for this type of battery, too.

Later, we will exemplarily present some members of the lithium battery family, that is, the lithium manganese dioxide battery, the lithium thionyl chloride battery and the lithium sulfur dioxide battery.

We start with the lithium manganese dioxide battery. Quite often, manganese(IV) oxide plays the role of the oxidizing agent in lithium batteries, too. This type of battery is also called a lithium manganese dioxide battery or $Li$–$MnO_2$ battery. It exhibits a nominal voltage of 3.0 V. The cathode of the $Li$–$MnO_2$ battery is made of tetravalent $MnO_2$ that is reduced during the discharging process or the corresponding redox reaction, respectively. A typical setup of a button-type $Li$–$MnO_2$ battery is shown in Figure 4.20. The anodic half-cell (metallic lithium) and the cathodic half-cell (paste of heat-treated manganese(IV) dioxide and a Teflon binder, for instance) are separated by an ion-conducting microporous separator that consists of a combination of polyethylene and polypropylene. The electrolyte is a lithium salt, for instance, lithium perchlorate ($LiClO_4$) in a mixture of organic solvents (e.g., propylene carbonate (PC) and dimethoxyethane). The electrolyte does not contain hydrogen due to the strong reactivity of lithium and hydrogen.

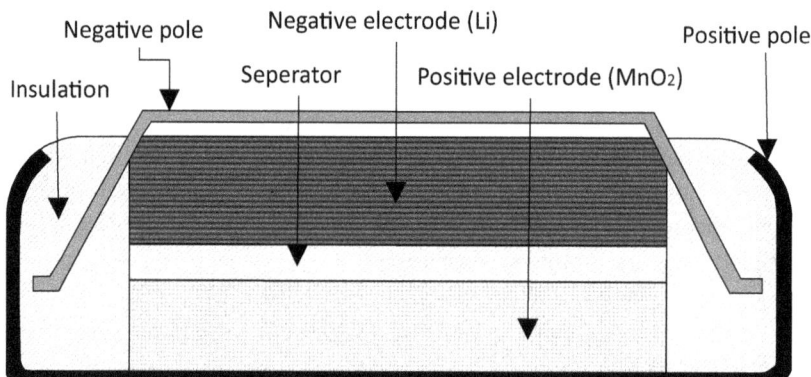

**Figure 4.20:** Cross-sectional layout of a bottom-type $Li$–$MnO_2$ battery.

At the anode of the $Li$–$MnO_2$ battery, the oxidation of metallic lithium occurs according to the following oxidation reaction:

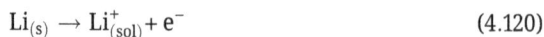

$$Li_{(s)} \rightarrow Li^+_{(sol)} + e^- \tag{4.120}$$

The subscript (sol) indicates that $Li^+$ is dissolved in organic solvents (and not in an aqueous solution). This oxidation reaction takes place in all lithium batteries.

At the cathode of the $Li$–$MnO_2$ battery, the reduction of tetravalent Mn(IV) in $MnO_2$ to trivalent Mn(III) occurs, where $Li^+$ is intercalated, too; hence, lithium–manganese dioxide is formed:

$$MnO_{2(s)} + Li^+_{(sol)} + e^- \rightarrow LiMnO_{2(s)} \qquad (4.121)$$

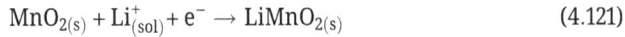

In total, the redox reaction that takes place in the Li–MnO$_2$ battery can be written as

$$Li_{(s)} + MnO_{2(s)} \rightarrow LiMnO_{2(s)} \qquad (4.122)$$

The Li–MnO$_2$ battery is a very common lithium battery. It can be used in the temperature range between about −30 and +60 °C; however, at higher temperatures, the self-discharge is rather high. For the fabrication of Li–MnO$_2$ batteries, relatively low-cost materials are employed. In general, the lifetime of Li–MnO$_2$ batteries is quite high, and they are suitable for low-cost applications. Moreover, the gravimetric and volumetric energy densities are also rather high; that is, gravimetric and volumetric energy densities of 280 Wh/kg and 580 Wh/L can be found, respectively. Altogether, these favorable properties make Li–MnO$_2$ batteries a common product for the consumer market.

The second example we want to look at is the lithium thionyl chloride battery, that provides a nominal cell voltage of 3.7 V. A typical setup of the Li–SOCl$_2$ battery is shown in Figure 4.21. A separator separates the metallic lithium anode from the cathode. The cathode is made from a mixture of liquid thionyl chloride (SOCl$_2$) and graphite. Graphite is employed to reduce the electrical resistance of the cathode. Thionyl chloride is not only the polarizer but also the solvent. Therefore, the electrolyte of the lithium thionyl chloride battery consists of lithium tetrachloroaluminate (LiAlCl$_4$) in liquid thionyl chloride, whereupon lithium tetrachloroaluminate dissociates according to the following dissociation reaction:

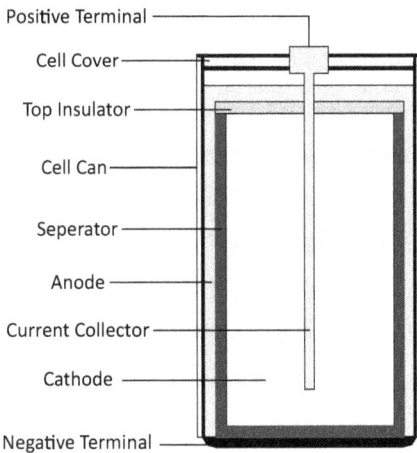

Positive Terminal

Cell Cover

Top Insulator

Cell Can

Seperator

Anode

Current Collector

Cathode

Negative Terminal

**Figure 4.21:** Typical setup of a lithium thionyl chloride battery.

$$LiAlCl_{4(sol)} \rightarrow Li^+_{(sol)} + AlCl^-_{4(aq)} \qquad (4.123)$$

At the anode, lithium oxidizes, and the $Li^+$ ions react with the $Cl^-$ ions in the electrolyte to form LiCl. The thionyl chloride ($SOCL_2$) in the cathode of the Li–$SOCl_2$ battery is reduced via several intermediate reaction steps to elemental sulfur (S). In parallel, sulfur dioxide ($SO_2$) is formed, too. Without going into much detail, the discharging process can be summarized by the redox reaction:

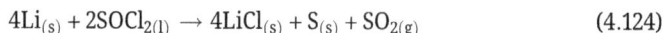

$$4Li_{(s)} + 2SOCl_{2(l)} \rightarrow 4LiCl_{(s)} + S_{(s)} + SO_{2(g)} \tag{4.124}$$

Li–$SOCl_2$ batteries exhibit some favorable properties. In particular, the gravimetric and volumetric energy densities are rather high, that is, 500 Wh/kg (or even higher) and 1,200 Wh/L, respectively. Moreover, they can operate over a wide temperature range; in particular, operation at temperatures down to −55 °C is possible. The shelf life is also quite high, and quite long storage is feasible. On the other hand, Li–$SOCl_2$ batteries contain highly toxic materials; and even they may explode, if they are shorted. Therefore, high costs, necessary safety features and special disposal procedures limit their use; that is, they are not used for consumer batteries.

The lithium sulfur dioxide battery was one of the early technically used lithium batteries. It was already developed in 1938. The Li–$SO_2$ battery exhibits a nominal voltage of 2.85 V. The gravimetric and volumetric energy densities have values of 250 Wh/kg and 400 Wh/L, respectively. It contains sulfur dioxide ($SO_2$) at high pressure. Therefore, $SO_2$ is in liquid form. Sulfur dioxide is the oxidizing agent (or oxidant); it is reduced during the redox reaction. Hence, the lithium sulfur dioxide battery has a liquid cathode. However, colloquially spoken, not the sulfur dioxide (i.e., the real cathode of the Li–$SO_2$ battery) is called the cathode, but the conducting structure of the electrode consisting of a mixture of Teflon and carbon black on a metal mesh. The electrolyte and solvent for the sulfur dioxide consist of lithium bromide (LiBr) and acetonitrile ($CH_3CN$). Lithium bromide plays the role of the conducting salt. A lithium foil (anode) is separated from the cathode by a microporous polypropylene separator.

For the Li–$SO_2$ battery, the discharging process can be described by the following simplified reactions. Again, metallic lithium is oxidized according to reaction (4.120). However, here we present the oxidation equation in the following form:

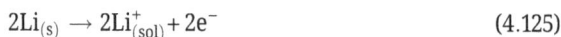

$$2Li_{(s)} \rightarrow 2Li^+_{(sol)} + 2e^- \tag{4.125}$$

In Li–$SO_2$ batteries, sulfur dioxide is reduced according to the reduction reaction:

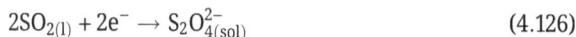

$$2SO_{2(l)} + 2e^- \rightarrow S_2O_{4(sol)}^{2-} \tag{4.126}$$

During this reduction reaction, dithionite ions ($S_2O_4^{2-}$) are formed. Finally, the oxidation and reduction reactions (4.125) and (4.126) can be merged into the following redox reaction:

$$2Li_{(s)} + 2SO_{2(l)} \rightarrow Li_2S_2O_{4(s)} \tag{4.127}$$

Lithium dithionite ($Li_2S_2O_4$) is hardly soluble. It is deposited in the pores of the cathode, and therefore, the cell resistance is enhanced.

The operation and disposal of Li–$SO_2$ batteries are rather problematic. They contain liquid $SO_2$ at high pressure and require safety vents. They can even explode under inappropriate conditions. Moreover, acetonitrile ($CH_3CN$) that is a component of the electrolyte, forms lithium cyanide (LiCN) and hydrogen cyanide (HCN) – the latter one at higher temperatures. Both substances are highly toxic. In total, Li–$SO_2$ batteries are expensive and face environmental and safety concerns. Hence, they are mainly used for military purposes, but not for consumer batteries.

A number of other types of lithium batteries are known; however, we will restrict ourselves to the examples discussed earlier.

## 4.4 Technically important secondary batteries

A secondary battery or secondary cell is a rechargeable galvanic cell. In contrast to the situation in primary cells, the electrochemical reactions in the secondary cells are reversible – at least to a large extent. During the discharging process, the electrochemical reactions in the secondary cell consume substances that are responsible for the electrochemical energy storage. At the negative electrode, an oxidation occurs, and at the positive electrode, a reduction takes place. During the discharging process at the electrodes, educts (the initial starting materials) are transformed into products (the final materials after the reaction). This is similar to the process in primary cells. However, if those substances (the educts) are exhausted and the electrochemical reactions stop, that is, if the secondary battery has finished its operation as a voltage source, the battery is not dead, since the reactions can be reversed and the initial charged state can be restored. During the charging process, at the positive electrode, a (forced) oxidation occurs and at the negative electrode, a (forced) reduction takes place, where electrons are consumed. These electrons are introduced into the (secondary) galvanic cell by applying an external current to the cell (i.e., to the electrodes) that is flowing in opposite direction as compared to the discharge process (Figure 4.22).

In comparison to primary batteries, secondary batteries are somewhat more environmentally friendly since they can be recharged and used for longer periods of operation. Nevertheless, secondary batteries also contain significant amounts of harmful substances, for instance, toxic heavy metals. Hence, the development and improvement of secondary batteries focus on a maximal number of charging/discharging cycles. The higher the number of charging/discharging cycles, the better the ecological footprint of secondary batteries. With this regard, rechargeable secondary batteries have a somewhat better reputation than primary cells, although they can also contain a variety of problematic materials.

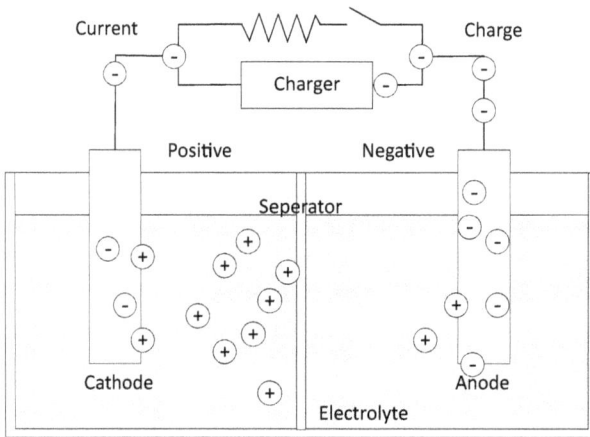

**Figure 4.22:** Principal setup for charging and discharging of secondary batteries.

In the following sections, we will describe important secondary batteries. We will start with the classic secondary cell, that is, the lead–acid battery that we encountered in Section 4.2 in the form of the Planté cell.

### 4.4.1 Lead–acid batteries

As mentioned in Section 4.2.3, the French scientist Gaston Planté invented the lead–acid battery in 1859. The so-called Planté cell was the first rechargeable battery. With this regard, the lead–acid battery is not only the oldest secondary battery but also the most successful one.

Although lead–acid batteries were invented more than one and a half centuries ago, even nowadays, they remain the technology of choice for a large variety of applications because they are robust, technically mature and low-priced. Although lead–acid batteries have rather low energy-to-weight and energy-to-volume ratios, respectively, they exhibit a rather large power-to-weight ratio. Moreover, they are able to supply and resist high surge currents. Hence, lead–acid batteries are attractive, for instance, in the automotive sector. They are used for ignition since they can provide the high currents necessary for automobile starter motors. On the other hand, lead–acid batteries are also widely used even when surge current is of lesser importance or when other battery designs and cell chemistries could provide higher energy densities. The reason is simple: lead–acid batteries are cheap and reliable.

Concerning higher power applications with intermittent loads, lead–acid batteries are often too big and heavy, and other types of rechargeable batteries are preferred. Moreover, lead–acid batteries exhibit rather short cycle lifetimes and typically feature usable power down to only 50% depth of discharge. Although lead–acid batteries feature a variety of shortcomings – for instance, with regard to energy densities, volume, mass and cell chemistry – they still serve a significant part of the high-power battery market.

As we already know, the single lead–acid galvanic cell is composed of porous, spongy metallic lead (Pb) and lead dioxide ($PbO_2$) electrodes. An aqueous sulfuric acid ($H_2SO_4$) solution acts as the electrolyte. During the discharging process, the negative lead electrode (negative plate) operates as the anode, and the positive lead dioxide electrode works as the cathode. Typical settings of lead–acid batteries – for instance, an automotive or car battery – are shown in Figure 4.23. The electrodes are formed in the shape of plates – or, better, comb-shaped plates. The comb-shaped negative and positive electrodes (plates) are fitted into each other, as shown in Figure 4.23. The structure of a single lead–acid cell is shown in Figure 4.24. An acid-resistant separator plate in between separates the negative plate (Pb) and the positive plate ($PbO_2$). The plates are immersed in an aqueous sulfuric acid solution. For the car battery, six galvanic lead–acid cells – with a cell voltage of about 2 V – are connected in series providing 12 V. A so-called cell divider separates single lead–acid cells in the car battery. Metallic lead or lead grids, respectively, at the outer boundaries of the battery realize electrical contact to the external electrical connections.

Electrochemical reactions that occur in modern lead–acid batteries are the same as those explained in Section 4.2.3 in the context of the historical Planté cell. Hence, during the discharging process, at the negative plate (anode) of the lead–acid battery, the following oxidation reaction occurs:

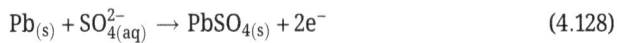

$$Pb_{(s)} + SO_{4(aq)}^{2-} \rightarrow PbSO_{4(s)} + 2e^- \tag{4.128}$$

At the positive plate (cathode), the following reduction reaction occurs during the discharging process:

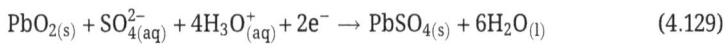

$$PbO_{2(s)} + SO_{4(aq)}^{2-} + 4H_3O_{(aq)}^+ + 2e^- \rightarrow PbSO_{4(s)} + 6H_2O_{(l)} \tag{4.129}$$

In these oxidation and reduction processes, the electrolyte – that is, sulfuric acid ($H_2SO_4$) in aqueous solution – is directly involved in the reactions. The dissociation of sulfuric acid occurs according to reactions (4.61a) or (4.61b), as mentioned earlier. The sulfate ions ($SO_4^-$) originating from the dissociated sulfuric acid are directly involved in the oxidation reaction (4.128). At the interface between the metallic lead electrode and the aqueous solution (electrolyte), lead reacts with the sulfate ions and forms electrically neutral lead sulfate ($PbSO_4$). $PbSO_4$ is deposited onto the negative electrode, and the remaining two electrons are delivered to the (negative) electrode. On the other hand, the hydrogen ions ($H^+$) that are produced according to the dissociation reactions (4.61a) or (4.61b) of the sulfuric acid enter the aqueous solution of the lead–acid cell. There, they immediately attach to water molecules and form hydronium ions ($H_3O^+$). At the positive plate, together with the sulfate ions, the $H_3O^+$ ions are subsequently consumed in the reduction reaction (4.129). During the reduction reaction, the tetravalent Pb(IV) in lead dioxide – Pb(IV)O$_2$ – is reduced to divalent Pb(II) in lead sulfate – Pb(II)SO$_4$. Since the sulfate ions (SO$_4^-$) are consumed within the oxidation and the reduction reactions at both the negative and the positive plates, respectively, lead sulfate (PbSO$_4$) is formed and deposited at both electrodes – the negative and the positive ones. During the discharging process, water is also generated, as shown in the redox reaction (4.129).

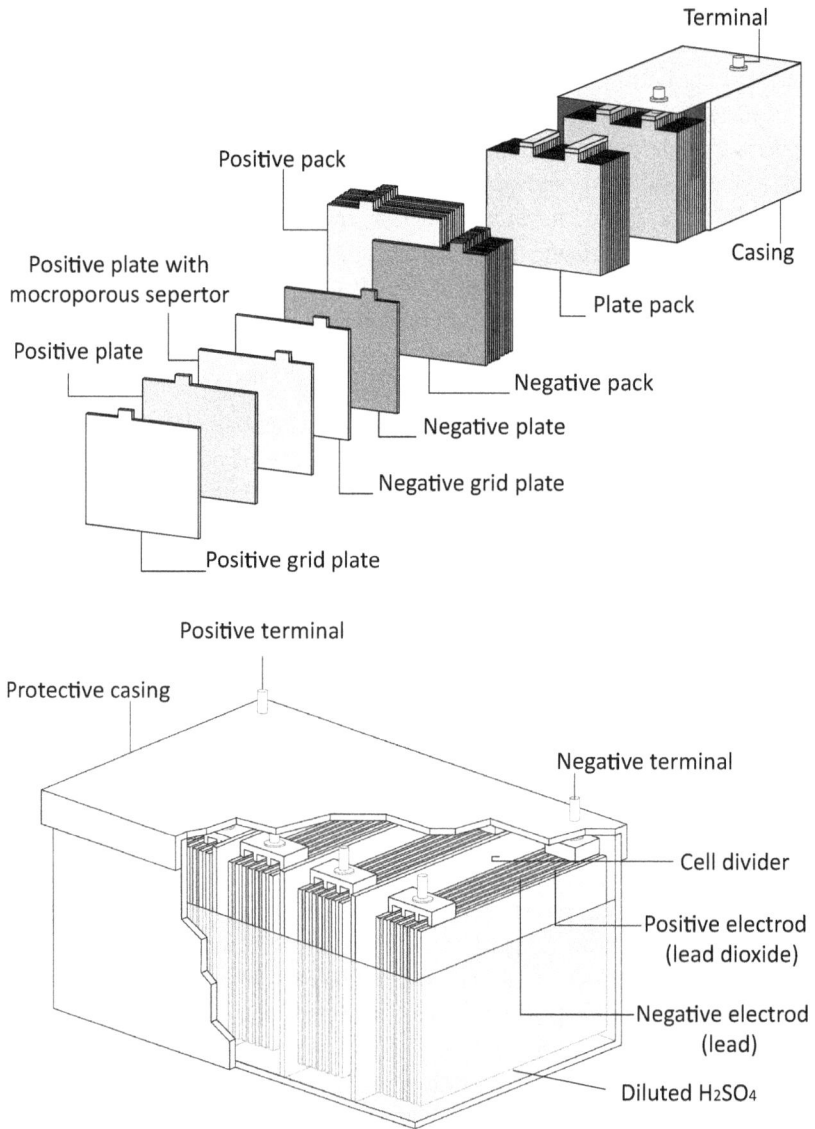

**Figure 4.23:** Typical setups of lead–acid batteries – for instance, automotive or car batteries.

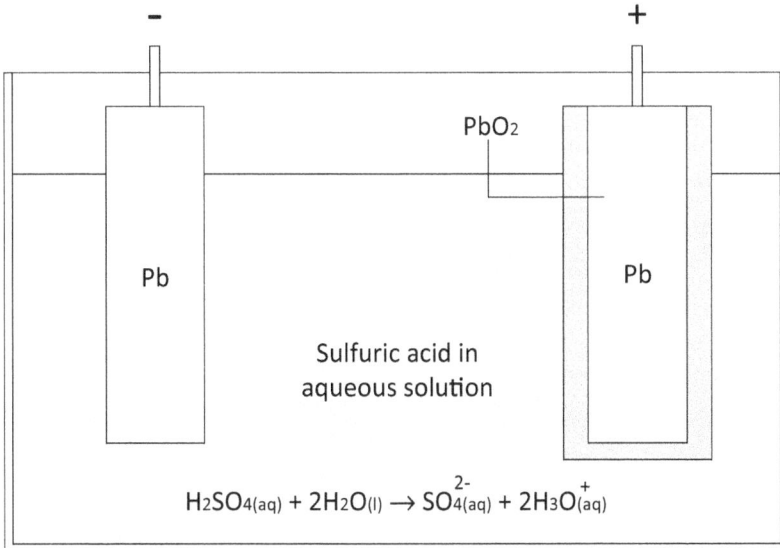

**Figure 4.24:** Principal structure of a single lead–acid cell. The figure shows the situation where the lead–acid battery is completely charged. The active material at the positive plate is $PbO_2$. Metallic lead (Pb) at the positive terminal only serves as an electrical contact.

In summary, the discharging process of lead–acid batteries can be described by the following redox reaction:

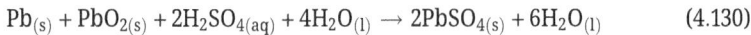

$$Pb_{(s)} + PbO_{2(s)} + 2H_2SO_{4(aq)} + 4H_2O_{(l)} \rightarrow 2PbSO_{4(s)} + 6H_2O_{(l)} \qquad (4.130)$$

Hereby, for the formation of $PbSO_4$ at the electrodes, the electrolyte has lost much of its dissolved sulfuric acid; therefore, if the lead–acid cell is completely discharged, the aqueous solution exhibits minimal sulfuric acid concentration. In addition, water is formed during the discharge redox reaction. Hence, during discharge, the acid density of the aqueous solution is steadily reduced; therefore, it is a measure of the charge state of the battery. Finally, in the discharged state, the negative plate (initially metallic Pb) and the positive plate (initially $PbO_2$) are both converted into lead sulfate (Figure 4.14).

In summary, during the discharging process of the lead–acid battery, both metallic lead at the negative plate and the lead dioxide at the positive plate react with the electrolyte of sulfuric acid and form lead sulfate, water and – finally – provide electric energy. During the charging process, this cycle is reversed. By an external electrical charging source, the lead sulfate at the electrodes and water in the aqueous solution are electrochemically converted into metallic lead, lead oxide and sulfuric acid.

By a forced dissociation of $PbSO_4$, and therefore, a forced reduction of divalent lead (Pb(II)) in lead sulfate at the electrode–electrolyte interface, metallic lead (Pb) is formed and deposited onto the electrode, so that finally the negative metallic Pb electrode is restored again:

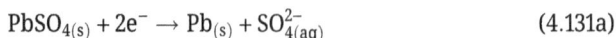

$$PbSO_{4(s)} + 2e^- \rightarrow Pb_{(s)} + SO_{4(aq)}^{2-} \quad (4.131a)$$

On the other side, by a forced dissociation of $PbSO_4$, and therefore, a forced oxidation of divalent Pb(II) in $PbSO_4$ at the electrode–electrolyte interface occurs; hence, $PbO_2$ with tetravalent Pb(IV) is formed and deposited onto the electrode so that, finally, the positive $PbO_2$ electrode is restored again, too:

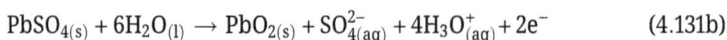

$$PbSO_{4(s)} + 6H_2O_{(l)} \rightarrow PbO_{2(s)} + SO_{4(aq)}^{2-} + 4H_3O_{(aq)}^+ + 2e^- \quad (4.131b)$$

In summary, as for the Planté cell, the charging process of the lead–acid battery can be described by the following redox reaction:

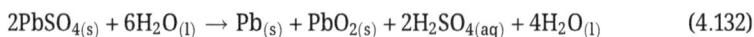

$$2PbSO_{4(s)} + 6H_2O_{(l)} \rightarrow Pb_{(s)} + PbO_{2(s)} + 2H_2SO_{4(aq)} + 4H_2O_{(l)} \quad (4.132)$$

During the charging process, besides the formation of metallic lead on one electrode and lead dioxide on the other electrode, water is consumed and sulfuric acid ($H_2SO_4$) is formed. Therefore, during the charging process, the acid density of the aqueous solution is steadily enhanced again. The completely charged lead–acid battery exhibits a maximal sulfuric acid concentration.

As already mentioned, the thermodynamic equilibrium of the lead–acid cell (i.e., the Planté cell) is the discharged state. Hence, the way toward the thermodynamic equilibrium points from the left to the right side of the redox reaction (4.130).

As mentioned in Section 4.2.3, when discussing the historic Planté cell, the lead–acid battery exhibits a significant tendency to self-discharge. Under open-circuit conditions, if the external electric circuit is not closed, at the negative plate, reaction (6.67) occurs that is similar to the oxidation reaction (4.60) upon the discharging process. Hence, under open-circuit conditions, at the negative plate, protons appear that directly combine to form $H_2$ gas molecules.

Going more into detail, the following reactions occur in the lead–acid battery under open-circuit conditions. At the negative electrode, metallic lead oxidizes according to the following reaction:

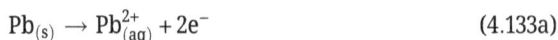

$$Pb_{(s)} \rightarrow Pb_{(aq)}^{2+} + 2e^- \quad (4.133a)$$

or rather:

$$2Pb_{(s)} \rightarrow 2Pb_{(aq)}^{2+} + 4e^- \quad (4.133b)$$

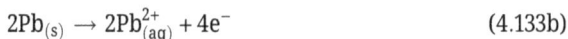

Moreover, the dissociation of sulfuric acid occurs according to the above-mentioned reaction (4.61a) – or, better, eq. (4.61b), that is,

$$2H_2SO_{4(aq)} + 4H_2O_{(l)} \rightarrow 2SO^{2-}_{4(aq)} + 4H_3O^+_{(aq)} \tag{4.134}$$

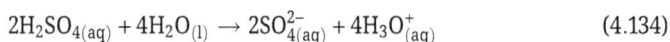

The hydronium ions in eq. (4.134) take over the electrons from the oxidation equation (4.133b), and we obtain:

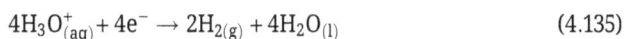

$$4H_3O^+_{(aq)} + 4e^- \rightarrow 2H_{2(g)} + 4H_2O_{(l)} \tag{4.135}$$

Combining reactions (4.133b), (4.134) and (4.135), we get, in total, for the self-discharge reaction at the negative electrode:

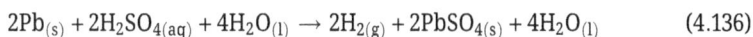

$$2Pb_{(s)} + 2H_2SO_{4(aq)} + 4H_2O_{(l)} \rightarrow 2H_{2(g)} + 2PbSO_{4(s)} + 4H_2O_{(l)} \tag{4.136}$$

or in brief:

$$Pb_{(s)} + H_2SO_{4(aq)} \rightarrow H_{2(g)} + PbSO_{4(s)} \tag{4.137}$$

This reaction was already mentioned in Section 4.2.3 for the Planté cell (see eq. (4.67)).

As shown in eq. (4.68), describing the self-discharge at the positive electrode of the Planté cell, lead dioxide reacts with sulfuric acid to form lead sulfate, while the remaining constituents produce water and oxygen gas. In detail, the following reactions occur under open-circuit conditions. Based on the self-ionization of water – eq. (4.39) – oxygen gas can be formed according to the following reaction:

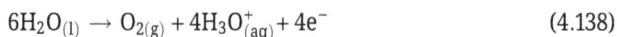

$$6H_2O_{(l)} \rightarrow O_{2(g)} + 4H_3O^+_{(aq)} + 4e^- \tag{4.138}$$

Subsequently, lead dioxide reacts with sulfuric acid and the hydronium ions, that is:

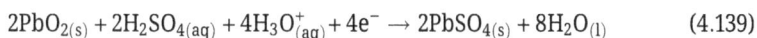

$$2PbO_{2(s)} + 2H_2SO_{4(aq)} + 4H_3O^+_{(aq)} + 4e^- \rightarrow 2PbSO_{4(s)} + 8H_2O_{(l)} \tag{4.139}$$

Hence, the overall chemical reaction causing self-discharge at the positive plate can be written as follows:

$$2PbO_{2(s)} + H_2SO_{4(aq)} \rightarrow O_{2(g)} + 2PbSO_{4(s)} + 2H_2O_{(l)} \tag{4.140}$$

This is identical to the above-mentioned equation (4.68).

As shown in reaction (4.137) occurring at the negative plate and eq. (4.140) that takes place at the positive plate, hydrogen gas ($H_2$) and oxygen gas ($O_2$) are formed under open-circuit conditions (self-discharge processes). Therefore, especially in large lead–acid batteries, the formation of an explosive oxyhydrogen gas can occur that can result in a strong exothermic reaction delivering water as the final product ($2H_2 + O_2 \rightarrow 2H_2O$).

The conventional lead–acid battery is a wet cell – it is also called a flooded lead–acid battery, where the electrolyte bathes the electrodes. In such a flooded lead–acid cell, the formed hydrogen and oxygen gases can escape. Moreover, if the charge current is too high, electrolysis will take place; hence, besides the other electrochemical reactions, a decomposition of water into hydrogen and oxygen gases is generated that can escape from the flooded lead–acid cell, too. The production and release of hydrogen and oxygen gas bubbles is called gassing. Gassing occurs, in particular, due to excessive charging. In large battery installations, gassing can cause explosive atmospheres in closed battery rooms. Moreover, gassing requires that water and/or electrolyte has to be added to refill the standard lead–acid wet cell.

The so-called valve-regulated lead–acid (VRLA) batteries – also known under the name sealed lead–acid battery – keep the formed gases within the battery as long as the pressure is not too high and remains within safety limits. Therefore, under normal operation conditions, they stay within the battery and can recombine (usually with the support by a catalyst). If the pressure is too high, a safety valve opens. It is usually not necessary to refill electrolyte and/or water. VRLA batteries are supposed to be "maintenance free"; however, this is somewhat misleading, since they require some frequent function testing anyway.

Three main types of VRLA batteries can be distinguished, that is, the sealed valve-regulated wet cell, the gel cell and the absorbent glass mat (AGM) battery, where the latter two can be mounted in any orientation due to the solidified and immobilized electrolyte. In AGM batteries, the electrolyte is kept in glass mats. These glass mats are woven from very thin glass fibers; therefore, the mats exhibit a strongly enhanced surface that holds so many electrolytes that it is sufficient for the lifetime of the AGM battery. The gel cell, or gel battery, is a VRLA battery with a jellied electrolyte that is composed of a mixture of the electrolyte – that is, sulfuric acid – and fumed silica. This mixture is quasi-immobile. Hence, the gel cell does not have to be kept in an upright position; it can be mounted as necessary in any orientation.

The electrochemical reaction processes in VRLA batteries are very similar to the reactions in the Planté cell or the standard lead–acid cell. However, the VRLA cell widely avoids the formation of the problematic oxyhydrogen gas.

Self-discharge processes in the VRLA battery are a little different as compared to the standard lead–acid battery. At the positive plate, oxygen is formed according to the following equation:

$$H_2O_{(l)} \longrightarrow \frac{1}{2}O_{2(g)} + 2H^+_{(aq)} + 2e^- \tag{4.141}$$

This is the equivalent brief notation of eq. (4.138) mentioned earlier. Through the gaseous space within the VRLA battery, the oxygen that was formed at the positive plate can reach the negative electrode very fast, where it can recombine with hydrogen in an exothermic reaction. However, further side reactions causing discharge can arise at the negative plate too; that is, under open-circuit conditions, a reduction of oxygen can occur. This is, in fact, the reversal of the reaction that could already be observed at the positive plate under open-circuit condition – that is, the reversal of eq. (4.38). Hence, the reduction of oxygen at the negative plate can be described by the following reaction:

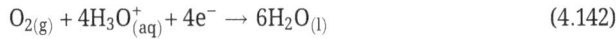

$$O_{2(g)} + 4H_3O^+_{(aq)} + 4e^- \rightarrow 6H_2O_{(l)} \tag{4.142}$$

or in brief:

$$\frac{1}{2}O_{2(g)} + 2H^+_{(aq)} + 2e^- \rightarrow H_2O_{(l)} \tag{4.143}$$

At the negative plate, together with the oxidation of lead according to eq. (4.133a) or (4.133b), the oxygen gas can also be involved in the following reaction:

$$Pb_{(s)} + \frac{1}{2}O_{2(g)} + H_2SO_{4(aq)} \rightarrow PbSO_{4(s)} + H_2O_{(l)} \tag{4.144}$$

or rather:

$$2Pb_{(s)} + O_{2(g)} + 2H_2SO_{4(aq)} \rightarrow 2PbSO_{4(s)} + 2H_2O_{(l)} \tag{4.145}$$

In this reaction at the negative plate, the electrolyte (dissociated sulfuric acid) is involved, and lead sulfate and water are the products. However, the produced lead sulfate will now be directly reduced to metallic lead again, consuming hydrogen $H^+_{(aq)}$ ions according to the following reaction:

$$PbSO_{4(s)} + H^+_{(aq)} + 2e^- \rightarrow Pb_{(s)} + HSO^-_{4(aq)} \tag{4.146}$$

We can summarize that, under the participation of sulfuric acid, the chemical reactions (4.137) occurring in the standard flooded electrolyte lead–acid battery and (4.142) that occur in the VRLA battery cause self-discharging processes, whereupon, in both cases, lead sulfate ($PbSO_4$) is the final product. However, in the case of the VRLA battery, the lead sulfate is instantly reduced according to reaction (4.146); therefore, no hydrogen gas is released.

Finally, self-discharge can also occur due to a corrosion process at the positive electrode consisting of a $PbO_2$ paste (i.e., the active material) that is smeared over a metallic lead lattice. The metallic lead lattice participates also in electrochemical processes and grid corrosion occurs. Such a grid corrosion can occur due to a series of reactions that are being discussed.

$Pb^{2+}$ can be found near the lead plates/grids due to an oxidation reaction (see also eq. (4.133a)):

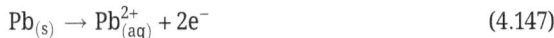

$$Pb_{(s)} \rightarrow Pb^{2+}_{(aq)} + 2e^- \tag{4.147}$$

Instantly, the $Pb^{2+}_{(aq)}$ ions pick up $O^{2-}$ ions, that is, the $O^{2-}$ ions neutralize the $Pb^{2+}_{(aq)}$ ions in the aqueous solution, and lead monoxide (PbO) is formed that is deposited onto the metallic lead grid:

$$Pb^{2+}_{(aq)} + O^{2-}_{(aq)} \rightarrow PbO_{(s)} \tag{4.148}$$

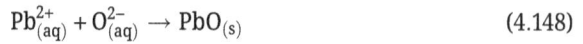

For the latter reaction, the active material $PbO_2$ donated the necessary $O^{2-}$ ions. Hence, in detail, we also have to consider the following discharge reaction of $PbO_2$ providing the $O^{2-}$ ions:

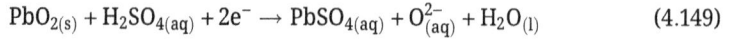

$$PbO_{2(s)} + H_2SO_{4(aq)} + 2e^- \rightarrow PbSO_{4(aq)} + O^{2-}_{(aq)} + H_2O_{(l)} \tag{4.149}$$

If we combine the involved reactions, we obtain:

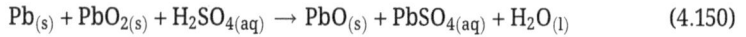

$$Pb_{(s)} + PbO_{2(s)} + H_2SO_{4(aq)} \rightarrow PbO_{(s)} + PbSO_{4(aq)} + H_2O_{(l)} \tag{4.150}$$

One can add that the lead monoxide created according to reaction (4.150) undergoes a further transformation to lead sulfate and water, if it gets in contact to sulfuric acid:

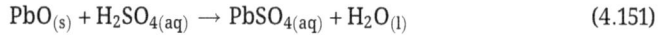

$$PbO_{(s)} + H_2SO_{4(aq)} \rightarrow PbSO_{4(aq)} + H_2O_{(l)} \tag{4.151}$$

To summarize, metallic lead is converted to lead monoxide (divalent Pb(II) oxide) with the help of the active material $PbO_2$, providing the necessary $O^{2-}$ ions. In addition, if $PbO_2$ has lost $O^{2-}$ ions according to the above-mentioned corrosion process, the remaining tetravalent Pb(IV) ions ($Pb^{4+}$) can take up again $O^{2-}$ ions that originate from self-ionized water:

$$3H_2O_{(l)} \rightarrow O^{2-}_{(aq)} + 2H_3O^+_{(aq)} \tag{4.152}$$

In this reaction, a water molecule is self-ionized, providing the $O^{2-}$ ion and two $H^+$ ions. The latter ones attach to two other water molecules and form the two hydronium ions ($H_3O^+$).

If the metallic lead grid is in contact with the electrolyte, the corrosion process can also follow the following reaction scheme:

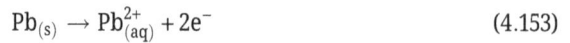

$$Pb_{(s)} \rightarrow Pb^{2+}_{(aq)} + 2e^- \tag{4.153}$$

and afterward

$$Pb^{2+}_{(aq)} + 2e^- + H_2SO_{4(aq)} + 2H_2O_{(l)} \rightarrow PbSO_{4(s)} + 2H_3O^+_{(aq)} + 2e^- \tag{4.154}$$

Immediately after oxidation (4.153), the $Pb^{2+}$ ions pick up the sulfate ions ($SO_4^{2-}$) from the dissociated sulfuric acid ($H_2SO_4$), whereas the remaining two $H^+$ ions are attached to water molecules (reaction (4.154)). Again, the discharge of $PbO_2$ has to be considered, now – as compared to eq. (4.149) – with a slightly modified reaction:

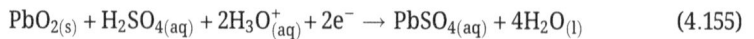

$$PbO_{2(s)} + H_2SO_{4(aq)} + 2H_3O^+_{(aq)} + 2e^- \rightarrow PbSO_{4(aq)} + 4H_2O_{(l)} \tag{4.155}$$

The combination of reactions (4.153), (4.154) and (4.155) yields

$$Pb_{(s)} + PbO_{2(s)} + 2H_2SO_{4(aq)} \rightarrow 2PbSO_{4(aq)} + 2H_2O_{(l)} \tag{4.156}$$

Supported by sulfuric acid, metallic lead and lead dioxide are converted to lead sulfate and water.

In general, one can state that the side reactions that occur in lead–acid batteries and that are responsible for the various self-discharge and corrosion processes at the positive and negative electrodes are independent of each other. Therefore, they occur with different time constants, and self-discharge rates are different for each specific process. Moreover, the various self-discharge rates or corrosion processes are dependent on a lot of parameters, like the temperature, the charge states of the battery, the acid density of the aqueous solution (or acidity) and so on. In addition, alloying of the metallic lead grid also has some influence. For instance, during a corrosion process at the positive plate, where metallic lead is finally converted into lead sulfate, alloyed additives – for example, antimony (Sb) – are set free and pollute the aqueous solution (electrolyte).

The advantages of lead–acid batteries are quite evident. Modern lead–acid batteries are based on a steadily developed and mature technology. Most reactions and side reactions are well-known and understood. Moreover, they are a mass product and fabricated in standardized manufacturing processes since decades. Therefore, lead–acid batteries are as well cheap and reliable. However, they also face some disadvantages that we have to keep in mind.

Despite the rather low energy density (as shown in the Ragone diagram of Figure 4.1), sulfation of the electrodes can be a problem that causes irreversible damage, and therefore, it is a disadvantage of lead–acid batteries. The sulfation is critical for the conversion of chemical energy into electrical energy. Essentially, sulfation is part of the electrochemical mechanism of lead–acid batteries, as shown in the redox reaction (4.130) of the discharge process. Hence, without sulfation, battery operation and electrical release are not possible. To illustrate this again, we emphasize the sulfation during the discharge process at the negative and positive plates of the lead–acid battery, respectively:

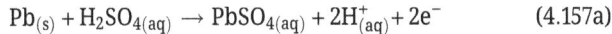

$$Pb_{(s)} + H_2SO_{4(aq)} \rightarrow PbSO_{4(aq)} + 2H^+_{(aq)} + 2e^- \tag{4.157a}$$

$$PbO_{2(s)} + H_2SO_{4(aq)} + 2H^+_{(aq)} + 2e^- \rightarrow PbSO_{4(aq)} + 2H_2O_{(l)} \tag{4.157b}$$

So again, as shown in eq. (4.156), metallic lead and lead dioxide are converted to lead sulfate and water. These sulfation reactions are completely reversible under charging conditions – but this holds, of course, only for ideal conditions.

In general, the formation of small sulfate crystals on the plates of lead–acid batteries is not very problematic. However, the formation of hardly soluble lead sulfate crystals can be significantly enhanced either if the battery is left in a discharged state for a long time or if the battery is only partly recharged. Under such disadvantageous circumstances, an accumulation of larger lead sulfate crystals on the negative and positive plates occurs and reduces the active material of the lead–acid battery; hence, its capacity is reduced. In the worst case, lead sulfate crystals can even cause electrical shorts.

Another disadvantage of lead–acid batteries is the risk of oxyhydrogen gas reactions due to the formation of $H_2$ and $O_2$ gas upon open-circuit conditions lasting for long periods. Such long periods without battery discharge may appear, for instance, in case of large lead–acid starter batteries in stand-alone emergency generators. As mentioned in Section 4.2.3 describing the historic Planté cell, lead–acid batteries tend to self-discharge under open-circuit conditions, since lead and lead dioxide react with the electrolyte according to the following reactions:

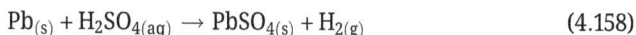

$$Pb_{(s)} + H_2SO_{4(aq)} \rightarrow PbSO_{4(s)} + H_{2(g)} \tag{4.158}$$

and

$$2PbO_{2(s)} + H_2SO_{4(aq)} \rightarrow 2PbSO_{4(s)} + 2H_2O_{(l)} + O_{2(g)} \tag{4.159}$$

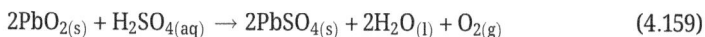

Thus, self-discharge occurs at both electrodes that are completely similar, as shown in the Planté cell (eqs. (4.67) and (4.68)). Under these self-discharge conditions, where the external electrical circuit is not closed, electron migration through the external wiring cannot occur. Then, the oxidation reaction at the negative electrode cannot occur, as shown in eq. (4.128) – or eq. (4.60) – for the discharge process; therefore, the $H^+$ ions and electrons combine and form $H_2$ gas molecules. At the positive electrode, under open-circuit conditions, lead dioxide reacts with the sulfuric acid to form lead sulfate; and the remaining constituents form water and oxygen gas ($O_2$).

Strong overcharging may also cause the formation of oxygen and hydrogen gas due to electrolysis (see also Section 4.4.3). This process is also known as gassing and implies the risk of explosion. Vents or valves that can release produced gases within batteries are integrated in wet cells or VRLA batteries, respectively. However, malfunction (failure of valves or plugging of vents by dirt) can cause a problematic accumulation of $H_2$ and $O_2$ gases within the batteries. Overheating may also cause an accumulation of gases and an enhancement of pressure within batteries if valves or vents fail.

In the following sections, we focus on nickel-based secondary batteries that have some advantages compared to lead–acid cells.

### 4.4.2 Nickel–cadmium batteries

If the same amount of energy is stored in nickel–cadmium (NiCd) and lead–acid batteries, the weight of the nickel–cadmium battery is lower than the weight of the lead–acid battery; that is, the specific energy density of the NiCd battery is higher. NiCd batteries are much more resilient and powerful than lead–acid batteries, especially concerning rapid charging and low ambient temperatures. Moreover, due to a rather low internal resistance, they can deliver significantly higher currents; that is, higher power densities. Overall, the rechargeable NiCd batteries exhibit better properties than the lead–acid cell. However, because of the toxicity of the heavy metal cadmium, they also cause serious problems and environmental impacts and are nowadays substituted by other battery systems, that is, nickel–metal hydride (NiMH) batteries that will be discussed somewhat later.

Nickel–cadmium cells provide a nominal voltage of 1.2 V. This value is somewhat lower than for standard 1.5 V batteries, for instance, AA-sized zinc–carbon or alkaline batteries for consumer applications; therefore, are NiCd cells sometimes could not be used for a replacement of those primary cells. However, in general this is not a real problem, since many devices are designed for an operation at about 1 V.

The fully charged nickel–cadmium battery exhibits a negative cadmium electrode – that is, the anode – and a positive trivalent nickel(III) oxide-hydroxide (NiO(OH)) electrode – that is, the cathode. NiCd batteries usually use a 20% solution of the alkaline electrolyte potassium hydroxide (KOH).

A typical setup for a NiCd battery is shown in Figure 4.25. The positive electrode is constructed using a polypropylene mat of fibers, where the powdered active material, that is, the trivalent nickel(III) oxide-hydroxide (NiO(OH)), is pressed in. For the enhancement of internal conductivity, powdered graphite is added, too. The negative electrode is constructed in a similar way. In this case, powered cadmium is pressed into a polypropylene mat of fibers and pressed onto a metal working as the current collector. The electrodes are isolated from each other by the separator soaked with the electrolyte, and they are curled inside the metallic case. Separators are made from polyamide, polyethylene or polypropylene mats of fiber. This spiral shape of the curled electrodes is favorable with regard to low internal resistance, and therefore, quite high currents can be delivered by the NiCd cell. The whole setup is integrated in a metal case with a sealing plate and often with a self-sealing safety valve. The negative electrode is in good electrical contact to the metallic case that provides an electrical contact to the outside world and functions as the negative terminal. If very high power densities are required, the electrodes are sintered.

During the discharging process at the negative plate (anode), the metallic cadmium is oxidized to cadmium(II) hydroxide ($Cd(OH)_2$) according to the following oxidation reaction:

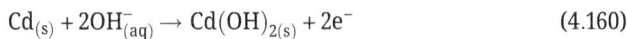

$$Cd_{(s)} + 2OH^-_{(aq)} \rightarrow Cd(OH)_{2(s)} + 2e^- \tag{4.160}$$

Cd in $Cd(OH)_2$ is divalent. At the positive plate (cathode), during the discharging process, the following reduction reaction occurs:

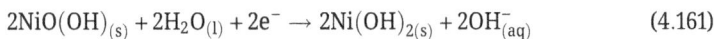

$$2NiO(OH)_{(s)} + 2H_2O_{(l)} + 2e^- \rightarrow 2Ni(OH)_{2(s)} + 2OH^-_{(aq)} \tag{4.161}$$

In this reduction reaction, trivalent nickel(III) oxide-hydroxide is reduced to divalent nickel(II) hydroxide. In summary, the discharging process of nickel–cadmium batteries can be described by the following redox reaction:

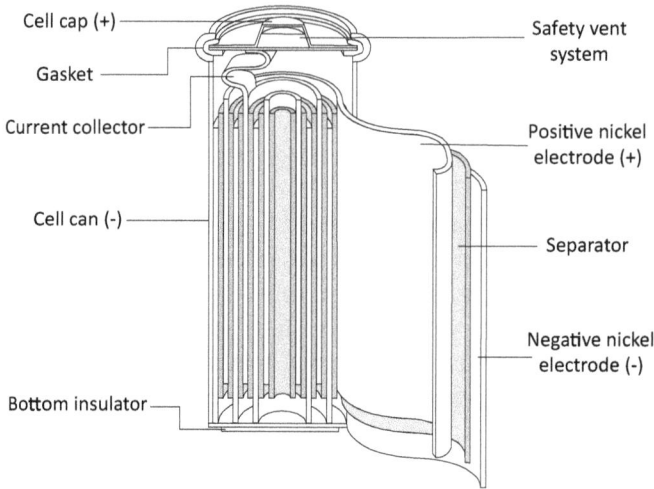

**Figure 4.25:** Typical setup of a NiCd battery. The spiral shape of the curled electrodes is favorable with regard to low internal resistance; hence, quite high currents can be delivered by the NiCd cell.

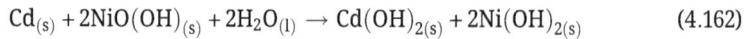

$$Cd_{(s)} + 2NiO(OH)_{(s)} + 2H_2O_{(l)} \rightarrow Cd(OH)_{2(s)} + 2Ni(OH)_{2(s)} \tag{4.162}$$

The electrolyte, that is, the aqueous KOH solution, is not consumed during this discharge redox reaction – the same amount of $OH^-$ ions consumed in the anodic half-cell during discharge is released in the cathodic half-cell of the battery; therefore, the overall concentration of the alkaline electrolyte (aqueous KOH solution) remains constant during the discharge process. Hence, the electrolyte only acts as an ion conductor. It does not really play a significant role in the electrochemical cell reactions. Therefore, the KOH density cannot be a measure of the charge state of the nickel–cadmium battery – unlike in lead–acid batteries, where the acid density provides a measure of its charge state.

The charging process is executed in the opposite direction, that is:

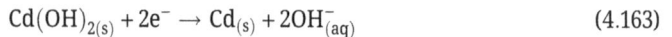

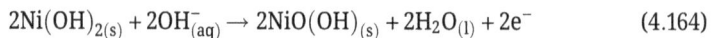

$$Cd(OH)_{2(s)} + 2e^- \rightarrow Cd_{(s)} + 2OH^-_{(aq)} \tag{4.163}$$

$$2Ni(OH)_{2(s)} + 2OH^-_{(aq)} \rightarrow 2NiO(OH)_{(s)} + 2H_2O_{(l)} + 2e^- \tag{4.164}$$

$$Cd(OH)_{2(s)} + 2Ni(OH)_{2(s)} \rightarrow Cd_{(s)} + 2NiO(OH)_{(s)} + 2H_2O_{(l)} \tag{4.165}$$

During the charging process, the cadmium electrode is still the negative terminal, but now it is the cathode, since during charging a (forced) reduction of cadmium hydroxide ($Cd(OH)_2$) occurs (eq. (4.163)). Then the nickel electrode ($Ni(OH)_2$) is the positive terminal, but now it is the anode, since during charging a (forced) oxidation of nickel(II) hydroxide occurs (eq. (4.164)). The overall redox reaction for the charging process is given by eq. (4.165).

The charging and discharging mechanisms in the NiCd battery, as shown in reactions (4.160)–(4.165), can be regarded as simplified descriptions of the real processes. For instance, the reactions at the negative electrode occur in two steps, that is, via dissolution and precipitation mechanisms. For the discharging reaction, the oxidation process can be described by the following two equations that proceed consecutively. First, cadmium enters the electrolyte solution, leaving two electrons in the electrode (oxidation process during discharge):

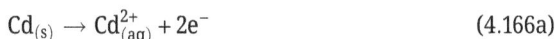

$$Cd_{(s)} \rightarrow Cd^{2+}_{(aq)} + 2e^- \qquad (4.166a)$$

Directly after having entered the aqueous solution, the positively charged cadmium ion reacts with two negatively charged hydroxide ions:

$$Cd^{2+}_{(aq)} + 2OH^-_{(aq)} \rightarrow Cd(OH)_{2(s)} \qquad (4.166b)$$

Thus, reaction (4.160) at the negative plate is the shortened description of the oxidation process described in detail by eqs. (4.166a) and (4.166b). During charging, the reactions run from the right to left side.

The situation can be even somewhat more complex: first, metallic cadmium enters the electrolyte solution and forms, together with three hydroxide ions, a negatively charged hydroxo complex, leaving two electrons in the electrode (oxidation process):

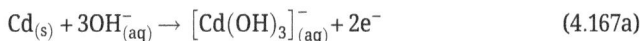

$$Cd_{(s)} + 3OH^-_{(aq)} \rightarrow \left[Cd(OH)_3\right]^-_{(aq)} + 2e^- \qquad (4.167a)$$

Immediately after this reaction, the hydroxo complex decomposes and releases a hydroxide ion, that is, the hydroxo complex transforms to solid cadmium(II) hydroxide – that precipitates onto the electrode – and a hydroxide ion that is left in the electrolyte solution:

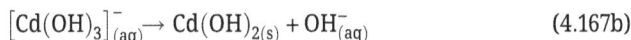

$$\left[Cd(OH)_3\right]^-_{(aq)} \rightarrow Cd(OH)_{2(s)} + OH^-_{(aq)} \qquad (4.167b)$$

In summary, the metallic cadmium is oxidized to cadmium(II) hydroxide, and we obtain the above-mentioned discharge oxidation reaction (4.160). In the case of charging, the two reactions (4.167a) and (4.167b) are running in the opposite direction. Finally, the charging and discharging reactions at the negative electrode are given by

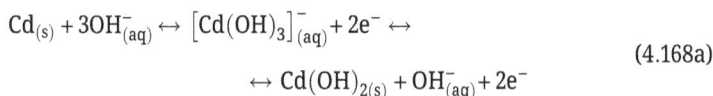

$$Cd_{(s)} + 3OH^-_{(aq)} \leftrightarrow \left[Cd(OH)_3\right]^-_{(aq)} + 2e^- \leftrightarrow$$
$$\leftrightarrow Cd(OH)_{2(s)} + OH^-_{(aq)} + 2e^- \qquad (4.168a)$$

or rather overall by

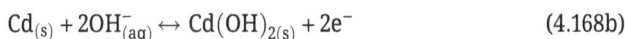

$$Cd_{(s)} + 2OH^-_{(aq)} \leftrightarrow Cd(OH)_{2(s)} + 2e^- \qquad (4.168b)$$

At the positive electrode, the reactions are more complex. For the completely charged state of the NiCd battery, the positive electrode consists not only of nickel(III) oxide-hydroxide (NiO(OH)), but it also contains other constituents. Hence, we can state that the positive plate consists mostly of NiO(OH), but it also contains tetravalent nickel(IV) dioxide ($NiO_2$), water and potassium hydroxide (KOH). Besides the above-mentioned discharging and charging reactions (4.161) and (4.164), the following side reaction can also occur at the positive electrode:

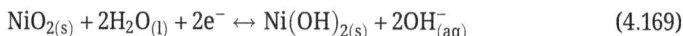

$$NiO_{2(s)} + 2H_2O_{(l)} + 2e^- \leftrightarrow Ni(OH)_{2(s)} + 2OH^-_{(aq)} \tag{4.169}$$

The discharge reaction is directed toward the right side, and the charging reaction toward the left side of eq. (4.169). The corresponding redox reaction can be written as

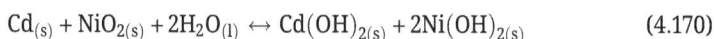

$$Cd_{(s)} + NiO_{2(s)} + 2H_2O_{(l)} \leftrightarrow Cd(OH)_{2(s)} + 2Ni(OH)_{2(s)} \tag{4.170}$$

Furthermore, during charging, the situation can become much more complex. Besides the main formation of trivalent nickel(III) oxide-hydroxide (NiO(OH)) during the charging process at the positive electrode, more complex hydroxides can be formed by side reactions, for instance, mixed oxyhydroxide [$Ni_4O_4(OH)_4$]$(OH)_2$K that includes potassium from the aqueous potassium hydroxide solution:

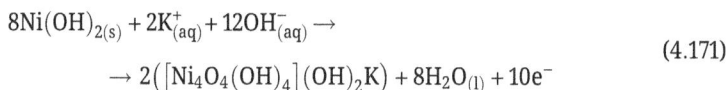

$$8Ni(OH)_{2(s)} + 2K^+_{(aq)} + 12OH^-_{(aq)} \rightarrow$$
$$\rightarrow 2\left(\left[Ni_4O_4(OH)_4\right](OH)_2K\right) + 8H_2O_{(l)} + 10e^- \tag{4.171}$$

Upon discharge, such oxyhydroxide compounds can also contribute to the current flow of the NiCd battery. As shown in eq. (4.171), during this charge reaction, water is formed. Hence, the hydroxide concentration is changed. However, this charging reaction is only a side reaction without a specified amount of reaction products. Therefore, the altered electrolyte density cannot be a measure of the charge state of the NiCd battery.

The advantages of NiCd batteries are their durability and rapid charging capability. They exhibit good low-temperature resistance. Even at temperatures of −40 °C, their performance and capacity are quite high. The energy density of NiCd batteries is better than that of lead–acid cells (specific energy density: 50–70 Wh/kg, volumetric energy densities: ca. 130 Wh/L); however, it is worth than the energy densities of NiMH or LIB. Typical power densities are 150–200 W/kg.

The lifetime (typical cycle lifetime of NiCd batteries: ca. 1,500) and storability of NiCd cells are also advantageous, and they are quite robust against deep discharge. Finally, the cost–benefit ratio is good as well. However, the self-discharge of NiCd batteries is not advantageous (ca. 20% per month); that is, within 3 months, self-discharge reduces the cell capacity down to a rest capacity of about 50%.

Moreover, with improper handling, NiCd batteries can exhibit a memory effect. The memory effect has been observed in NiCd batteries with sintered electrodes. It occurs when the batteries are repeatedly only partially discharged and then recharged. In such cases, the NiCd battery seems to remember that only a small fraction of its capacity was used, and it holds less charge than it initially held before the repeated partial discharge and recharge procedure began. Due to the memory effect, the discharge curve of a NiCd battery can be significantly altered (Figure 4.26).

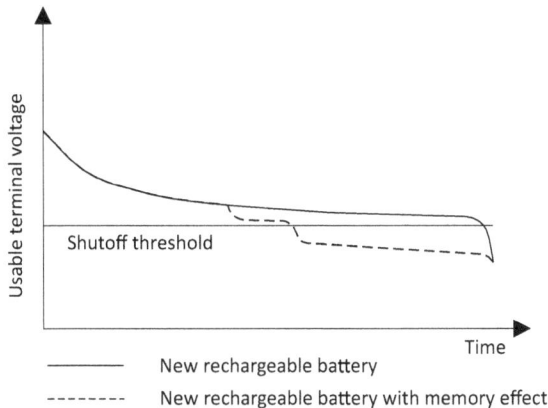

Figure 4.26: Memory effect.

The memory effect can be explained by the formation of cadmium microcrystals during recharging. If the NiCd battery was only partially discharged, upon recharge, the formation of Cd crystals can occur in regions of the battery that were not discharged yet. Repeated partial discharge supports the formation of larger Cd crystals. Since – at constant mass in total – larger crystals exhibit a smaller surface area than smaller ones, they react less efficiently during discharge than the smaller microcrystals. Therefore, eventually, the NiCd battery degrades. In the worst case, such effects can cause serious damage to the cell. If NiCd cells are continuously charged at low currents, at the cathode metallic cadmium crystals can grow (whiskers) that – beside a reduction of the active surface of the electrode – can even cause short circuits within the cell. Modern chargers that control the charge state of the battery largely avoid those memory effect problems.

Nickel–cadmium batteries were used for decades in portable devices. Until the end of the twentieth century, they were the rechargeable battery of choice for home electronic applications. But because of the disadvantage of environmental incompatibility due to the toxic heavy metal cadmium, in the 1990s NiMH batteries took over the domination due to the toxicity problem of the NiCd cells. Although in general recycling is possible, a lot of used NiCd cells are taken out of the recycling loop and simply thrown away. This causes problems for the environment and for the health of man and beast.

Many of the favorable characteristics of nickel–cadmium cells can also be found in NiMH batteries. This made the replacement of NiCd cells with the NiMH battery system easy. The latter combines similar favorable technical properties with much higher environmental friendliness. Because of its environmental problems and stricter regulations, NiCd batteries are nowadays limited to some special applications (e.g., emergency lighting, medical equipment). Since 2009, NiCd batteries with a cadmium content of more than 0.002 wt% are prohibited within the European Union.

Before we continue to discuss other types of prominent secondary batteries, we want to make a digression and look at overcharging and overdischarging processes.

### 4.4.3 Overcharging and overdischarging processes in NiCd batteries

A common problem of secondary batteries is overcharging and overdischarging processes. We will exemplarily discuss such processes for the case of the NiCd batteries, and we will start with overcharging of NiCd batteries.

As mentioned in the last section, during the usual discharging process at the negative plate (anode) of the NiCd battery, metallic cadmium is oxidized to divalent cadmium(II) hydroxide ($Cd(OH)_2$) according to the oxidation reaction (4.160). At the positive plate (cathode), during discharging, the reduction reaction (4.161) occurs, where trivalent nickel(III) oxide-hydroxide ($NiO(OH)$) is reduced to divalent nickel(II) hydroxide ($Ni(OH)_2$). On the other hand, during the normal charging process, the cadmium electrode of the NiCd battery is still the negative terminal, but now it functions as a cathode, since during charging a (forced) reduction of cadmium hydroxide ($Cd(OH)_2$) occurs (eq. (4.163)). Hence, the nickel electrode ($Ni(OH)_2$) is the positive terminal, but now it is the anode, since during charging a (forced) oxidation of nickel(II) hydroxide occurs (eq. (4.164)).

If more than the necessary charging is applied, that is, if we overcharge the NiCd battery, other chemical reactions occur than those for the "normal" charging process that was presented in the last section in reactions (4.163) and (4.164) for the negative and positive electrodes of the NiCd battery, respectively. $H_2$ and $O_2$ gas generation can occur either at overcharge and overdischarge conditions. Likewise, during rapid charging and discharging processes, $H_2$ and $O_2$ gas generation can occur. An overview can be achieved from Figure 4.27. The details will be described now.

By the end of the charging process, that is, if the normal load cycle of the NiCd battery is nearly finished, the cell voltage increases. If we enter the overcharging regime, gassing can occur, since the decomposition voltage of water (ca. 1.6 V) can be exceeded that allows for electrolysis. Therefore, during overcharging, at the negative electrode, hydrogen gas ($H_2$) can be formed and at the positive electrode, oxygen gas ($O_2$), respectively. In detail, at the negative electrode, the following reaction causing hydrogen gas formation can be observed:

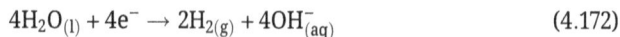

$$4H_2O_{(l)} + 4e^- \rightarrow 2H_{2(g)} + 4OH^-_{(aq)} \tag{4.172}$$

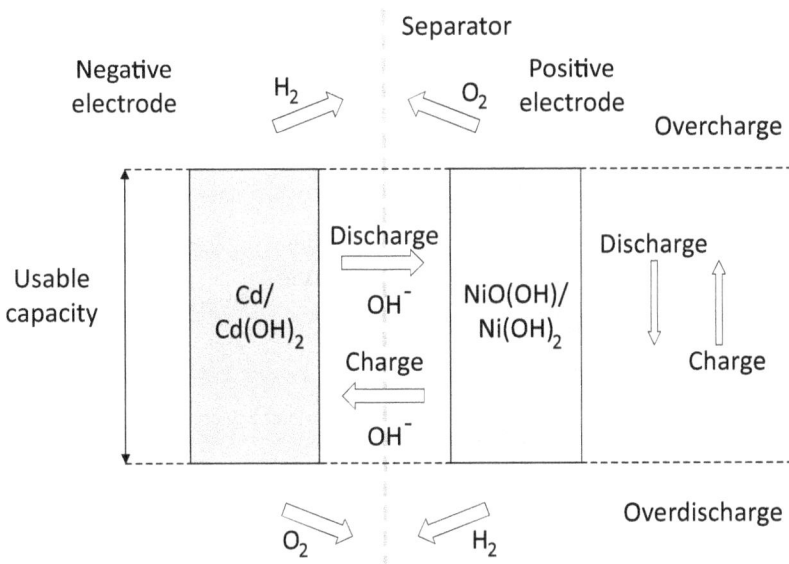

**Figure 4.27:** $H_2$ and $O_2$ gas formation during overcharging and overdischarging processes in NiCd batteries.

while at the positive electrode, oxygen gas is formed:

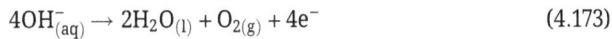

$$4OH^-_{(aq)} \rightarrow 2H_2O_{(l)} + O_{2(g)} + 4e^- \tag{4.173}$$

that is, in total, electrolysis of water occurs:

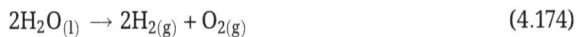

$$2H_2O_{(l)} \rightarrow 2H_{2(g)} + O_{2(g)} \tag{4.174}$$

If the NiCd battery has a safety valve or if it is vented, the $H_2$ and $O_2$ gases can escape from the cell container; that is, $H_2$ and $O_2$ gases – that is, water in the form of its gas constituents – are lost. Hence, the amount of electrolyte and its composition are changed, and the capacity and proper operation of the cell are negatively affected. Otherwise, if the container is gas-tight, an internal overpressure might occur as long as the pressure limit of the safety valve of the container is not exceeded. If the internal pressure within the NiCd battery exceeds the limit of the safety valve, again $H_2$ and $O_2$ gases – that is, water – are lost, and the amount and composition of the electrolyte are changed, altering the capacity and properties of the cell. Sealed NiCd batteries are constructed with a pressure vessel that should retain the generated oxygen and/or hydrogen gases until they recombine back into water (reversal of eq. (4.174)).

Under normal operation conditions, the $H_2$ gas formation should be suppressed – or at least, $H_2$ formation should be much lower than the formation of $O_2$ gas. This can be achieved by a proper design of the NiCd battery: The negative electrode is oversized compared to the positive electrode (Figure 4.28). Therefore, under normal operation conditions, the negative electrode does not reach a potential where hydrogen generation can occur; that is, the smaller positive plate is designed in a way that it goes into overcharge first. Hence, $O_2$ gas is already generated at the positive plate due to overcharging conditions, while the formation of $H_2$ gas – at the negative plate – has not yet started.

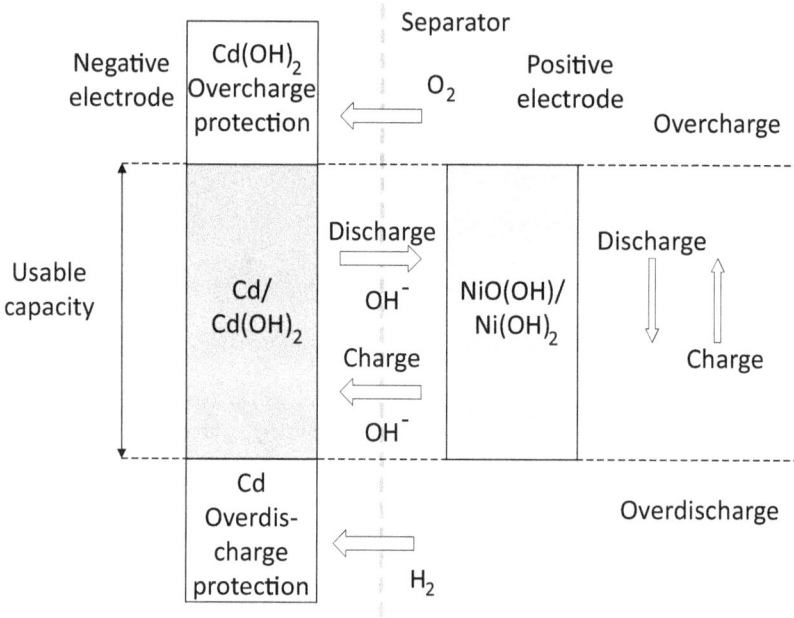

**Figure 4.28:** $H_2$ and $O_2$ gas formation during overcharging and overdischarging processes in NiCd batteries with oversized negative electrode.

In a sealed NiCd battery, the generated $O_2$ molecules diffuse to the negative cadmium electrode, where the following recombination reaction occurs:

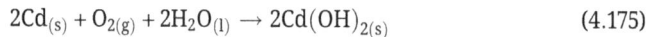

$$2Cd_{(s)} + O_{2(g)} + 2H_2O_{(l)} \rightarrow 2Cd(OH)_{2(s)} \tag{4.175}$$

In detail, this recombination reaction occurs in two steps. At the negative electrode, the arriving oxygen molecules are reduced:

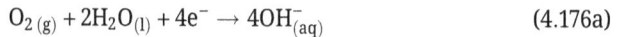

$$O_{2\,(g)} + 2H_2O_{(l)} + 4e^- \rightarrow 4OH^-_{(aq)} \tag{4.176a}$$

These hydroxide ions are joined with cadmium and form cadmium(II) hydroxide:

$$2Cd_{(s)} + 4OH^-_{(aq)} \rightarrow 2Cd(OH)_{2(s)} \tag{4.176b}$$

In total, the oxygen gas and the metallic cadmium from the negative electrode react to form cadmium(II) hydroxide according to the recombination reaction (4.175). Hence, finally the generated oxygen is immediately consumed again. Since we are still dealing with charging conditions, at the negative electrode, the cadmium(II) hydroxide (on the right side of the recombination reaction (4.175)) is again converted back to metallic cadmium via the intermediate formation of a dissolved negative hydroxo complex – [Cd(OH)$_3$]$^-$ – according to the following two-step mechanism that we already know:

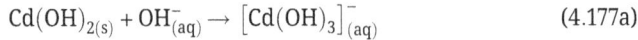

$$Cd(OH)_{2(s)} + OH^-_{(aq)} \rightarrow \left[Cd(OH)_3\right]^-_{(aq)} \qquad (4.177a)$$

Then, the divalent cadmium in the [Cd(OH)$_3$]$^-$ complex is immediately reduced to metallic Cd:

$$\left[Cd(OH)_3\right]^-_{(aq)} + 2e^- \rightarrow Cd_{(s)} + 3OH^-_{(aq)} \qquad (4.177b)$$

In a vented NiCd battery, the oxygen – formed upon overcharging – can be released to the external environment. Therefore, as mentioned earlier, a loss of electrolyte occurs and alters the cell performance. Vented NiCd batteries are used if large capacities are required. They are also called wet or flooded NiCd cells. Such larger cells need maintenance because of the possible loss of electrolyte.

In sealed NiCd batteries, gas formation can also occur if deep discharge or overdischarging occurs. We discuss overdischarge again for the case of an oversized negative electrode (Figure 4.28). In case of an overdischarge, a polarity reversal can occur at the positive electrode, and H$_2$ gas can be generated according to the following reaction:

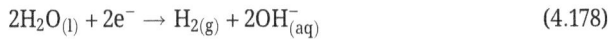

$$2H_2O_{(l)} + 2e^- \rightarrow H_{2(g)} + 2OH^-_{(aq)} \qquad (4.178)$$

that is, at the positive electrode, water molecules from the aqueous KOH solution are reduced. The H$_2$ molecules from this reaction diffuse toward the negative electrode, where they react with cadmium(II) hydroxide:

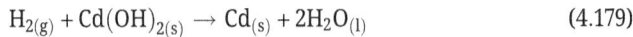

$$H_{2(g)} + Cd(OH)_{2(s)} \rightarrow Cd_{(s)} + 2H_2O_{(l)} \qquad (4.179)$$

Finally, the created H$_2$ gas is consumed again and – for the ideal case – cannot cause problems in the NiCd battery setup. In the ideal case, the electrolyte is not altered during this reaction. However, since hydrogen usually cannot be completely recombined again, further discharge might even cause a polarity reversal of the negative electrode that causes oxygen gas formation. Due to H$_2$ and O$_2$ gas formation, in the worst case, a pressure increase might cause a breaking of the NiCd battery.

By an appropriate construction, the positive electrode of the NiCd battery can also be protected against gassing upon deep discharge (Figure 4.29). For this purpose, powdered cadmium(II) hydroxide is added to the nickel electrode. Hereby, Cd(OH)$_2$ works as an antipolar mass.

Due to the oversized negative electrode, the positive electrode is exhausted first upon deep discharge, and the voltage breaks down. At this stage, the divalent Cd in

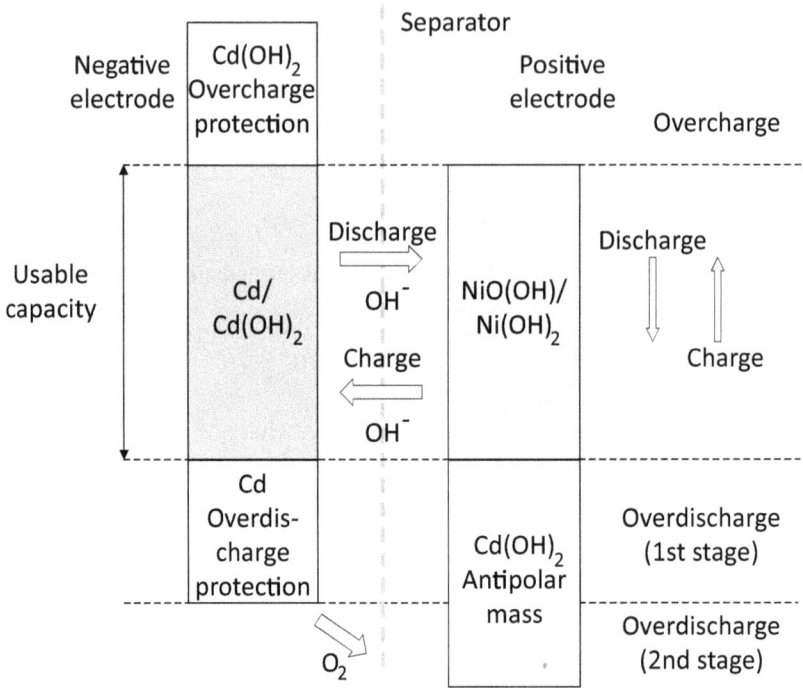

**Figure 4.29:** $H_2$ and $O_2$ gas formation during overcharging and overdischarging processes in NiCd batteries with oversized negative electrode and antipolar mass at the positive electrode.

cadmium(II) hydroxide (i.e., the antipolar mass) is reduced at the positive plate – that is, at the "nickel plate" – to metallic Cd:

$$Cd(OH)_{2(s)} + 2e^- \rightarrow Cd_{(s)} + 2OH^-_{(aq)} \tag{4.180}$$

At the negative plate, the following reaction occurs, and the cycle is closed:

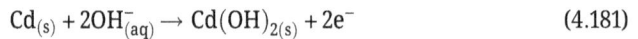

$$Cd_{(s)} + 2OH^-_{(aq)} \rightarrow Cd(OH)_{2(s)} + 2e^- \tag{4.181}$$

At the negative electrode, discharge continues due to the provided discharge reserve. A weak negative cell voltage appears. Based on reactions (4.180) and (4.181), this process is called the first stage of overdischarge.

If the discharge reserve at the negative electrode is exhausted, the second stage of the overdischarge begins. Now, upon discharge at the (initially negative) cadmium electrode, the following reaction occurs:

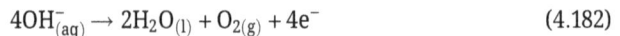

$$4OH^-_{(aq)} \rightarrow 2H_2O_{(l)} + O_{2(g)} + 4e^- \tag{4.182}$$

Oxygen gas is formed at the cadmium electrode (at positive voltage!). Note that this is the same reaction that occurs in the case of overcharging at the negative electrode (see reaction (4.173)) – but now with swapped roles. The oxygen gas that was formed according to reaction (4.182) diffuses to the (initially positive) nickel electrode, where it is reduced again (at negative voltage!):

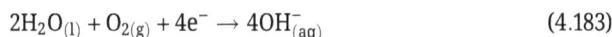

$$2H_2O_{(l)} + O_{2(g)} + 4e^- \rightarrow 4OH^-_{(aq)} \tag{4.183}$$

The battery has faced a pole change, and we observe an oxygen cycle. Electrical energy that is introduced into the NiCd battery is now nearly completely transformed into thermal energy (heat).

### 4.4.4 Nickel–metal hydride batteries

In this section, we want to discuss the peculiarities of NiMH batteries. The NiMH battery is a well-developed rechargeable battery with a potassium hydroxide (KOH) electrolyte. It is closely related to the NiCd battery. The electrochemical reactions are rather similar to that of the NiCd battery. But now, instead of the somewhat problematic Cd, the negative electrode employs a metal alloy that absorbs hydrogen, that is, the aqueous electrolyte solution.

The typical setup of a NiMH battery is similar to that of the NiCd batteries shown in Figure 4.25. The electrodes are isolated from each other by the separator. They are curled inside the metallic case. The spiral shape of the curled electrode setup is favorable with regard to low internal resistance; therefore, NiMH batteries deliver quite high currents. Separators are made from a synthetic polymer or cellulose layer; polyamide, polyethylene or polypropylene mats of fiber can be used. The separator is soaked with electrolyte, that is, an aqueous solution of potassium hydroxide (20% KOH with a pH value of 14).

The positive electrode is constructed in a similar way than in case of the NiCd battery. It consists of a polypropylene mat of fibers, where the powdered active material, that is, the trivalent nickel(III) oxide-hydroxide ($NiO(OH)$) is pressed in. For the enhancement of the internal conductivity, powdered graphite is added, too. Alternatively, it can also consist of a divalent nickel(II) hydroxide ($Ni(OH)_2$) layer that is covered by the active material, that is, the trivalent nickel(III) oxide-hydroxide ($NiO(OH)$). In this case, the divalent nickel(II) hydroxide ($Ni(OH)_2$) sheet at the positive electrode can also act as an antipolar mass (see Section 4.4.3).

The negative electrode is constructed in a quite similar way. In this case, a powered intermetallic metal alloy is pressed into a metal grid that works as current collector and that is connected to the outer metal case. Various intermetallic metal alloys can be used for the negative electrode of NiMH batteries. For instance, so-called $AB_5$ alloys are employed, where A can be a rare earth metal or a mixture of rare earth metals (e.g., lanthanum (La), cerium (Ce), neodymium (Ne) and praseodymium (Pr)). B can be nickel (Ni), manganese (Mn) or cobalt (CO), where mostly Ni is used. A typical example would be $LaNi_5$. In addition, the so-called $AB_\alpha$ alloys can be used. In this case, A can be titanium (Ti) or zirconium (Zr), and B can be again nickel or manganese; $\alpha$ is the fraction of B. Moreover, also more complex alloys might be used for the negative electrode, for example, $La_{0.8}Nd_{0.2}Ni_{2.5}Co_{2.4}Si_{0.1}$. As can be easily seen, this latter complex alloy is related to the $AB_5$ alloys, at which a mixture of lanthanum (La) and neodymium (Nd) are related to A and a mixture of nickel (Ni), cobalt (Co), and silicon (Si) to B. In some cases, $AB_2$ alloys are used, where A is titanium (Ti) or vanadium (V), and B is nickel or zirconium that contains fractions of chromium, cobalt, manganese, or iron. A negative electrode based on an $AB_2$ alloy provides a higher capacity.

The energy densities of NiMH batteries can be somewhat higher than those of NiCd batteries; that is, 60–120 Wh/kg or 140–300 Wh/L are possible, respectively. Specific power densities up to 1,000/kg can be obtained. NiMH batteries provide a nominal cell voltage of 1.2 V – similar to the NiCd cells. The ideal value is given by $E^0_{cel} = 1.32$ V; that is, $E^0_{anode} = -0.83$ V at the negative electrode and $E^0_{cathode} = 0.49$ V at the positive electrode. Again – like for the NiCd – we can state that this value is somewhat lower than for standard 1.5 V zinc–carbon or alkaline batteries for consumer applications. Therefore, NiMH cells sometimes might not be used as a replacement for those primary cells. However, in general, this is not a real problem, since many devices are designed for operation at about 1 V.

The fully charged NiMH battery consists of a negative MH electrode – that is, the anode – and a positive trivalent nickel(III) oxide-hydroxide (NiO(OH)) electrode – that is, the cathode. NiMH batteries usually use a 20% solution of the alkaline electrolyte potassium hydroxide (KOH).

Upon charging, at the negative electrode, a forced reduction of hydrogen ions ($H^+$) occurs:

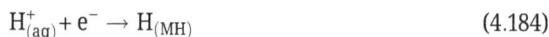

$$H^+_{(aq)} + e^- \rightarrow H_{(MH)} \tag{4.184}$$

Those $H^+$ ions result from the self-dissociation of water, as explained earlier (see eq. (4.39)). If the $H^+$ ion is reduced – that is, neutralized – at the negative electrode, the neutral hydrogen atom is bonded to the intermetallic metal alloy (M), and a metal hydride (MH) is formed. The metal in the metal hydride exhibits an oxidation number of +1 ($M^+H^-$). The charging reaction at the negative electrode can be described in more detail by the following (forced) reduction equation:

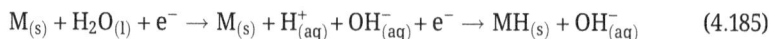

$$M_{(s)} + H_2O_{(l)} + e^- \rightarrow M_{(s)} + H^+_{(aq)} + OH^-_{(aq)} + e^- \rightarrow MH_{(s)} + OH^-_{(aq)} \tag{4.185}$$

This reaction is reversible upon discharge.

At the positive electrode, upon charging, the same (forced) oxidation of divalent nickel(II) hydroxide ($Ni(OH)_2$) to trivalent nickel(III) oxide hydroxide (NiO(OH)) occurs, as was already explained for the NiCd battery (see eq. (4.164)), that is:

$$Ni(OH)_{2(s)} + OH^-_{(aq)} \rightarrow NiO(OH)_{(s)} + H_2O_{(l)} + e^- \tag{4.186}$$

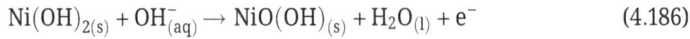

Hence, for the charging process, we finally obtain the following redox reaction:

$$M_{(s)} + Ni(OH)_{2(s)} \rightarrow MH_{(s)} + NiO(OH)_{(s)} \tag{4.187}$$

We can summarize that during the charging process, the intermetallic metal alloy (M) electrode is the negative terminal, but now it is the cathode, since here, during charging, a (forced) reduction of $H^+$ ions occurs (eqs. (4.184) and (4.185)). Then, the nickel electrode ($Ni(OH)_2$) is the positive terminal, but now it is the anode, since during charging, a (forced) oxidation of nickel(II) hydroxide occurs (eq. (4.186)).

During discharge at the negative electrode, hydrogen atoms from metal hydride ($H_{(MH)}$) ions are oxidized:

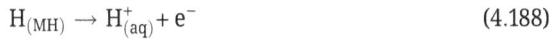

$$H_{(MH)} \rightarrow H^+_{(aq)} + e^- \tag{4.188}$$

and immediately afterward:

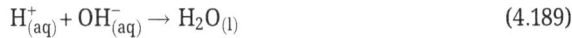

$$H^+_{(aq)} + OH^-_{(aq)} \rightarrow H_2O_{(l)} \tag{4.189}$$

These discharging reactions at the negative electrode can be described in more detail by the following oxidation equation:

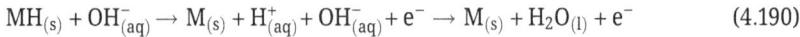

$$MH_{(s)} + OH^-_{(aq)} \rightarrow M_{(s)} + H^+_{(aq)} + OH^-_{(aq)} + e^- \rightarrow M_{(s)} + H_2O_{(l)} + e^- \tag{4.190}$$

The hydrogen atoms from the metal hydride (MH) are oxidized, and a metal with an oxidation number of 0 is formed ($M^0$).

During the discharge process, at the positive electrode, trivalent nickel(III) oxide hydroxide (NiO(OH)) is reduced to divalent nickel(II) hydroxide $Ni(OH)_2$:

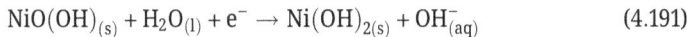

$$NiO(OH)_{(s)} + H_2O_{(l)} + e^- \rightarrow Ni(OH)_{2(s)} + OH^-_{(aq)} \tag{4.191}$$

Hence, for the discharge process, we finally obtain the following redox reaction:

$$MH_{(s)} + NiO(OH)_{(s)} \rightarrow M_{(s)} + Ni(OH)_{2(s)} \tag{4.192}$$

We can summarize the electrochemical discharge and charging reactions for the NiMH batteries for the negative electrode, the positive electrode, and the corresponding redox equation, respectively, as follows:

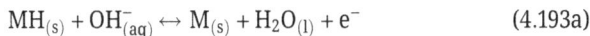

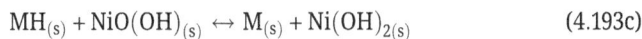

$$MH_{(s)} + OH^-_{(aq)} \leftrightarrow M_{(s)} + H_2O_{(l)} + e^- \tag{4.193a}$$

$$NiO(OH)_{(s)} + H_2O_{(l)} + e^- \leftrightarrow Ni(OH)_{2(s)} + OH^-_{(aq)} \tag{4.193b}$$

$$MH_{(s)} + NiO(OH)_{(s)} \leftrightarrow M_{(s)} + Ni(OH)_{2(s)} \tag{4.193c}$$

In these equations, discharge reactions point toward the right side, and charge reactions toward the left side (as usual).

Similar to the reactions occurring in NiCd cells, in NiMH batteries, the electrolyte is not altered or consumed during the discharge and charge redox reactions. The same amount of $OH^-$ ions consumed in the anodic half-cell during discharge is released in the cathodic half-cell of the battery; therefore, the overall concentration of the alkaline electrolyte (aqueous KOH solution) remains constant during the discharge process. For charging reactions, a similar statement can be given, only vice versa. Hence, the electrolyte only plays an intermediate role in the electrochemical cell reactions and simply acts as an ion conductor. Therefore, the KOH density cannot be a measure of the charge state of the NiMH battery – unlike in lead–acid batteries, where the acid density gives a measure of its charge state.

The electrochemical reactions (4.193a)–(4.193b) occurring in the NiMH battery upon charging and discharging only present a simplified version of the more complicated situation in real life. The real situation can be more complex, and, in particular, side reactions can occur too.

For the negative electrode, one can expect similar reactions as can be observed in NiCd cells. However, the specific reactions in NiMH battery depend on the composition of the intermetallic metal alloys that are used and will not be discussed in detail here.

At the positive electrode, the same reactions than observed in NiCd batteries can be found. Also for the case of the completely charged NiMH battery, the positive electrode consists not only of nickel(III) oxide-hydroxide (NiO(OH)) but also contains other constituents, for instance, tetravalent nickel(IV) dioxide $NiO_2$), water and potassium hydroxide (KOH). Hence, as with the NiCd battery, for NiMH batteries as well, besides the above-mentioned discharging and charging reactions (4.186) and (4.191), the following side reaction can also occur at the positive electrode:

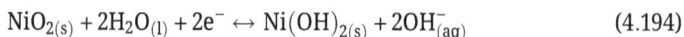

$$NiO_{2(s)} + 2H_2O_{(l)} + 2e^- \leftrightarrow Ni(OH)_{2(s)} + 2OH^-_{(aq)} \tag{4.194}$$

The discharge reaction directs toward the right side, and the charging reaction toward the left side of eq. (4.194). Keeping in mind the discharge and charge reactions at the negative electrode – that is, eq. (4.193a) – we can write the corresponding redox reaction for the side reaction (4.194) as follows:

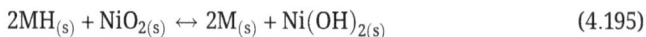

$$2MH_{(s)} + NiO_{2(s)} \leftrightarrow 2M_{(s)} + Ni(OH)_{2(s)} \tag{4.195}$$

Furthermore, during charging, the situation can be more complicated. Beside the main formation of trivalent nickel(III) oxide-hydroxide (NiO(OH)) during charging at the positive electrode, more complex hydroxides can be formed by various side reactions, for instance, mixed oxyhydroxide $[Ni_4O_4(OH)_4](OH)_2K$ that includes potassium from the aqueous potassium hydroxide solution:

$$8Ni(OH)_{2(s)} + 2K^+_{(aq)} + 12OH^-_{(aq)} \longrightarrow$$
$$\longrightarrow 2\left([Ni_4O_4(OH)_4](OH)_2K\right) + 8H_2O_{(l)} + 10e^- \qquad (4.196)$$

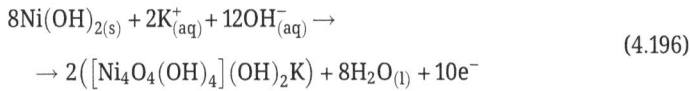

This reaction also occurs in NiCd batteries (eq. (4.171)). Upon discharge, such oxyhydroxide compounds also contribute to the current flow of the NiMH battery. As shown in eq. (4.196), during this charge reaction, water is formed. Hence, the hydroxide concentration is changed. As mentioned earlier for the NiCd battery, this charging reaction is only a side reaction without a specified amount of reaction products. Hence, the altered electrolyte density cannot be a measure of the charge state of the NiMH battery.

Although NiMH battery exhibits some similar peculiarities as NiCd batteries, they show significant advantages. First, without cadmium, they are much more environmentally friendly than NiCd batteries. Their energy densities are about twice as high than those of NiCd cells. It is advantageous that NiMH batteries do not exhibit a memory effect.

However, a so-called lazy effect can occur. The lazy effect (or lazy battery effect) means that the NiMH battery exhibits power degradation and a small voltage reduction of about 50–100 mV upon discharge. However, this lazy effect is not as dramatic as the memory effect (Figure 4.30). For a NiCd cell with a memory effect, the discharge voltage drops significantly well before the nominal capacity is reached (Figure 4.26). This is not the case with the lazy effect. The cell voltage is only slightly reduced over the whole discharge process. The capacity that can be used is nearly the same as that of a NiMH battery without the lazy effect. The lazy effect can appear after permanent charging or long storage. To a lesser extent, it appears after permanent partial discharge. The lazy effect can be healed by running through several complete discharge and charge cycles; that is, it is reversible.

Disadvantages of NiMH batteries – compared to NiCd cells – are a somewhat lower cycle lifetime and lower maximum charge and discharge currents. Moreover, the favorable temperature window, where NiMH battery can be used, is smaller than for NiCd batteries. NiMH batteries are not suitable for an operation at temperatures below 0 °C. The resilience against deep discharge or poor charging methods is also somewhat lower. With this regard, NiCd cells are more robust.

NiMH batteries are sensitive to overcharging and overdischarging (deep discharge). We start with the overcharging process. By the end of the charging process, the cell voltage increases. If the decomposition voltage of water, at about 1.6 V, is exceeded, the NiMH battery enters the overcharging regime. At the negative (M/MH) electrode, hydrogen gas ($H_2$) can be formed according to the following reaction:

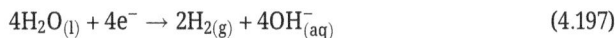

$$4H_2O_{(l)} + 4e^- \longrightarrow 2H_{2(g)} + 4OH^-_{(aq)} \qquad (4.197)$$

At the positive (nickel) electrode, oxygen gas ($O_2$) is formed, that is,

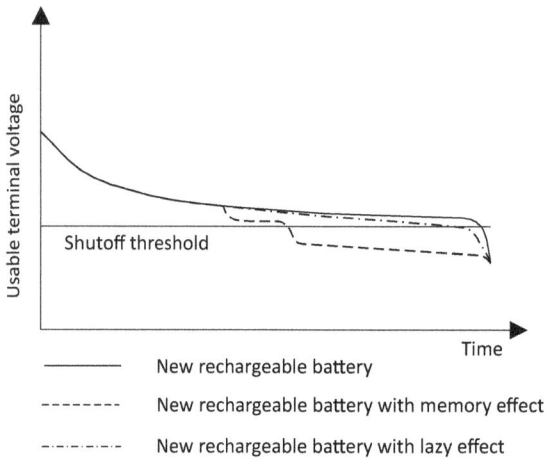

**Figure 4.30:** Lazy effect: the lazy effect is much weaker than the memory effect (Figure 4.26).

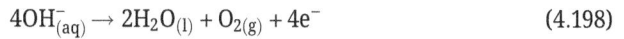

$$4OH^-_{(aq)} \rightarrow 2H_2O_{(l)} + O_{2(g)} + 4e^- \tag{4.198}$$

In total, electrolysis occurs:

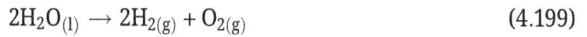

$$2H_2O_{(l)} \rightarrow 2H_{2(g)} + O_{2(g)} \tag{4.199}$$

Gassing in NiMH batteries upon overcharging runs in the same way than for NiCd batteries. By gassing, the electrolyte is altered, and the cell performance is changed. Due to the formation of $H_2$ and $O_2$ gases, in closed systems the internal pressure increases, even an exothermal oxyhydrogen gas reaction might be possible in larger battery setups. Vented battery systems are somewhat better with this regard; however, the electrolyte is altered anyway.

Gassing can also occur upon overdischarge (or deep discharge). In this case, a polarity change can occur, and water molecules in the aqueous KOH solution face electrolysis, too. Now, at the nickel electrode (initially the positive one) $H_2$ gas is generated, and at the M/MH electrode (initially the negative one) $O_2$ gas. Again, the electrolyte is altered, and the cell performance is changed.

Gassing upon overcharging and overdischarging processes is schematically shown in Figure 4.31. In the actual discussion up to now, we assumed that both electrodes have the same size.

Under normal operation conditions, gassing should be avoided – or at least, the formation of hydrogen gas should be much less than the oxygen gas formation. With this regard, the construction of NiMH batteries with an oversized negative electrode is helpful. If the negative electrode is larger than the positive one, the battery has a charge reserve in case that overcharging occurs, and it has a discharge reserve, in case of overdischarge.

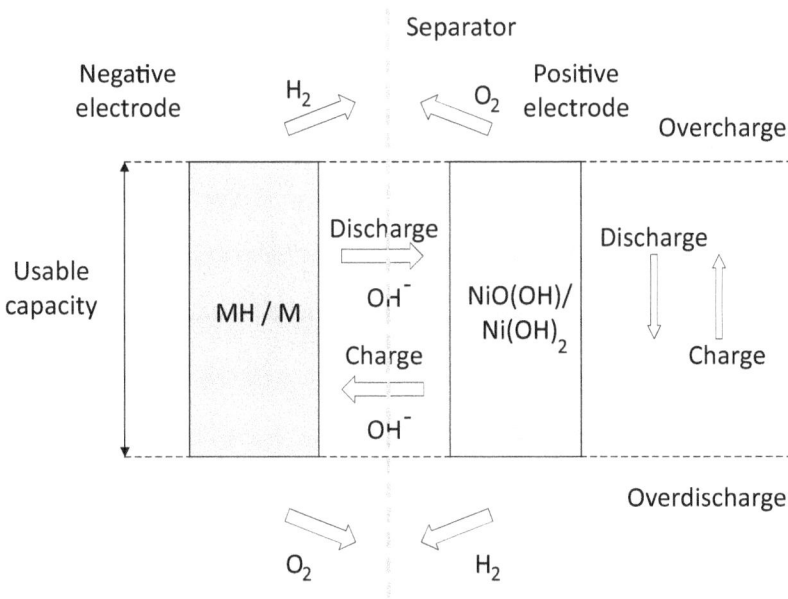

**Figure 4.31:** $H_2$ and $O_2$ gas formation during overcharging and overdischarging processes in NiMH batteries.

We first take a look at the overcharging procedure with charge reserve. In this case, the NiMH battery has a negative electrode available that contains more intermetallic alloys and metal hydrides, respectively, than are needed for complete charge and discharge processes. Under these circumstances, upon overdischarge, at the negative electrode, a forced reduction occurs according to the following reaction:

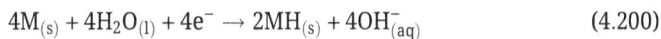

$$4M_{(s)} + 4H_2O_{(l)} + 4e^- \rightarrow 2MH_{(s)} + 4OH^-_{(aq)} \tag{4.200}$$

Due to the charge reserve in the oversized negative, the normal charge reaction continues without the formation of $H_2$ gas.

On the other hand, if the smaller positive electrode is exhausted and goes into the overcharging regime first, here $O_2$ gas formation occurs:

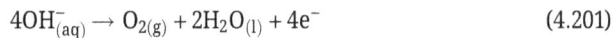

$$4OH^-_{(aq)} \rightarrow O_{2(g)} + 2H_2O_{(l)} + 4e^- \tag{4.201}$$

This forced oxidation is the same overcharging reaction that can be observed for the situation with two electrodes of the same size (eq. (4.198)). The oxygen molecules that were formed at the positive electrode diffuse to the negative electrode, where they are directly consumed again, that is,

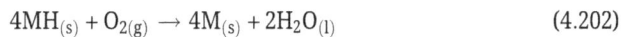

$$4MH_{(s)} + O_{2(g)} \rightarrow 4M_{(s)} + 2H_2O_{(l)} \tag{4.202}$$

Due to this recombination reaction, no increase of pressure can be faced even in sealed NiMH batteries upon overdischarge, if an oversized negative electrode is employed.

In case of overdischarging with an oversized negative electrode, at the negative electrode the normal discharge reaction continues to proceed (compare with reaction (4.190)):

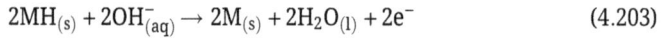

$$2MH_{(s)} + 2OH^-_{(aq)} \rightarrow 2M_{(s)} + 2H_2O_{(l)} + 2e^- \tag{4.203}$$

Now, at the positive electrode, $H_2$ formation occurs because (again) the positive electrode is exhausted first and enters the overdischarge regime:

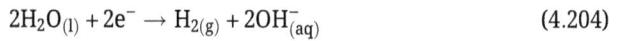

$$2H_2O_{(l)} + 2e^- \rightarrow H_{2(g)} + 2OH^-_{(aq)} \tag{4.204}$$

The $H_2$ molecules diffuse to the negative electrode, where they are directly consumed again:

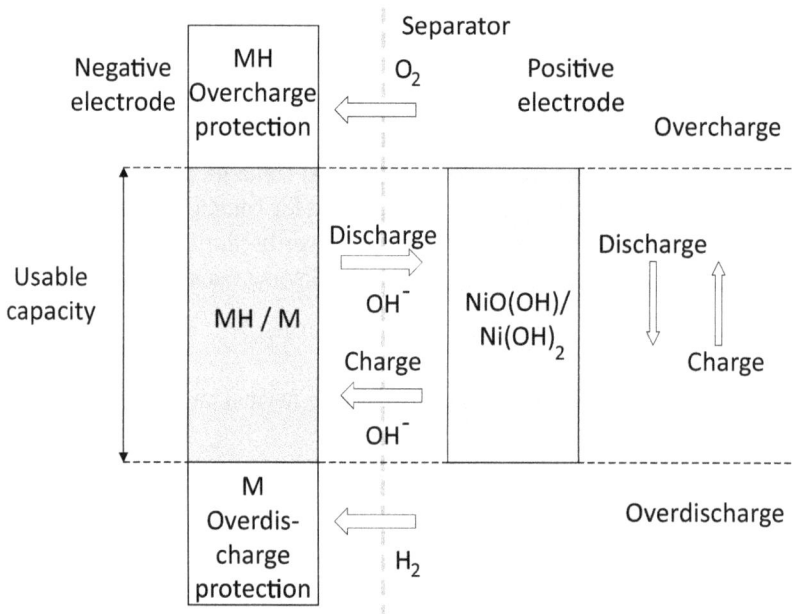

$$4M_{(s)} + 2H_{2(g)} \rightarrow 2MH_{(s)} \tag{4.205}$$

**Figure 4.32:** $H_2$ and $O_2$ gas formation during overcharging and overdischarging processes in NiMH batteries with oversized negative electrode.

that is, metal hydrides are formed. Hence, the cycle is closed, in total no gassing occurs, and the internal pressure within the NiMH battery does not increase.

Gassing upon overcharging and overdischarging processes in NiMH batteries with oversized negative electrode is schematically summarized in Figure 4.32. It can be seen that gassing occurs only at the smaller positive electrode, that is, $O_2$ formation upon overcharging, when the positive electrode is exhausted and goes into the overcharging regime, and $H_2$ formation upon overdischarge, when the positive electrode is exhausted and goes into the overdischarge regime. Both gases are consumed at the oversized negative electrode again. Therefore, internal pressure enhancement is avoided.

### 4.4.5 Lithium-ion batteries

One of the most important and powerful secondary batteries is the LIB. However, this is not a specific type of battery but, in fact, a quite large family of batteries with a rather wide variation in its specific components and, therefore, a large variation in corresponding chemical and electrochemical reactions, too. Hence, the electrochemical reactions – including especially the various side reactions – of LIBs and their detailed properties, safety characteristics and costs vary across the various LIB types.

In general, the LIB family exhibits many attractive properties, for instance, beside long lifetimes and cycle lifetimes, high energy densities and high power densities are most attractive, as shown in the Ragone diagram (Figure 4.1), where the performance of various batteries and other energy storage devices is compared. Moreover, LIBs have a very small memory effect; and they feature good thermal stability as well as low self-discharge. These are the strong advantages of this battery type.

However, the general tendency to move toward higher energy densities in battery technology and applications represents an increased risk because of general physical and chemical principles. This holds, of course, for LIB too. Therefore, the operation and handling of LIB require some precautions, since unwanted strong exothermic reactions can occur. In particular, lithium (Li) is a very reactive material that strongly reacts with water. Therefore, LIB practically have to be free of water; that is, water or humidity concentrations within the battery should be very low ($[H_2O] < 20$ ppm), since water causes an intense exothermic reaction with lithium, resulting in deflagration, explosion or fire. Moreover, the electrolytes can be highly reactive too. Highly toxic and corrosive hydrofluoric (HF) acid can be formed upon the reaction of water with organic electrolytes that contain fluorine. Such reactions are strongly supported by water. Even very small amounts of moisture within LIB can result in the formation of problematic HF acid, as will be shown later. However, these problems can be handled if LIB are carefully constructed and precautions are considered. In general, the operation of LIB should occur under careful consideration of the technical instructions.

The main components of LIB are – as always – the negative and the positive electrodes, and in between, the ion-conducting electrolyte. The electrodes and the electrolyte are the distinguishing components within the LIB family.

LIBs use intercalated lithium for the negative electrode. Hence, for this electrode, materials have to be utilized that exhibit layered, open, or porous crystal structures, allowing for lithium intercalation. The electrical properties of the electrode materials have to be good too (i.e., with low resistivity). A good example of an appropriate material for the negative electrodes of conventional LIBs is carbon, that is, graphite with its layered crystal structure that allows for lithium intercalation (Figure 4.33); graphite exhibits good electrical properties as well. In the case of a negative graphite electrode, charge transfer occurs between the lithium and carbon atoms.

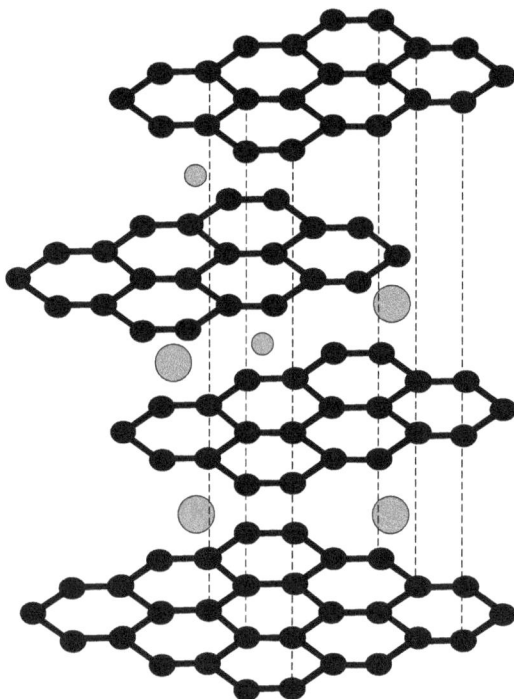

**Figure 4.33:** Layered structure of graphite with intercalated atoms (gray).

Another material for the negative electrode can be titanium(IV) disulfide ($TiS_2$) that also has a layered crystal structure (Figure 4.34) and allows for the intercalation of lithium atoms. Each of the tetravalent titanium atoms is bond to a sulfur atom. The $TiS_2$ structure is hexagonal close-packed. The titanium ($Ti^{4+}$) is surrounded by six sulfide ($S^{2-}$) ligands in an octahedral structure, and each sulfide is connected to three titanium centers. While covalent bonds exist within the $TiS_2$ layers, adjacent $TiS_2$ layers are bonded together by van der Waals forces. In this case, charge transfer occurs between the lithium and titanium atoms.

We will take a closer look at the intercalation/deintercalation processes using the example of the $TiS_2$ electrode. In the case where lithium is the intercalating element,

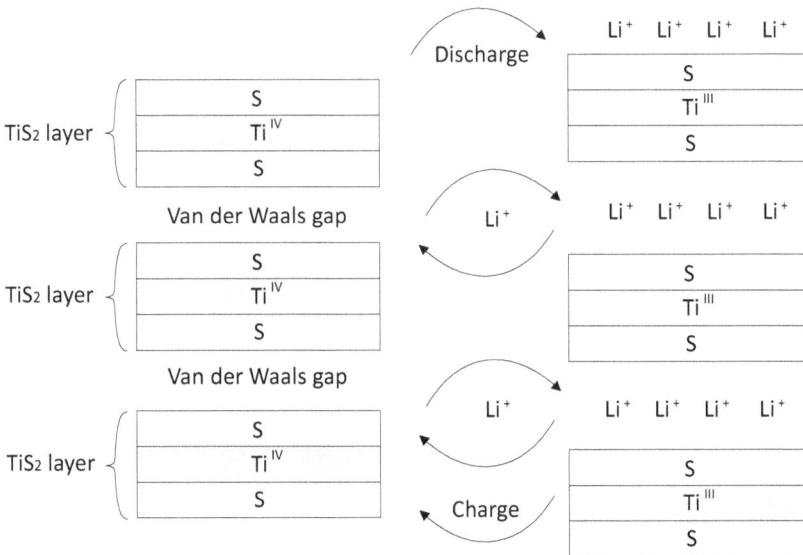

**Figure 4.34:** Layered structure of titanium(IV) disulfide ($TiS_2$). Intercalation of $Li^+$ ions during discharge causes an expansion of the vertical *c*-axis of $TiS_2$. Lithium deintercalation occurs upon charging.

the intercalation interaction is a redox process and can be described by the following reaction:

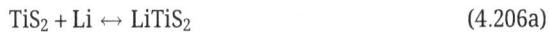

$$TiS_2 + Li \leftrightarrow LiTiS_2 \qquad\qquad (4.206a)$$

or rather:

$$TiS_2 + Li \leftrightarrow Li^+[TiS_2]^- \qquad\qquad (4.206b)$$

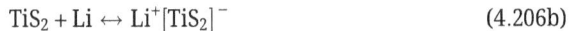

The intercalation process points toward the right, where lithium enters the layered structure of titanium(IV) disulfide. Deintercalation, where lithium leaves the layered $TiS_2$ structure, points toward the left. During the intercalation/deintercalation processes, a continuous variation of stoichiometries occurs that can be described by the formula $Li_xTiS_2$ with $x < 1$. Usually, intercalation causes an expansion of the spacing between the adjacent $TiS_2$ layers; that is, the *c*-axis of the structure expands. Upon deintercalation, the *c*-axis becomes smaller again. Intercalation/deintercalation processes occurring periodically cause stress within the negative electrode that even might cause cracks within the structure.

Although titanium(IV) disulfide electrodes are nowadays usually replaced by other materials like manganese or cobalt oxides, the use of $TiS_2$ cathodes still faces some interest for various applications.

It is to add that in case of negative electrodes made from graphite, similar processes occur upon intercalation and deintercalation processes, respectively, that is,

$$C_n + Li_x \leftrightarrow Li_x C_n \qquad\qquad (4.207)$$

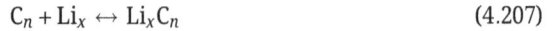

Again, the intercalation process points toward the right, and the deintercalation process to the left. The intercalation/deintercalation processes occurring periodically will cause stress also within the negative graphite electrode – and again, even cracks might occur in the electrode. Intercalation causes swelling; upon deintercalation, swelling ebbs away again.

Other materials with appropriate open or porous structures are, for instance, alloys based on amorphous or nanocrystalline silicon (i.e., a-Si or nc-Si, respectively) that form alloys with lithium. Using such a negative electrode, the corresponding LIB is sometimes called a lithium–silicon battery.

The openly structured lithium titanate ($Li_4Ti_5O_{12}$), that is, a mixed oxide, is also a reasonable material for negative electrodes in LIBs. Such a mixed oxide is composed of several oxides, and the crystal lattice consists of oxygen and the cations of different elements ($Li_2O$ or $TiO_2$). Therefore, the crystal structure of $Li_4Ti_5O_{12}$ contains $Li^+/Ti^{4+}$ ions, $O^{2-}$ ions and $Li^+$ ions.

For the charged LIB, the negative electrode (anode) contains lithium, for instance, intercalated lithium atoms in the graphite layers, that is, the anode is the lithium-based electrode, where oxidation occurs upon discharge.

Now we want to look a little deeper at the various kinds of positive electrodes (or cathodes) of LIBs and their properties. As was already mentioned, there is not only one type of LIB existing. Instead, we are dealing with a rather large LIB family that is mainly characterized by quite a variation of different kinds of positive electrodes (which are the cathodes upon discharge). The positive electrodes consist of lithium metal oxides (or rather lithium transition-metal oxides), for instance, lithium cobalt dioxide ($LiCoO_2$); this is the most prominent one. A variety of similar lithium metal oxides can be used for the positive electrode too, for instance, lithium manganese oxide ($LiMn_2O_4$) or lithium nickel dioxide ($LiNiO_2$). By the way, these chemical formulas reflect the totally discharged state of the positive electrode in the LIB.

In these positive electrodes consisting of lithium metal oxides, the cobalt, nickel or manganese atoms are formally in the trivalent oxidation state (+3). Thus, at this point, we speak about lithium cobalt(III) oxide, lithium nickel(III) oxide or lithium manganese(III) oxide. As we will see somewhat later in more detail, the oxidation states of either cobalt, nickel or manganese are cycling between the trivalent and tetravalent states upon battery operation under charging and discharging conditions, respectively. In particular, upon discharge, tetravalent cobalt(IV), nickel(IV) or manganese(IV) are reduced to trivalent cobalt(III), nickel(III) or manganese(III), respectively, and vice versa upon charging.

$LiCoO_2$ has a layered hexagonal crystallographic structure belonging to the trigonal crystal system. It is usually described by a rhombohedral symmetry with the space group R$\bar{3}$m. The layered structure of $LiCoO_2$ is characterized by alternating layers of lithium and cobalt ions that are each separated by layers of oxygen ions, as

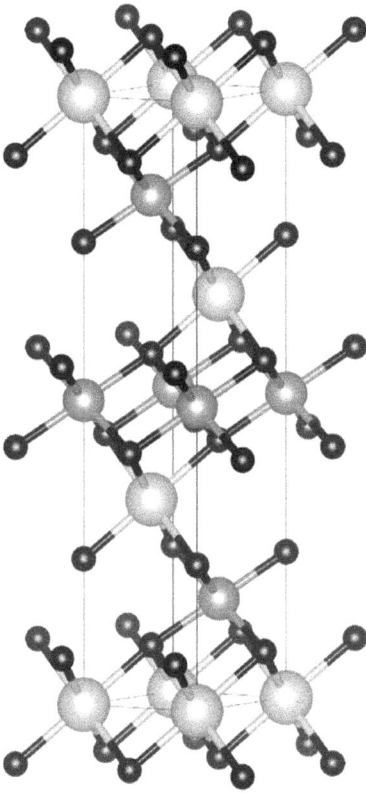

**Figure 4.35:** Layered structure of lithium cobalt dioxide (LiCoO$_2$). The small black spheres are oxygen ions, the medium-sized gray spheres are cobalt ions and the big light gray spheres are lithium ions.

can be seen in Figure 4.35. The monovalent lithium cations (Li$^+$) occupy octahedral sites in the interstitial spaces between the cobalt and oxygen layers, where Li$^+$ ions are embedded between anionic sheets of cobalt and oxygen ions. The cobalt atoms are octahedrally surrounded by oxygen atoms, forming CoO$_6$ octahedra, where these CoO$_6$ octahedra are edge-sharing. In other words, within the LiCoO$_2$ structure, the CoO$_2$ sheets are arranged as edge-sharing octahedra, with two faces parallel to the sheet plane. Cobalt exists as Co(III) in the fully lithiated state (LiCoO$_2$) after complete discharge of the LIB. Thus, the cobalt atoms in the trivalent state are laying between layers of divalent oxygen atoms (O$^{2-}$). Since the lithium layers are positioned between the cobalt-oxygen layers, well-defined pathways for lithium-ion intercalation and de-intercalation during charge and discharge cycles are provided. During charging (delithiation), trivalent Co(III) is partially oxidized to tetravalent Co(IV).

The layered crystallographic structure of the LiCoO$_2$ electrode is highly ordered and provides excellent structural stability under moderate cycling conditions. It allows for high ionic conductivity and facile lithium-ion migration within the electrode during the charging and discharging processes, respectively. The LiCoO$_2$ electrode provides a high energy density and good cycling performance. Therefore, it was the dominant cathode material in the early development of LIBs. LIBs with $LiCoO_2$ electrodes typically operate at 3.0–4.2 V. They exhibit a specific capacity of about 140–160 mAh/g. Due to the high operating voltage and specific capacity, LIBs with $LiCoO_2$ electrodes also have high energy densities of about 550 Wh/kg.

However, if too much lithium ions have left the electrode (upon charging, when $Li^+$ ions migrate toward the graphite anode), structural changes are possible; for instance, phase transitions to spinel structures or disordered phases might arise. Thus, overcharging has to be avoided. $LiCoO_2$ decomposes at elevated temperatures (at about 180–200 °C), releasing oxygen and posing safety risks. Thermal runaway is a concern, especially at high states of charge. Structural degradation can also occur at high voltages due to the destabilization of the cobalt-oxygen bonds. In general, the layered structure is quite stable but undergoes significant stress and strain during lithium intercalation and deintercalation, which can lead to microcracking. Prolonged cycling causes structural collapse and capacity fading.

Cobalt is more toxic compared to other cathode materials, and it poses environmental hazards during mining and disposal. Thus, the usage of cobalt is environmentally questionable. $LiCoO_2$ electrodes are susceptible to thermal runaway; thus, advanced battery management systems are required to ensure safety. Moreover, cobalt is a relatively expensive raw material, and it has supply chain challenges due to geopolitical issues. Therefore, alternative cathode materials are considered.

Anyway, $LiCoO_2$ electrodes have significant advantages. $LiCoO_2$ remains a cornerstone of LIB technology due to its high energy density and reliability. High energy densities make it ideal for portable electronic devices. Moreover, the usage of $LiCoO_2$ electrodes is a mature technology, and extensive research and development have optimized its performance over decades. Especially, synthesis is easy since conventional solid-state methods can be employed. However, the search for alternatives is driven by the need for improved safety, lower cost and sustainability.

$LiNiO_2$ adopts a layered structure with rhombohedral symmetry belonging to the space group $R\overline{3}m$. The structure consists of alternating layers of lithium ions and nickel ions, separated by oxygen planes, which allows for facile lithium-ion intercalation and deintercalation. Thus, the crystal structure of $LiNiO_2$ looks very similar to the structure of $LiCoO_2$. Looking at Figure 4.35, one simply has to replace the cobalt atoms with nickel atoms to get the crystallographic structure of $LiNiO_2$. The layered structure of $LiNiO_2$ is characterized by alternating layers of lithium and nickel ions that are each separated by layers of oxygen ions (similar to the $LiCoO_2$ structure shown in Figure 4.35). A common issue with $LiNiO_2$ electrodes is the partial mixing of $Ni^{2+}$ into the lithium layer due to the similar ionic radii of both ions, that is, the ionic radii of $Ni^{2+}$ and $Li^+$ are 0.69 Å and 0.76 Å, respectively. This cation mixing can reduce lithium mobility and capacity.

The theoretical specific capacity of $LiNiO_2$ is about 275 mAh/g; however, practical capacities are often lower due to structural instability at high delithiation levels. LIBs with $LiNiO_2$ electrodes operate at average voltages of about 3.6–4.0 V, depending on the degree of delithiation and cycling conditions. $LiNiO_2$ exhibits a moderate cycle life, which can degrade due to structural changes, electrolyte decomposition and surface reactions during cycling. Overcharging might cause phase transitions (e.g., from layered to spinel or rock-salt phases), which reduce performance, as the penetration of lithium ions into the electrode is hindered. Moreover, especially in highly delithiated states, $LiNiO_2$ is less thermally stable compared to materials like $LiCoO_2$ or $LiFePO_4$ (the latter one will be discussed in the next chapter). The decomposition temperature decreases as the state of charge increases, releasing oxygen and potentially leading to thermal runaway. During cycling, heat is generated due to resistive losses and side reactions, necessitating careful thermal management in battery systems.

The advantages of $LiNiO_2$ electrodes – that is, high energy densities and rather high specific capacities – face some disadvantages or challenges, respectively, that is, safety concerns due to thermal runaway in overcharged or overheated conditions, and the reduction of the mobility of lithium ions and overall electrochemical performance due to cation mixing. Also, environmental concerns have to be mentioned, since nickel mining and processing have significant environmental impacts; and also, the recycling of LIBs with $LiNiO_2$ electrodes requires careful handling.

$LiMn_2O_4$ has a crystallographic cubic spinel structure with the space group $Fd\bar{3}m$ (Figure 4.36). The cubic spinel structure can be described by the general chemical formula $AB_2O_4$. The structure forms a three-dimensional network of edge-sharing $MnO_6$ octahedra, where the manganese ions are located at the octahedral sites, meaning in the center of the octahedra. The oxygen ions are located around the manganese ions at the tips of the octahedra. Mn exists as a mixture of trivalent Mn(III) and tetravalent Mn(IV) oxidation states, facilitating the redox reactions for lithium intercalation and deintercalation in the $LiMn_2O_4$ electrode during discharge and charging processes, respectively (the corresponding redox reaction will be shown somewhat later (eq. (4.218)). Lithium ions occupy tetrahedral sites within this network, meaning that these sites are located within the interstitial spaces of the Mn-O framework, thus, providing Li-ion channels, where a facile migration of $Li^+$ ions during the charging and discharging processes is ensured.

In total, one can state that the cubic symmetry of $LiMn_2O_4$ and the open channels within the crystallographic structure allow for high ionic conductivity and facile lithium-ion migration with the $LiMn_2O_4$ electrode. Moreover, the spinel structure ensures structural stability during highly repeated lithium insertion and extraction during charging and discharging, respectively. However, $LiMn_2O_4$ can suffer from structural degradation upon prolonged cycling. This is partly due to manganese dissolution in the electrolyte.

The theoretical specific capacity of LIBs with $LiMn_2O_4$ is about 148 mAh/g. However, the practical capacity is usually slightly lower due to side reactions and structural degradation over longer cycling periods; for example, capacity degradation occurs due to the dissolution of manganese into the electrolyte. Side reactions with the electrolyte can occur, especially at elevated temperatures. Moreover, the presence of

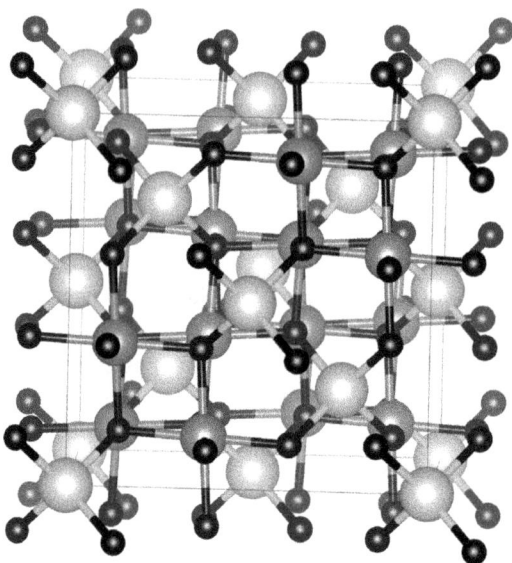

**Figure 4.36:** Layered structure of lithium manganese oxide ($LiMn_2O_4$). The small black spheres are the oxygen ions, the medium-sized gray spheres are manganese ions, and the big light grey spheres are lithium ions.

$Mn^{3+}$ ions leads to structural distortions that reduce cycling stability (this distortion is based on the Jahn–Teller effect).

The Jahn–Teller effect is a structural distortion phenomenon that occurs in certain transition metal complexes or compounds due to electronic degeneracy in their molecular or crystal orbitals. In the context of $LiMn_2O_4$ electrodes in LIBs, this effect significantly impacts the structural stability and the electrochemical performance of the electrode material. Manganese in $LiMn_2O_4$ exists in mixed valence states of Mn(III) and Mn(IV). The trivalent $Mn^{3+}$ ions exhibit a $3d^4$ electronic configuration, where the fourth electron occupies the $e_g$ orbital. The $e_g$ orbitals ($d_{z^2}$ and $d_{z^2-y^2}$) are degenerated in a perfect octahedral environment, and this degeneracy makes $MnO_6$ octahedra unstable. To lower the system's energy, the $MnO_6$ octahedra distort and lift the degeneracy. Typically, this distortion involves the elongation or compression of Mn–O bonds, and it depends on the occupancy of the $e_g$ orbitals. In $LiMn_2O_4$, elongation along one axis is common and results in a tetragonal distortion.

The Jahn–Teller distortion is particularly pronounced during lithium extraction and reinsertion. As lithium ions are removed, the proportion of $Mn^{3+}$ increases and intensifies the distortion. This structural instability can lead to phase transitions from cubic to tetragonal symmetry. Localized stress from the distortion can cause fractures in the electrode material (microcracking). Moreover, poor lithium-ion diffusion due to structural degradation results in capacity fading. In summary, the structural instability caused by the Jahn–Teller effect leads to a gradual loss of capacity over cycling.

Distortion increases the resistance to lithium-ion transport; thus, the electrochemical performance is reduced. Finally, distorted structures may react more readily with the electrolyte, increasing the risk of thermal runaway.

Beside the above-mentioned three lithium metal oxides ($LiCoO_2$, $LiNiO_2$, and $LiMn_2O_4$), the usage of mixed metal oxides can also be employed for positive electrodes (cathodes) of LIBs. Examples of such electrode materials can be materials like lithium nickel cobalt dioxide compositions ($LiNi_{1-x}Co_xO_2$) with a varying fraction $x$ of cobalt. Such materials $LiNi_{1-x}Co_xO_2$ materials are also called NC-electrode materials, where N stands for nickel and C for cobalt. One advantage of such materials is that, on one hand, cost intensive cobalt can be partly reduced by nickel, and on the other hand, the electrodes can be tailored with regard to an optimization of the LIB.

LIBs with positive $LiCoO_2$ electrodes (cathodes) exhibit stable capacities. However, in comparison to nickel-rich cathodes, the capacities and power are lower (for instance, if nickel cobalt manganese oxides or nickel cobalt aluminum oxides are used). With regard to thermal stability, $LiCoO_2$ cathodes are somewhat better than the nickel-rich ones. But the drawback is that they are more sensitive to thermal runaway, for instance, if they operated at higher temperatures above about 130 °C. Thermal runaway of $LiCoO_2$ electrodes can also occur upon overcharging.

At elevated temperatures, $LiCoO_2$ can decompose and release oxygen that might, for instance, react with the organic electrolyte in the battery cell. The decomposition of $LiCoO_2$ is problematic with regard to safety issues, since highly exothermic reactions might be initiated, leading to destruction or causing a fire hazard.

Moreover, following the argumentation of tailoring electrode properties, so-called NMC electrodes are also employed, where manganese (M) is involved in the formation of the mixed metal oxide. The NMC materials have layered structures similar to the individual $LiCoO_2$ metal oxide compound. Thus, all these mixed lithium metal oxide structures allow for easy penetration of lithium ions into the electrode material.

In Figure 4.37, the ternary phase diagram of the mixed lithium transition metal oxide system is shown, where the considered transition metals are nickel, manganese and cobalt (Li-NMC). By the modification of the stoichiometry of the transition metals, the characteristics of the mixed compounds change, and thus, cathode performances of the electrode, respective the cathodes can be changed. Not all possible variations of Li-NMC are realized for practical or commercial NMC cathode applications. But some concern around cobalt with regard to availability, costs, environmentally problematic mining, etc. and the need of extremely high number of high-performance batteries for the increasing electric vehicle market drive further Li-NMC research. In general, Li-NMC materials ($LiNi_xMn_yCo_zO_2$, where $x + y + z = 1$) are a versatile and high-performing option for cathodes in LIBs. Their tunable composition allows a balance between energy density, stability, cycle life, thermal stability and finally, costs. Therefore, Li-NMC electrodes are ideal for a wide range of applications, such as electric vehicles, portable electronics, etc.

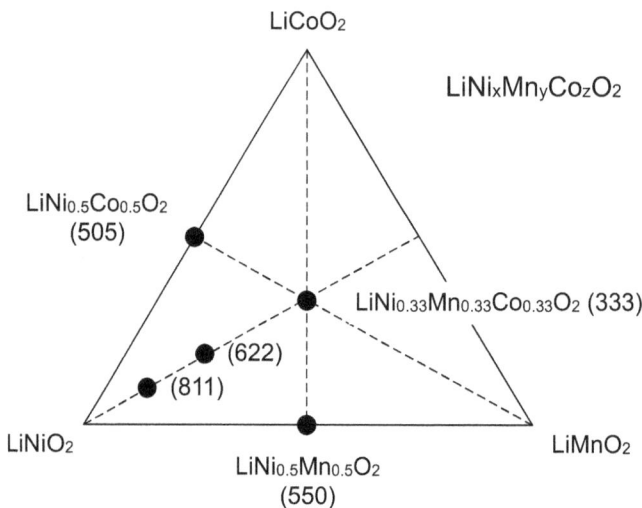

**Figure 4.37:** Ternary phase diagram of the mixed transition metal oxides LiNi$_x$Mn$_y$Co$_z$O$_2$ ($x + y + z = 1$).

Li-NMC materials with the general formula LiNi$_x$Mn$_y$Co$_z$O$_2$ ($x + y + z = 1$) exhibit a layered hexagonal crystal structure with a rhombohedral symmetry belonging to the R$\bar{3}$m space group. The structure is similar to the layered hexagonal LiCoO$_2$ crystallographic structure (Figure 4.35). Lithium ions (Li$^+$) occupy interstitial sites between layers of transition metals (Ni, Mn, Co) and oxygen. Now we give a brief overview about the properties of Li-NMC materials, summarizing their role as electrode material in LIBs.

Nickel improves the layered structure and facilitates the mobility of lithium during intercalation (i.e., discharge) and deintercalation (i.e., charging) in the Li-NMC cathode. Nickel also participates in the redox reactions during charging and discharging, switching its oxidation states (Ni(II) ↔ Ni(III) ↔ Ni(IV)). Higher nickel content increases capacity but reduces thermal stability and can result in chemical degradation of the electrode.

Manganese enhances thermal and structural stability by suppressing unwanted phase transitions and oxygen release. However, manganese does not actively contribute to capacity, as it remains in the Mn(IV) oxidation state during cycling between charging and discharging.

Cobalt (like nickel) improves the layered structure and facilitates lithium-ion mobility during intercalation (i.e., discharge) and deintercalation (i.e., charging) in the Li-NMC cathode. Cobalt also participates in the redox reactions during charging and discharging, switching its oxidation states (Co(III) ↔ Co(IV)). As was mentioned already above, the disadvantages of cobalt are that it increases costs and raises ethical concerns due to its limited supply and often problematic mining issues.

LiNi$_{0.33}$Mn$_{0.33}$Co$_{0.33}$O$_2$ (NMC 333) is a very good material that can be found in electric cars. The ideal oxidation states of the Li-NMC 333 material with regard to charge distribution are divalent nickel (Ni(II)), tetravalent manganese (Mn(IV)) and trivalent cobalt (Co(III)). During the charging process, cobalt and nickel partially oxidize to Co(IV) and Ni(IV); however, Mn(IV) remains inactive and maintains structural stability. Thus, manganese plays an important role with regard to mechanical electrode stability.

Li-NMC materials with higher nickel content provide higher capacities; however, they exhibit decreasing cycling stability with higher nickel content. Examples of such nickel-rich Li-NMC cathodes are, for instance, LiNi$_{0.6}$Mn$_{0.2}$Co$_{0.2}$O$_2$ (NMC 622) and LiNi$_{0.8}$Mn$_{0.1}$Co$_{0.1}$O$_2$ (NMC 811). It is assumed that with higher nickel concentrations, higher nickel oxidation states (Ni(III), Ni(IV)) are arising. Nickel ions at higher oxidation states are unstable, and they tend to chemical reactions with the electrolyte, and thus, leading to oxygen release and phase transitions. Therefore, it is not astonishing that for the Li-NMC 622 and Li-NMC 811 electrodes, degradation is a more serious problem than for NMC 333 electrodes. Various undesired processes can occur.

If the battery is completely discharged, oxygen can be released from the metal oxide at about 300 °C. This process degrades the lattice structure. Higher nickel content in the electrode results in the reduction of the temperature at which oxygen is released from the structure, thus, it supports degradation. Moreover, higher nickel content increases also the heat generation during the battery operation; thus, careful thermal management is necessary. In the case of higher nickel content in the electrode, a so-called cation mixing can occur, where Li$^+$ ions can substitute Ni$^{2+}$ ions, since the ions' diameters are rather similar (Ni$^{2+}$ has a diameter of 0.69 Å, in case of Li$^{2+}$, we have 0.76 Å). Cation mixing results in a displacement of nickel ions from the layered structure, altering the bonding characteristics of the material. Undesired impurity phases can be formed, and the capacity is lowered.

Before we look at the charging and discharging redox reactions in LIBs, we will have a brief look at the electrolytes.

With regard to the electrolyte, LIBs have various options. For instance, aprotic solvents can be used. Aprotic solvents cannot donate H$^+$ ions, in contradiction to protic solvents that have a labile hydrogen atom that is bound to an oxygen atom, and therefore, protic solvents donate H$^+$ ions. Due to the reactivity of lithium, protic solvents have to be strictly avoided. Aprotic electrolytes that can be used are, for instance, various carbonates like ethylene carbonate (C$_3$H$_4$O$_3$, EC), propylene carbonate (C$_4$H$_6$O$_3$) or dimethyl carbonate (C$_3$H$_6$O$_3$). The structures of these aprotic carbonates are shown in Figure 4.38.

Another option for electrolytes in LIB can be lithium salts, for instance, lithium hexafluorophosphate (LiPF$_6$). LiPF$_6$ dissociates into a positive Li$^+$ ion and a negatively charged hexafluorophosphate [PF$_6$]$^-$ anion with an octahedral structure. Another lithium salt that can be used as an electrolyte for LIB is lithium tetrafluorooxalatophosphate (LiPF$_4$(C$_2$O$_4$)); however, it is not used as often. In dissociated lithium tetrafluorooxalatophosphate, the double negatively charged dianion is an oxalate [C$_2$O$_4$]$^{2-}$.

C₃H₄O₃

Ethylene carbonate

C₄H₆O₃

Propylene carbonate

C₃H₆O₃

Dimenthyl carbonate

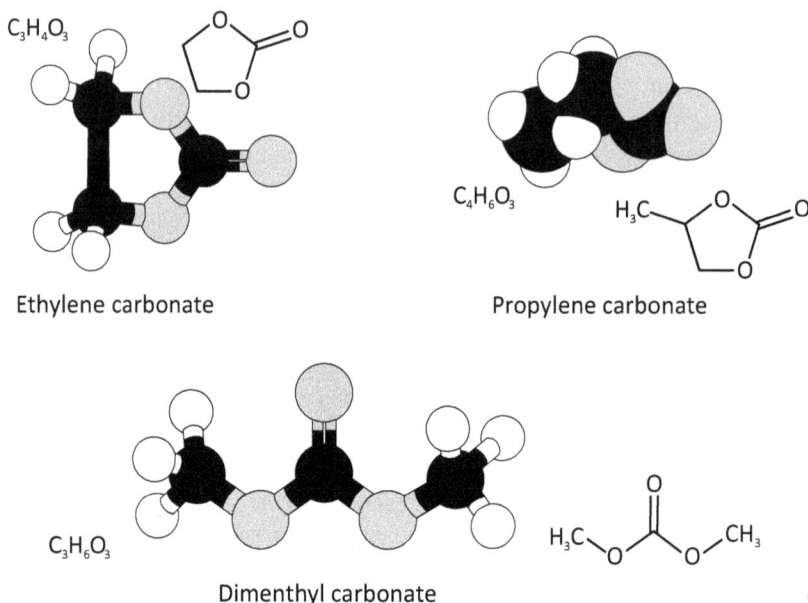

**Figure 4.38:** Structures of various aprotic carbonates (hydrogen: white balls; carbon: black balls; oxygen: gray balls).

Moreover, polymers can also be used as electrolytes in LIBs. An example is polyvinylidene fluoride, with the chemical formula $(C_2H_2F_2)_n$, where the monomer is vinylidene fluoride $(C_2H_2F_2)$. Another example is polyvinylidene fluoride hexafluoropropylene, with the formula $(C_2H_2F_2 + C_3F_6)_n$ and the monomer $(C_2H_2F_2 + C_3F_6)$. Using polymers as electrolytes for lithium, we speak about lithium-ion polymer batteries. However, the general principles of the cell chemistry do not differ in both battery types. While standard LIBs use nonaqueous liquid electrolyte solutions, lithium-ion polymer batteries employ solid or gel-like electrolytes.

The general setup of a LIB is shown in Figure 4.39. The negative electrode consists of a layered structure – mostly graphite. The graphite is deposited onto a copper sheet that forms the electrical contact with the negative terminal of the battery. The positive electrode consists of a lithium metal oxide with a porous structure that allows for the penetration of $Li^+$ ions, such as lithium cobalt dioxide $(LiCoO_2)$. The lithium metal oxide is deposited onto an aluminum sheet that provides the electrical contact with the positive terminal of the battery. A separator separates both half-cells of the LIB, and a nonaqueous electrolyte solution is used. Lithium ions can migrate through the separator from one half-cell to the other.

The main principles of the electrochemical reactions and the energy storage mechanism in LIB can be summarized as follows. The electrical energy is stored in the intercalated lithium atoms at the negative electrode and (mostly) in transition

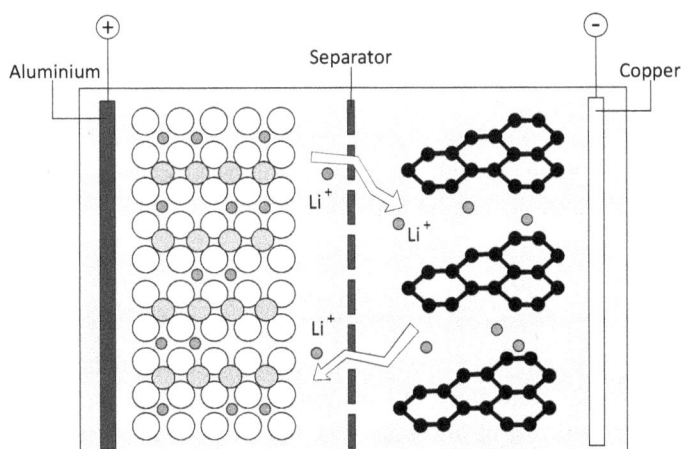

**Figure 4.39:** General setup of a lithium-ion battery (lithium: small gray balls, oxygen: white balls, metal (e.g., cobalt): bigger gray balls, carbon (graphite): black balls). The charging process occurs from the left to the right side; the discharge process occurs from the right side to the left side.

metal ions at the positive electrode. Energy storage and release in LIB are chemical processes with a conversion of the participating materials by chemical reactions. The lithium ions are mobile, while the transition metal ions in the negative electrode are not mobile (Figure 4.39). Involved transition metals can be cobalt (Co), nickel (Ni) or manganese (Mn); very often cobalt is used. The migration of the lithium ions ($Li^+$) through the battery upon discharge and charging processes is balanced by the external electrical current ($e^-$).

If the LIB is fully charged, the lithium atoms are intercalated within the layered graphite structure. In the ideal case, this should hold true for all lithium atoms. Upon discharge, lithium is oxidized; that is, electrons are donated to the graphite electrode. Those electrons migrate via external wiring to the positive electrode. The lithium ions migrate internally to the positive electrode. The ionized transition metals (e.g., Co) accept the arriving electrons ($e^-$) – not the $Li^+$ ions that had penetrated the porous positive electrode structure.

Upon discharge, the following processes and electrochemical reactions occur in the LIB (Figure 4.39). At the negative electrode, intercalated lithium is oxidized according to the oxidation reaction:

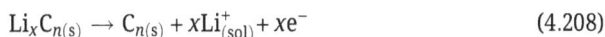

$$Li_xC_{n(s)} \rightarrow C_{n(s)} + xLi^+_{(sol)} + xe^- \qquad (4.208)$$

The electrons migrate to the positive electrode via external wiring. The positive electrode consists of lithium cobalt dioxide ($Li_{1-x}CoO_2$) with a deficit of lithium ($Li_{1-x}$) in the case of the fully charged LIB. The oxidized lithium ions in solution ($Li^+_{(sol)}$) internally migrate to the positive electrode, enter the porous structure of the electrode and fill up the lithium deficit. Then, at the positive electrode, the arriving lithium ions are reduced according to the following reduction reaction:

$$Li_{1-x}CoO_{2(s)} + xLi^+_{(sol)} + xe^- \rightarrow LiCoO_{2(s)} \tag{4.209}$$

In total, the discharge process of the LIB with a negative graphite electrode and a positive lithium cobalt dioxide electrode can be described by the following redox reaction:

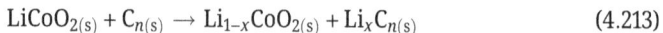

$$Li_{1-x}CoO_{2(s)} + Li_xC_{n(s)} \rightarrow LiCoO_{2(s)} + C_{n(s)} \tag{4.210}$$

If the LIB is completely discharged, the $Li^+$ ions are bound within the lithium cobalt dioxide. Hence, in the discharged state of the LIB, neutral Li atoms are not located in the open pores of the crystal structure or intercalated within the positive electrode as they were in the graphite electrode in the charged state of the LIB. If recharging of the LIB gets under way, $Li^+$ ions start to migrate to the negative electrode, where they are intercalated (Figure 4.39).

In ideal circumstances, the fully discharged lithium cobalt dioxide electrode ($LiCoO_2$) has no lithium deficit. Then, upon charging, the following electrochemical reaction is initiated at the positive electrode – reversal of eq. (4.209):

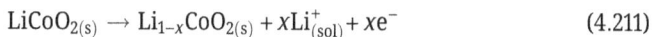

$$LiCoO_{2(s)} \rightarrow Li_{1-x}CoO_{2(s)} + xLi^+_{(sol)} + xe^- \tag{4.211}$$

Hence, at the positive electrode, trivalent Co(III) in $LiCoO_2$ is transformed to tetravalent cobalt(IV). This means that a forced oxidation of the transition metal (Co) occurs during charging. The released (donated) electrons migrate to the negative graphite electrode via external wiring. There, the arriving $Li^+$ ions are reduced according to the following (forced) reduction reaction – reversal of eq. (4.208):

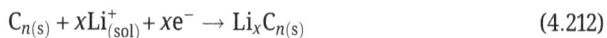

$$C_{n(s)} + xLi^+_{(sol)} + xe^- \rightarrow Li_xC_{n(s)} \tag{4.212}$$

Lithium remains in the graphite as neutral, intercalated lithium atoms. In total, the charging process of the LIB, with a negative graphite electrode and a positive lithium cobalt dioxide electrode, can be described by the following redox reaction:

$$LiCoO_{2(s)} + C_{n(s)} \rightarrow Li_{1-x}CoO_{2(s)} + Li_xC_{n(s)} \tag{4.213}$$

We can subsume that at the positive electrode during discharge, a reduction of tetravalent Co(IV) to trivalent Co(III) occurs within the lithium cobalt dioxide electrode with lithium in deficit ($Li_{1-x}CoO_2$). Upon charging, Co(III) is oxidized to Co(IV) within the lithium cobalt dioxide electrode ($LiCoO_2$). The fully discharged electrode has no lithium deficit. The lithium deficit emerges during the charging process. It is important to underline that the reactions at the positive lithium cobalt dioxide electrode are only reversible if the lithium deficit becomes not too large, that is, $x < 0.5$. Hence, $x < 0.5$

is a limiting condition for the allowed depth of discharge. This means that the overall redox reaction describing the discharging and charging processes for this type of LIB, that is,

$$Li_{1-x}CoO_{2(s)} + Li_xC_{n(s)} \leftrightarrow LiCoO_{2(s)} + C_{n(s)} \tag{4.214}$$

is valid only for $x < 0.5$. The discharge process points toward the right side of the redox reaction (4.214), and charging toward the left side.

In the case of a positive electrode made from $LiNiO_2$ that is combined with a graphite electrode, we get a reaction equation that is similar to eqs. (4.209) and (4.213), that is, for both discharge and charge directions:

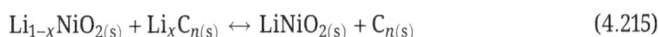

$$Li_{1-x}NiO_{2(s)} + Li_xC_{n(s)} \leftrightarrow LiNiO_{2(s)} + C_{n(s)} \tag{4.215}$$

Upon discharge, a reduction of tetravalent Ni(IV) to trivalent Ni(III) occurs within the positive lithium nickel dioxide electrode with lithium in deficit ($Li_{1-x}NiO_2$). Upon charging, Ni(III) is oxidized to Ni(IV) within the lithium nickel dioxide electrode ($LiNiO_2$); and, in succession, $Li^+$ ions can leave the positive electrode and migrate toward the graphite electrode, leaving a nickel dioxide electrode with lithium in deficit behind.

Thus, nickel can substitute cobalt in the positive electrode. A lithium nickel dioxide electrode provides some advantages. First, nickel is a cheaper material than cobalt. The capacity of the LIB is somewhat increased, since the lithium deficit in the fully charged battery can be somewhat higher than $x = 0.5$, allowing for a higher number of lithium ions to migrate through the battery upon discharge and charging, respectively. On the other hand, the voltage is somewhat lower. With regard to security aspects, $LiNiO_2$ electrodes are somewhat more problematic than $LiCoO_2$ electrodes.

Using a positive electrode that consists of lithium manganese oxide ($LiMn_2O_4$) with a spinel structure and a negative graphite electrode, the following discharging (pointing toward the right side) and charging reactions (pointing toward the left side) occur. For the negative electrode, we obtain the same reaction that was discussed earlier:

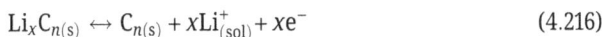

$$Li_xC_{n(s)} \leftrightarrow C_{n(s)} + xLi^+_{(sol)} + xe^- \tag{4.216}$$

At the positive electrode, we obtain

$$Li_{1-x}Mn_2O_{4(s)} + xLi^+_{(sol)} + xe^- \leftrightarrow LiMn_2O_{4(s)} \tag{4.217}$$

Upon discharge, a reduction of tetravalent Mn(IV) to trivalent Mn(III) occurs within the lithium manganese oxide electrode with lithium in deficit ($Li_{1-x}Mn_2O_4$). Upon charging, a forced oxidation of Mn(III) to Mn(IV) occurs within the lithium manganese oxide electrode ($LiMn_2O_4$). Again, in ideal circumstances, the fully discharged electrode has no lithium deficit. The total redox equation for a LIB with a graphite and a lithium manganese oxide electrode is given by

$$\text{Li}_{1-x}\text{Mn}_2\text{O}_{4(s)} + \text{Li}_x\text{C}_{n(s)} \leftrightarrow \text{LiMn}_2\text{O}_{4(s)} + \text{C}_{n(s)} \tag{4.218}$$

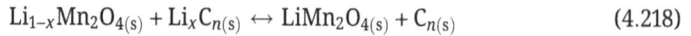

The cycle stability of lithium manganese oxide electrodes is not as good as the cycle stability of lithium cobalt dioxide electrodes. Moreover, it is sensitive to deep discharge, since the spinel structure gets unstable, and the electrode alters. Nevertheless, concerning costs, $\text{LiMn}_2\text{O}_4$ electrodes are attractive. Combining $\text{LiMn}_2\text{O}_4$ with $\text{LiCoO}_2$ in one electrode, electrodes with positive properties of both materials can be constructed that overbalance the negative aspects of the pure electrode materials.

The redox reactions in LIBs with Li-NMC cathodes ($\text{LiNi}_x\text{Mn}_y\text{Co}_z\text{O}_2$, where $x + y + z = 1$) involve the intercalation and deintercalation of $\text{Li}^+$ ions between the positive electrode (Li-NMC) and the negative electrode (graphite). During discharge, intercalated lithium atoms in the graphite electrode are oxidized, deintercalated and move as $\text{Li}^+$ ions from the graphite electrode to the Li-NMC electrode. The electrons flow through an external circuit to the Li-NMC electrode and provide power. The redox reactions for discharge can be expressed as follows.

Upon discharge, at the graphite electrode (anode), we get:

$$\text{Li}_x\text{C}_{n(s)} \rightarrow \text{C}_{n(s)} + x\text{Li}^+_{(sol)} + xe^- \tag{4.219}$$

Sometimes, a more simplified discharge reaction formula is used:

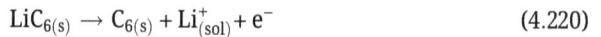

$$\text{LiC}_{6(s)} \rightarrow \text{C}_{6(s)} + \text{Li}^+_{(sol)} + e^- \tag{4.220}$$

At the Li-NMC electrode (cathode), $\text{Li}^+$ ions are intercalated into the NMC electrode structure, and the transition metals undergo a reduction. Since nickel is the primary redox-active species in the Li-NMC electrode, and cobalt only contributes partially to the redox reaction, the redox reaction at the cathode can be written as follows in a somewhat simplified expression:

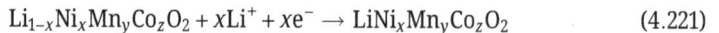

$$\text{Li}_{1-x}\text{Ni}_x\text{Mn}_y\text{Co}_z\text{O}_2 + x\text{Li}^+ + xe^- \rightarrow \text{LiNi}_x\text{Mn}_y\text{Co}_z\text{O}_2 \tag{4.221}$$

Upon charging, the directions of the reactions at the anode and cathode are reversed. The overall redox reaction formula can be written as:

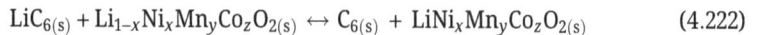

$$\text{LiC}_{6(s)} + \text{Li}_{1-x}\text{Ni}_x\text{Mn}_y\text{Co}_z\text{O}_{2(s)} \leftrightarrow \text{C}_{6(s)} + \text{LiNi}_x\text{Mn}_y\text{Co}_z\text{O}_{2(s)} \tag{4.222}$$

In the case of the fully charged battery (left side of the reaction formula), $\text{LiC}_6$ is the lithium-intercalated graphite (anode material), and $\text{Li}_{1-x}\text{Ni}_x\text{Mn}_y\text{Co}_z\text{O}_2$ is the partially delithiated NMC cathode. In the case of the fully discharged battery (right side of the reaction formula), $\text{C}_6$ is the delithiated graphite anode after all lithium has been released, and $\text{LiNi}_x\text{Mn}_y\text{Co}_z\text{O}_2$ is the fully lithiated NMC electrode. Nickel, as the primary redox-active species, is cycling between Ni(IV), Ni(III) and Ni(II) oxidation states. Cobalt only partially contributes to the redox reactions; it is cycling between Co(III) and Co(IV) oxidation states. Manganese mostly remains in the Mn(IV) oxidation state. In the Li-NMC electrodes manganese rather acts as a structural stabilizer than as a redox-active species.

The properties of LIB can be significantly altered upon overdischarging and overcharging processes, resulting in strong cell degradation and electrode damage if the cell voltage falls below about 2 V due overdischarging. Hereinafter, we will briefly discuss some examples and aspects of overdischarging and overcharging processes in LIB. We start with some overdischarging reactions.

Upon overdischarge, at the negative electrode (anode), copper can be dissolved into the electrolyte resulting in an enhanced self-discharge rate. The copper originates from the current collector foil that realizes the electrical contact to the negative terminal of the battery. Upon recharging, the dissolved copper is deposited as metallic copper, for instance, at the electrodes or other internal components of the LIB. Finally, in the worst case, short circuits between the electrodes arise, and the LIB is broken.

Overdischarging strongly affects the positive electrode as well. For example, when using a positive lithium cobalt dioxide electrode, an irreversible side reaction occurs upon overdischarging, when $Li^+$ ions oversaturate the $LiCoO_2$ electrode material, that is, if the lithium deficit in the $Li_{1-x}CoO_2$ electrode is filled up ($x = 0$). This oversaturation leads to the following side reaction, where lithium oxide ($Li_2O$) and divalent cobalt(II) oxide (CoO) are formed, that is,

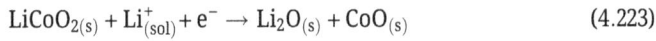

$$LiCoO_{2(s)} + Li^+_{(sol)} + e^- \rightarrow Li_2O_{(s)} + CoO_{(s)} \tag{4.223}$$

Hence, if overdischarging occurs, reaction (4.223) significantly alters the positive lithium cobalt dioxide electrode; and as a consequence, the LIB faces a loss of capacity.

Now we discuss some unwanted side reactions in LIB that occur upon overcharging. Those side reactions alter and degrade the properties of LIB, too. Moreover, side reactions can also lead to the formation of problematic impurities within the battery, for instance, hydrogen fluoride (HF), which will be mentioned later.

During overcharging, metallic lithium can be deposited onto the surface of the negative electrode through a reduction process:

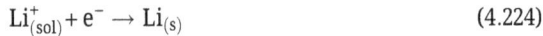

$$Li^+_{(sol)} + e^- \rightarrow Li_{(s)} \tag{4.224}$$

This can happen, in particular, if the LIB contains excess lithium that cannot be completely intercalated within the electrode. The deposition of metallic lithium onto the surface of the electrode is not reversible. Hence, a loss of mobile lithium ions occurs in the system. Moreover, due to the high reactivity of metallic lithium, side reactions with the solvent or electrolyte can occur, too. In particular, lithium carbonate ($Li_2CO_3$) can be formed, that in turn – for example – can react with a lithium hexafluorophosphate ($LiPF_6$)-based electrolyte:

$$Li_2CO_{3(s)} + LiPF_{6(sol)} \rightarrow POF_{3(g)} + CO_{2(g)} + 3LiF_{(s)} \tag{4.225}$$

Besides carbon dioxide ($CO_2$) and lithium fluoride (LiF), another product of this side reaction is phosphorus trifluoride oxide ($POF_3$) that is also called phosphoryl fluoride or phosphorus oxyfluoride. $POF_3$ is a colorless gas with a pungent smell. Phosphorus oxyfluoride reacts very quickly with water (hydrolysis) and forms difluorophosphoric acid ($HPO_2F_2$) and – especially – hydrogen fluoride (HF):

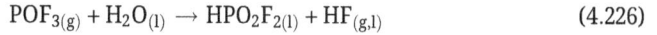

$$POF_{3(g)} + H_2O_{(l)} \rightarrow HPO_2F_{2(l)} + HF_{(g,l)} \tag{4.226}$$

Hence, we can list another reason for a very low water content in LIB, that is, to avoid the formation of hydrogen fluoride. Hydrogen fluoride has a boiling point of 19.5 °C. Hence, depending on the ambient temperature, it can be either in the liquid or gaseous state.

Anyway, as shown in reaction (4.225), gassing occurs ($CO_2$), which increases the internal pressure within the LIB. Gassing is always problematic, since, in the worst case, gassing can ultimately cause cracks in the casing of the battery that allow for moisture penetration; therefore, the formation of unwanted, highly problematic chemical compounds like hydrogen fluoride becomes possible.

Upon overcharging, metallic lithium that is deposited onto the surface of the negative electrode (lithium plating) and the subsequent formation of side products can block the pores and impede the further mobility of lithium ions during battery operation. Moreover, active material like lithium is lost and/or the electrolyte is altered during the irreversible side reactions. Thus, the properties of LIBs are degraded.

Overcharging can also alter the positive electrodes. A rather wide variety of chemical and electrochemical reactions can occur depending on the composition of the positive electrode. In particular, overcharging can result in the formation of stable materials that are lost for further cell reactions. Two examples are shown.

First, we look at a LIB with a positive lithium cobalt dioxide electrode and a negative graphite electrode. As already mentioned, upon charging, at the positive electrode the electrochemical reaction (4.211) is initiated. For the ideal case, the completely discharged positive electrode has a composition without lithium deficit, that is, $LiCoO_2$. During charging, a lithium deficit emerges, since $Li^+$ ions migrate to the negative graphite electrode, where they are intercalated. By this means, during the charging process, $LiCoO_2$ transforms to $Li_yCoO_2$; in this process, $y < 1$, and $y$ steadily decreases. If the value of the lithium deficit gets too high, that is, $y < 0.4$, $Li_yCoO_2$ becomes unstable. The battery enters the overcharging regime, and the positive $Li_yCoO_2$ electrode decomposes according to the following reaction:

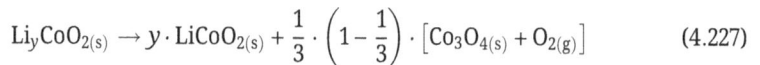

$$Li_yCoO_{2(s)} \rightarrow y \cdot LiCoO_{2(s)} + \frac{1}{3} \cdot \left(1 - \frac{1}{3}\right) \cdot \left[Co_3O_{4(s)} + O_{2(g)}\right] \tag{4.227}$$

Thus, upon overcharging, $LiCoO_2$ without lithium deficit, $O_2$ gas and cobalt(II, III) oxide ($Co_3O_4$) are formed. The crystal structure of the inorganic $Co_3O_4$ compound is composed of divalent and trivalent cobalt.

Second, in the case of a positive lithium nickel dioxide electrode, upon overcharging, a related decomposition of the electrode can be observed if $y < 0.3$; that is, if the electrode exhibits a high lithium deficit:

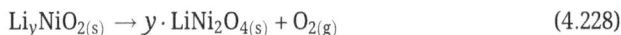

$$Li_yNiO_{2(s)} \rightarrow y \cdot LiNi_2O_{4(s)} + O_{2(g)} \qquad (4.228)$$

In both cases, the overcharge reactions (4.227) and (4.228) at the positive lithium cobalt dioxide and lithium nickel dioxide electrode, respectively, result in an $O_2$ gassing. Hence, pressure enhancement occurs that in the worst case can result in cracking of the casing, and strong exothermal reactions arise if ambient moisture gets in contact with the highly reactive battery components (lithium, electrolyte, etc.).

Approaching the end of this chapter, we also want to briefly describe the so-called solid electrolyte interface (SEI) in LIBs. Deeper insight can be obtained from the literature mentioned in the further reading publication list at the end of the book.

The SEI is a thin, protective passivating layer that forms on the surface of the anode (usually graphite) in LIBs. The formation of the SEI occurs during the first few charge–discharge cycles, and it plays a crucial role in the performance, stability and lifespan of the LIB. The SEI forms due to chemical reactions between the electrolyte and the anode material in the battery. This layer is essential because it prevents further electrolyte decomposition, stabilizes the anode, and improves battery performance and longevity. The SEI layer is electronically insulating but allows $Li^+$ ions to pass through while blocking electrons and thus enables battery operation.

When a LIB is first charged, the electrolyte undergoes electrochemical decomposition due to reactions with the anode. This results in the formation of a thin, protective SEI layer that is composed of organic and inorganic compounds such as lithium carbonate ($Li_2CO_3$), LiF, or organic lithium salts.

An example of such an organic lithium salt is lithium ethylene dicarbonate, with the chemical formula $(CH_2OCO_2)_2Li_2$ (abbreviated as $Li_2EDC$). $Li_2EDC$ is formed from the reduction of ethylene carbonate, that has the chemical formula $(CH_2O)_2CO$ (abbreviated as EC). The chemical formula of EC can also be written in brief $C_3H_4O_3$. EC is one of the main solvents in the electrolytes of Li-ion batteries. Lithium ethylene dicarbonate forms a soft, organic-rich outer layer in the SEI, which supports the transport of lithium ions. However, it can break down over time, leading to further SEI growth and capacity loss.

Another example is lithium alkyl carbonates, for instance, lithium methyl carbonate (LMC) or lithium ethyl carbonate (LEC). The general chemical formula for lithium alkyl carbonates is given by $LiOCO_2R$, where R stands for $CH_3$ in the case of LMC, and in the case of LEC, R stands for $C_2H_5$. Lithium alkyl carbonates help to stabilize the SEI structure by improving its elasticity. Another positive effect is that lithium alkyl carbonates enhance the conductivity of the $Li^+$ ions, too. However, lithium alkyl carbonates can decompose at high temperatures, leading to the formation of carbon dioxide or methane gases ($CO_2$, $CH_4$).

Several key chemical reactions leading to SEI formation can be mentioned. SEI formation primarily occurs due to electrolyte decomposition when the anode potential drops below the reduction potential of certain electrolyte components. The most common electrolytes in Li-ion batteries typically contain lithium salts (e.g., $LiPF_6$) that are dissolved in carbonate-based organic solvents like EC with the chemical formula $(CH_2O)_2CO$ or dimethyl carbonate (DMC) with the chemical formula $OC(OCH_3)_2$. During the initial charging cycle, lithium ions move from the cathode to the anode, and the surface of the anode is reduced (electrons are added). This leads to subsequent chemical reactions, where the reduction of the electrolyte solvents occurs.

For example, we look at ethylene carbonate (EC) with the chemical formula $C_3H_4O_3$. The chemical formula of EC can be written in a more precise way that better reflects the EC structure, that is, $(CH_2O)_2CO$ (see Figure 4.38). EC is a key solvent in the electrolyte of Li-ion batteries that has a high reduction potential. This property makes it one of the first species to decompose at the anode. The reduction of EC leads to the formation of lithium ethylene dicarbonate ($Li_2EDC$) with the chemical formula $(CH_2OCO_2)_2Li_2$. $Li_2EDC$ is a key SEI component.

$$2(CH_2O)_2CO + 2e^- + 2Li^+ \rightarrow (CH_2OCO_2)_2Li_2 + C_2H_4 \tag{4.229}$$

As a byproduct, ethylene gas ($C_2H_4$) is formed, and this can lead to an accumulation of gas inside the battery.

EC and other carbonate solvents can also break down into lithium carbonate ($Li_2CO_3$). Lithium carbonate ($Li_2CO_3$) is highly stable, and thus, contributes to the overall stability of the SEI. In fact, lithium carbonate is one of the most important components of the SEI. The formation of lithium carbonate can occur according to the following reaction:

$$C_3H_4O_3 + 2e^- + 2Li^+ \rightarrow Li_2CO_3 + C_2H_4 \tag{4.230}$$

Lithium hexafluorophosphate salt ($LiPF_6$) is used in the electrolytes of LIBs. Lithium hexafluorophosphate can decompose, leading to the formation of LiF and phosphorus pentafluoride ($PF_5$):

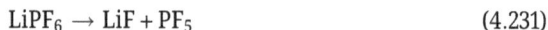

$$LiPF_6 \rightarrow LiF + PF_5 \tag{4.231}$$

LiF is another important component of the SEI. In particular, LiF is crucial for the formation of a solid and stable SEI layer. On the other hand, phosphorus pentafluoride is a strong Lewis acid and reacts with trace moisture ($H_2O$) to form hydrofluoric acid (HF) and gaseous phosphoryl fluoride ($POF_3$).

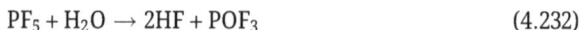

$$PF_5 + H_2O \rightarrow 2HF + POF_3 \tag{4.232}$$

HF is highly reactive and can further degrade the SEI or corrode battery components. Thus, the last reaction has to be avoided as far as possible; that is, moisture in the Li-ion batteries must be strongly reduced to nearly zero concentrations.

Several organic compounds can also form as part of the SEI, including lithium alkyl carbonates and lithium oxide ($Li_2O$). The first ones are products of solvent reduction and polymerization reactions, such as dimethyl carbonate (DMC) with the chemical formula $C_3H_6O_3$ (see Figure 4.38). The second one may also form under certain conditions and contribute to the SEI structure.

The final composition of the SEI typically includes the following organic and inorganic compounds. In the case of the main organic compounds, we speak about lithium ethylene dicarbonate ($Li_2EDC$) and polymeric species from EC. Inorganic compounds in the SEI that have to be mentioned are lithium carbonate ($Li_2CO_3$), LiF, lithium oxide ($Li_2O$) and phosphorus oxyfluoride ($POF_3$).

To ensure long-lasting and robust SEI layers, several approaches have been developed. Electrolyte additives improve SEI stability by modifying its composition, thickness, or formation mechanism. Some of the most effective additives include vinylene carbonate (VC, $C_3H_4O_3$), fluoroethylene carbonate (FEC, $C_3H_3FO_3$), lithium bis(oxalato)borate (LiBOB, $C_4BLiO_8$), or phosphates, for example, trimethyl phosphate (TMSP) with the chemical formula $C_3H_9O_4P$, or rather $(CH_3O)_3PO$.

It is to annotate that vinylene carbonate (VC) and EC are both organic carbonates with the same chemical formula, $C_3H_4O_3$. Both are used in LIB electrolytes, but they serve different roles due to their distinct chemical structures and electrochemical properties.

We can finally conclude that the SEI is very important in Li-ion batteries, since it prevents continuous electrolyte decomposition by stabilizing the anode–electrolyte interface while still allowing lithium-ion transport. Besides this prevention of decomposition of the electrolyte, a stable SEI layer reduces unwanted side reactions, and improves the cycle lifetime, and thus, the efficiency of Li-ion batteries. In general, the composition and thickness of the SEI influence battery capacity, efficiency and safety, thus affecting the overall battery performance. Irregular or unstable SEI can lead to capacity loss and performance degradation. With repeated cycling, the SEI can crack. It also can be reconstructed, and by this, it can consume more lithium, and therefore, reduces the battery efficiency. Moreover, an unstable SEI can also lead to lithium plating on the anode reducing the amount of available lithium ions for efficient battery operation and resulting in capacity fade. At elevated temperatures, the SEI can degrade, causing side reactions and safety issues, since even thermal runaway can occur in extreme cases. Electrolyte additives – for example, VC ($C_2H_2O_3$) – can help form a more stable SEI.

In the next paragraph, we briefly want to describe thermal runaway effects in Li-ion batteries.

### 4.4.6 Thermal runaway in lithium-ion batteries

Thermal runaway is a dangerous and self-sustaining failure mechanism in LIBs, where heat generation exceeds heat dissipation, leading to an uncontrollable temperature rise and potentially catastrophic failure (fire or explosion). Thermal runaway is not a single chemical reaction. In fact, it is a chain of interlinked chemical and physical processes. The critical dangers arise from self-heating reactions, the generation of combustible gases, oxygen release from the cathode (positive electrode), and, finally, the combustion of gases once ignited. Thermal runaway events follow several subsequent key stages that are partially overlapping.

The first stage of a thermal runaway event is the initiation by a triggering event. Usually, this initiation is caused by a single or a combination of the following incidents. External heating from high ambient temperatures or fire can be such a triggering event. Internal short circuits due to mechanical damage, dendrite growth inside the battery, physical damage by crushing or puncture, or simply manufacturing defects can be other reasons for the triggering of thermal runaway. Finally, overcharging or overdischarging can also trigger thermal runaway events. All of these triggering events result in a localized heating with the LIB.

As the internal temperature rises in the Li-ion battery, the second stage of a thermal runaway event sets in. This is the breakdown of the SEI layer on the anode of the Li-ion battery that occurs at temperatures in the range between about 80 and 150 °C. As described above, the SEI is a passivation layer that normally protects the anode. Due to one or several triggering events and the corresponding rise in the internal cell temperature, the SEI layer decomposes. The SEI layer decomposition is accompanied by a further increase in the internal cell temperature, since the decomposition mechanisms are (at least partly) exothermic. Thus, the second stage of the thermal runaway boosts the increase in the internal temperature, and, in addition, increased internal pressure occurs within the battery cell due to the release of gases.

The third stage is the decomposition of the electrolyte arising at temperatures between about 120 and 250 °C. The organic electrolytes are typically carbon-based. Upon decomposition in this temperature range, flammable gases like carbon monoxide (CO), methane ($CH_4$) and ethylene ($C_2H_4$) are released, effecting further heat generation, increase of the cell pressure and the risk of venting. Ethylene ($C_2H_4$) is one of the main decomposition products of the solid electrolyte interphase (SEI) layer, which was formed on the anode (typically graphite) during initial battery operation. At the beginning of battery operation, when lithium reacts with electrolyte solvents like EC, the SEI forms; but at elevated temperature upon unwanted heating, the SEI decomposes and releases ethylene gas again. Ethylene also can be formed under electrolyte breakdown. Ethylene is a highly flammable hydrocarbon gas. Very low energy is required to ignite ethylene; for instance, a spark or a hot surface can be sufficient for ignition. The autoignition temperature of ethylene is about 450 °C.

It is to underline that ethylene is important in the context of the thermal runaway event in LIBs, since it is highly flammable. Before venting, it causes a rapid pressure build-up inside the battery cell. Thus, an increased flammable gas concentration within the cell occurs, finally resulting in a violent combustion once oxygen becomes available from the cathode decomposition or air ingress. The combustion of ethylene is an exothermic reaction that delivers $CO_2$, $H_2O$ and heat, thus an amplification of the thermal runaway event occurs.

If the internal temperatures in the Li-ion batteries are in the range between about 200 and 300 °C, the materials of the positive electrode (cathode) start to release oxygen (especially metal oxides like $LiCoO_2$ release $O_2$). This fourth stage results in an exothermic breakdown of the cathode material, since the released oxygen reacts with the electrolyte and intensifies the combustion. Therefore, temperature rapidly increases, and flammable gas generation occurs that in turn undergoes exothermal reaction with the oxygen.

Above internal cell temperatures of about 300 °C, the full thermal runaway breaks through. Temperatures are exponentially increasing, and exothermic combustion reactions of released gases occur in the presence of oxygen; und thus, flammable gases ignite. Finally, cell rupture or explosion of the Li-ion battery occurs due to overpressure; and fire propagation to adjacent cells can occur (thermal propagation).

After this brief description, we now want to look a little deeper at the stages of the thermal runaway event and reaction sequences. Within the framework of this book, this can only be done qualitatively, since the reactions can be very complex, often arising in parallel, and are sometimes not completely understood and still a matter of research. We start our description at lower elevated temperatures, with the decomposition of the solid electrolyte interphase layer.

The SEI layer decomposes first because it is thermally unstable. As we mentioned already above, the SEI layer on the anode is made of a mixture of organic and inorganic compounds, like, for instance, lithium carbonate ($Li_2CO_3$), LiF and/or lithium oxide ($Li_2O$). Under heating at temperatures between about 80 and 150 °C, it breaks down and releases gases (e.g., ethylene ($C_2H_4$)) and, in particular, more heat. For example, three main reactions can be mentioned. Organic lithium ethylene dicarbonate (abbreviated $Li_2EDC$) with the chemical formula $(CH_2OCO_2)_2Li_2$ (in brief, $Li_2C_4H_4O_6$) is the first substance that decomposes at increasing temperatures in the mentioned temperature range. $Li_2EDC$ decomposes according to the following formula:

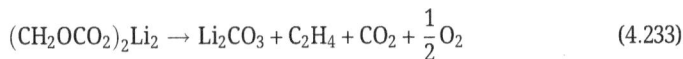

$$(CH_2OCO_2)_2Li_2 \rightarrow Li_2CO_3 + C_2H_4 + CO_2 + \frac{1}{2}O_2 \qquad (4.233)$$

By this reaction, solid lithium carbonate ($Li_2CO_3$) and ethylene gas ($C_2H_4$) are formed. This is the main part of the reaction. Keeping in mind the atomic balance, one carbon atom and three oxygen atoms are remaining in addition. Thus, it might be expected that also carbon dioxide ($CO_2$) and oxygen molecules ($^1/_2 O_2$) are formed in the decomposition reaction (4.233). Alternatively, also carbon monoxide (CO) and oxygen gas ($O_2$)

might be formed. More reasonable is that in a real decomposition process, we get a mixture of CO, $CO_2$ and oxygen upon SEI decomposition, depending on the actual situation. The mechanisms are rather complicated and not fully resolved experimentally.

Moreover, a subsequent decomposition of lithium carbonate can occur at a higher temperature:

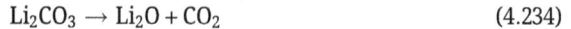

$$Li_2CO_3 \rightarrow Li_2O + CO_2 \tag{4.234}$$

Thus, lithium oxide ($Li_2O$) and carbon dioxide ($CO_2$) are formed. $CO_2$ is an inert gas, thus non-flammable and, in this regard, not problematic; however, it contributes to an increase in the internal cell pressure. Lithium oxide ($Li_2O$) is a highly reactive solid that easily reacts with traces of moisture ($H_2O$) and electrolyte solvents like ethylene carbonate (EC, $C_3H_4O_3$) and ethyl methyl carbonate (EMC, $C_4H_8O_3$). However, one can again state that these mechanisms are also rather complicated and not fully resolved experimentally.

In the case of the reaction between lithium oxide and EC, the following reaction is realistic:

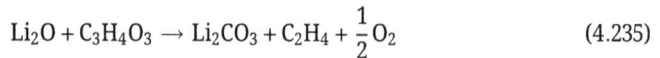

$$Li_2O + C_3H_4O_3 \rightarrow Li_2CO_3 + C_2H_4 + \frac{1}{2}O_2 \tag{4.235}$$

or rather:

$$2Li_2O + 2C_3H_4O_3 \rightarrow Li_2CO_3 + 2C_2H_4 + O_2 \tag{4.236}$$

The oxygen atom or molecule can combine with carbon atoms, forming carbon monoxide or carbon dioxide (CO or $CO_2$), respectively, where the former (CO formation) is the more realistic one under reducing or low oxygen concentrations in the degrading SEI layer. Thus, besides the formation of ethylene gas, further gas formation occurs in the battery cell upon the decomposition of the SEI layer. In general, one has to consider that several reactions occur in parallel via various pathways. Thermal decomposition of the SEI layer could involve, for instance, ring opening of the EC (by breaking C–O bonds), or the formation of lithium alkoxides might occur. Alkoxides are the conjugate base of alcohols and consist of an organic group bonded to a negatively charged oxygen atom.

Ethyl methyl carbonate is a major linear carbonate solvent used in typical Li-ion battery electrolytes. During SEI layer decomposition in the temperature range between about 80 and 150 °C, the reaction of lithium oxide ($Li_2O$) with ethyl methyl carbonate (EMC, $C_4H_8O_3$) can occur, and the situation gets even a little more complex compared to its reaction with EC. The reaction of $Li_2O$ and $C_4H_8O_3$ results in the formation of lithium carbonate ($Li_2CO_3$) and hydrocarbons like $C_2H_4$, $C_2H_6$ and $CH_4$; most probably, carbon monoxide and carbon dioxide balance the oxygen atoms. The exact distribution of products depends on temperature, oxygen availability and other SEI components. Details of the reaction sequences are complex and still part of ongoing research. A reasonable formal reaction could be given as follows:

$$Li_2O + C_4H_8O_3 \rightarrow Li_2CO_3 + 2CH_4 + CO \tag{4.237}$$

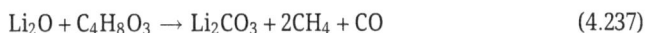

including the formation of methane and carbon monoxide.

Now, we want to look a little deeper into the electrolyte decomposition during a thermal runaway event that occurs in the temperature range between about 120 and 250 °C.

Organic solvents like ethylene carbonate (EC, $C_3H_4O_3$), ethyl methyl carbonate (EMC, $C_4H_8O_3$) or diethyl carbonate (DEC, $C_5H_{10}O_3$) are key carbonate solvents in LIBs. In the considered temperature range between 120 and 250 °C, they thermally decompose. Such decomposition can be catalyzed by $Li_2O$. The main products of the electrolyte decomposition are ethylene ($C_2H_4$), methane ($CH_4$), carbon monoxide (CO), carbon dioxide ($CO_2$) and hydrogen ($H_2$).

For example, the decomposition of EC can be described by the following reaction formula that delivers flammable gases ($CH_4$, CO):

$$C_3H_4O_3 \rightarrow CH_4 + CO_2 + CO \tag{4.238}$$

Ethyl methyl carbonate decomposes according to:

$$C_4H_8O_3 \rightarrow CH_4 + C_2H_4 + CO + O_2 \tag{4.239}$$

It is also reasonable that, in real electrolyte decomposition processes, we get a mixture of CO, $CO_2$ instead of the mentioned CO and $O_2$ formation in the formula.

The decomposition of diethyl carbonate ($C_5H_{10}O_3$) is rather complex. Under appropriate thermal conditions, DEC can react with $Li_2O$ or it decomposes thermally. Common decomposition products are ethylene ($C_2H_4$) or ethane ($C_2H_6$), carbon monoxide (CO) or carbon dioxide ($CO_2$), or sometimes methane ($CH_4$). Lithium carbonate is formed, if the reactions occur with lithium-containing compounds.

Concerning pure thermal decomposition of DEC in the temperature range about 120 and 250 °C, a realistic reaction is given by:

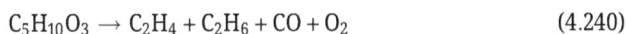

$$C_5H_{10}O_3 \rightarrow C_2H_4 + C_2H_6 + CO + O_2 \tag{4.240}$$

Again, one can state that in real electrolyte decomposition processes, it is reasonable that we get a mixture of CO, $CO_2$ of the mentioned CO and $O_2$ formation in the formula. Concerning a reaction where lithium oxide is involved, we get:

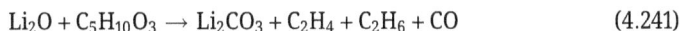

$$Li_2O + C_5H_{10}O_3 \rightarrow Li_2CO_3 + C_2H_4 + C_2H_6 + CO \tag{4.241}$$

We can summarize that, in the case of electrolyte decomposition in the temperature range between about 120 and 250 °C, significant amounts of flammable gases are formed that strongly support the further ongoing of the thermal runaway event.

Lithium hexafluorophosphate salt ($LiPF_6$) is often used in the electrolytes of LIBs. $LiPF_6$ is thermally unstable and decomposes at temperatures between 200 and 300 °C:

$$LiPF_6 \rightarrow LiF + PF_5 \qquad (4.242)$$

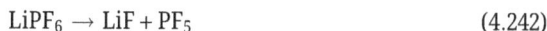

LiF is a solid, and phosphorus pentafluoride ($PF_5$) is a gas. The latter is a strong Lewis acid that reacts with trace moisture ($H_2O$), resulting in the formation of two gases:

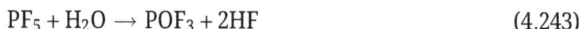

$$PF_5 + H_2O \rightarrow POF_3 + 2HF \qquad (4.243)$$

Hydrogen fluoride (HF) is the more problematic one. Beside that it is very poisonous, it attacks the separator and electrode binders, and thus, it accelerates the degradation of the LIB. Phosphoryl fluoride ($POF_3$) – commonly also called phosphorus oxyfluoride – is a toxic gas that also reacts with water and acts as the precursor of fluorophosphoric acids by hydrolysis. The reaction sequence starts with the formation of difluorophosphoric acid ($HPO_2F_2$) and hydrogen fluoride:

$$POF_3 + H_2O \rightarrow HPO_2F_2 + HF \qquad (4.244)$$

followed by the formation of monofluorophosphoric acid ($H_2PO_3F$) and hydrogen fluoride:

$$HPO_2F_2 + H_2O \rightarrow H_2PO_3F + HF \qquad (4.245)$$

And moreover, monofluorophosphoric acid reacts with water and forms phosphoric acid and hydrogen fluoride:

$$H_2PO_3F + H_2O \rightarrow H_3PO_4 + HF \qquad (4.246)$$

Thus, fluorophosphoric acids are formed by hydrolysis, accompanied by the formation of very problematic hydrogen fluoride. All these gases are corrosive and very toxic.

In the temperature range above 200 °C, the decomposition of the cathode occurs. Cathode materials in LIBs are commonly lithium metal oxides (abbreviated $LiMO_2$ or $LiM_2O_4$), like, for instance, $LiCoO_2$, $LiNiO_2$, $LnMn_2O_4$, or Li-NMC cathodes ($LiNi_xMn_yCo_zO_2$, where $x + y + z = 1$). Such cathode decomposition at temperatures between about 200 and 300 °C occurs by partial delithiation and oxygen release. These are very complex reactions depending on the specific cathode material. In general, cathode breakdown involves exothermic reactions, where transition metal oxides decompose and release oxygen, which can further react with the anode (usually graphite) or the electrolyte, releasing flammable gases. This intensifies the thermal runaway. The following key processes for the cathode decomposition at elevated temperatures can be listed. First of all, delithiated oxides ($Li_{1-x}MO_2$) become thermally unstable. They decompose and release oxygen gas ($O_2$). Then, exothermic reactions between the transition metals and oxygen can occur. Moreover, exothermic reactions can also occur between oxygen and the electrolyte or between oxygen and the flammable gases (especially ethylene). As a consequence, we face a strong generation of heat, the release of more oxygen, and possibly also the release of carbon monoxide and carbon dioxide, where the latter is not problematic since $CO_2$ is inert. In the following consideration, we mainly look at the $LiCoO_2$ and Li-NMC cathodes that exhibit the most dramatic oxygen release.

Above 200 °C, $LiCoO_2$ starts to decompose. This is a rather complex process and cannot be described by a simple reaction formula with exact stoichiometry and balance of the involved atoms. Several reactions occur in parallel and/or subsequently, where the real decomposition often proceeds through non-stoichiometric, mixed oxides. Thus, one can describe the decomposition of $LiCoO_2$ in phase terms rather than molecules or atoms. In solid-state materials like cathodes, a "phase" refers to a specific crystal structure or solid composition (like $LiCoO_2$, $Co_3O_4$, and $Li_2O$). During battery operation, that is, during discharge or charging, the $LiCoO_2$ electrode is mostly partially delithiated (see also reaction (4.214)):

$$Li_{1-x}CoO_{2(s)} + Li_xC_{n(s)} \leftrightarrow LiCoO_{2(s)} + C_{n(s)} \tag{4.247}$$

If we heat $LiCoO_2$, we face various peculiarities that happen over a broad range of temperatures and various compositions, depending on how much lithium has already been removed, how much oxygen is available, etc. Structural changes can occur, releasing $Li^+$ ions and oxygen. Moreover, a change of the oxidation state of the transition metals (e.g., cobalt) can occur. Therefore, new phases can be formed ($Co_3O_4$, CoO). $Co_3O_4$ is an inorganic compound that is denominated cobalt(II, III) oxide. It is a mixed-valence compound, where cobalt is divalent and trivalent; it is sometimes also written as $Co^{II}Co^{III}_2O_4$, or alternatively, $CoO·Co_2O_3$. Cobalt monoxide (CoO) is also an inorganic compound; it is denominated cobalt(II) oxide. Here, cobalt is divalent. At very high temperatures (950 °C), cobalt(II, III) oxide decomposes to cobalt(II) oxide and releases oxygen:

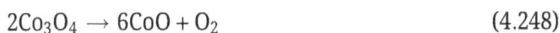

$$2Co_3O_4 \rightarrow 6CoO + O_2 \tag{4.248}$$

This can happen at the later final stages of the thermal runaway event, after the temperatures increase exponentially.

Considering partial delithiation, the decomposition of $LiCoO_2$ can be best described by the following qualitative (not exact) decomposition formula:

$$Li_{1-x}CoO_2 \rightarrow Co_3O_4 + O_2 + Li_2O \tag{4.249}$$

This is not necessarily a perfect atom-by-atom count! The decomposition formula does not describe a single, sharp reaction with precise stoichiometry, but more of a structural decomposition scheme depending on the value of $x$ (that describes the state delithiation). Thus, formula (4.249) is an approximated (or empirical) formula that represents phases that are present before and after the change. It is based on thermal studies and experimentally observed phase transitions.

Now, we look at the decomposition of a $LiNiO_2$ electrode. At high temperatures, $LiNiO_2$ decomposes to NiO, $Li_2O$ and $O_2$. The oxygen loss from the $LiNiO_2$ electrode starts at temperatures between around 200 and 300 °C. A transition from the layered

LiNiO$_2$ structure to nickel(II) oxide (NiO) with a cubic rock salt structure occurs. The decomposition can be described by the formula:

$$2LiNiO_2 \rightarrow 2NiO + \frac{1}{2}O_2 + Li_2O \qquad (4.250)$$

However, this single and sharp reaction with precise stoichiometry is only the ideal case. Depending on the grade of delithiation, the change in the oxidation state of nickel, various more complicated reactions can occur in parallel. Like in the case of a LiCoO$_2$ electrode, an approximated (or empirical) formula can be given that represents phases that are present before and after the change, and not necessarily a perfect atom-by-atom count:

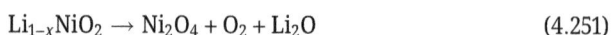

$$Li_{1-x}NiO_2 \rightarrow Ni_2O_4 + O_2 + Li_2O \qquad (4.251)$$

Again, this formula is not exact and is more of a structural decomposition scheme depending on the value of $x$ that describes the state of delithiation. It is not a perfect atom-by-atom count; however, it is based on thermal studies and observed phase transitions.

The decomposition of LiCoO$_2$ and LiNiO$_2$ electrodes causes a release of oxygen. The released oxygen can react highly exothermic with electrolyte solvents like diethyl carbonate (DEC, C$_5$H$_{10}$O$_3$) or ethyl methyl carbonate (EMC, C$_4$H$_8$O$_3$). The processes are autocatalytic, especially if the battery is already damaged or overcharged. This means that there is a positive feedback on the thermal reactions; thus, the thermal runaway event is strongly enhanced. Moreover, it is to mention that released NiO can catalyze further the decomposition of organic substances in the battery cell.

Now we look at the situation that occurs at the decomposition of the LiMn$_2$O$_4$ electrode. LiMn$_2$O$_4$ has a spinel structure, where Li$^+$ ions occupy tetrahedral sites and Mn$^{3+}$ and Mn$^{4+}$ ions occupy octahedral sites. The decomposition of the LiMn$_2$O$_4$ electrode starts at somewhat higher temperatures than the LiCoO$_2$ and LiNiO$_2$ electrodes, that is, above 250 °C. At increasing temperatures, LiMn$_2$O$_4$ undergoes a stepwise decomposition involving oxygen release and phase change:

$$2LiMn_2O_4 \rightarrow 2LiMn_2O_2 + O_2 \qquad (4.252)$$

Thus, oxygen gas is released, and LiMn$_2$O$_4$ transforms from a spinel structure to a distorted spinel phase in LiMn$_2$O$_2$. In the spinel structure of LiMn$_2$O$_4$ the valence of the two manganese atoms is equally mixed; thus, we have LiMn(III)Mn(IV)O$_4$ (or rather Li$^{1+}$Mn$^{3+}$Mn$^{4+}$O$^{2-}_4$). The LiMn$_2$O$_2$ phase is less well-defined. However, from thermal decomposition studies, it is assumed that both manganese atoms are in the trivalent Mn(III) state because oxygen is being released, meaning a reduction of tetravalent Mn(IV) to trivalent Mn(III) occurs. But formula (4.252) does not reflect this, since, according to (4.252), the oxidation state per manganese atom should be 1.5, which is chemically impossible. Thus, one can conclude that formula (4.252) is simplified and not correctly formulated. What actually happens in an early thermal decomposition of the $LiMn_2O_4$ electrode can be better described by:

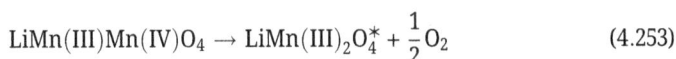

$$LiMn(III)Mn(IV)O_4 \rightarrow LiMn(III)_2O_4^* + \frac{1}{2}O_2 \qquad (4.253)$$

The $*$ at the right side of the formula represent the distorted oxygen-deficient spinel structure. A partial reduction of Mn(IV) to Mn(III) occurs, and oxygen ($O_2$) is released. Thus, the oxygen release must be balanced by the reduction of manganese, shifting the average oxidation state from +3.5 (in $LiMn_2O_4$) toward +3.

Trivalent Mn(III) is unstable at high temperatures; it is transformed into divalent Mn(II) and tetravalent Mn(IV) according to:

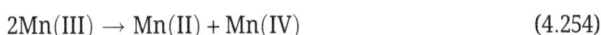

$$2Mn(III) \rightarrow Mn(II) + Mn(IV) \qquad (4.254)$$

Divalent Mn(II) can dissolve into the electrolyte, and it catalyzes electrolyte decomposition. In particular, it catalyzes the decomposition of carbonate solvents. This reaction is autocatalytic. At increasing temperatures, more divalent Mn(II) is formed, resulting in a faster degradation of the electrolyte; this provides more even heat and, thus, more trivalent Mn(III) that transforms according to the formula (4.254), and finally provides even more Mn(II).

We continue with our qualitative considerations. Upon continued heating and, according to the simplified formula (4.252), $LiMnO_2$ decomposes at temperatures between about 300 and 400 °C, according to the following reaction:

$$4LiMnO_2 \rightarrow 2Mn_2O_3 + 2Li_2O \qquad (4.255)$$

However, because the simplified reaction schemes are phase-based, they are often modeled in steps with various intermediate phases. Combining the possible reaction steps, a practical net decomposition might be described by the following estimated (not exact) decomposition formula of $LiMn_2O_4$:

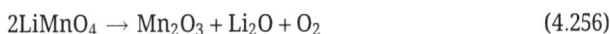

$$2LiMnO_4 \rightarrow Mn_2O_3 + Li_2O + O_2 \qquad (4.256)$$

Again, this approximated (or empirical) formula represents the phases that are present before and after the change, and it is not necessarily a perfect atom-by-atom count. However, the formed phases after the decomposition were experimentally proved. Again, oxygen generation occurs and increases the internal pressure in the battery cell and enhances flammability of the system. The dissolution of Mn(II) catalyzes the breakdown of the SEI layer and the electrolyte. The overall process becomes autocatalytic and exothermic. Moreover, the degradation of the spinel structure of $LiMn_2O_4$ loses electrochemical reversibility, and the battery dies permanently.

If we now look at the decomposition of Li-NMC cathodes ($LiNi_xMn_yCo_zO_2$, where $x + y + z = 1$), the situation becomes not easier. The breakdown is even more complex due to the participation of multiple transition metals. Keeping in mind that we are dealing with approximated formulas, the decomposition of Li-NMC cathodes ($LiNi_xMn_yCo_zO_2$, where $x + y + z = 1$) at temperatures above 200 °C can be illustrated by the following qualitative (not exact) decomposition formula:

$$LiNi_xMn_yCo_zO_2 \rightarrow NiO + MnO + CoO + O_2 + Li_2O \tag{4.257}$$

All these decomposition reactions of $LiMO_2$ (or $LiM_2O_4$) and Li-NMC electrodes are highly exothermic, releasing oxygen ($O_2$), and sometimes CO and/or $CO_2$ from reactions with the electrolyte.

### 4.4.7 Lithium iron phosphate batteries

As usual, the redox reactions in lithium iron phosphate (LFP) batteries involve the migration of lithium ions ($Li^+$) and electrons ($e^-$) during the discharging and charging processes. Similar to LIBs, the reactions occur between the cathode (i.e., the $LiFePO_4$ electrode) and the anode (usually a graphite electrode is used, represented as $C_6$), facilitated by an electrolyte. At the cathode, i.e., in the $LiFePO_4$ (LFP) material, the $Li^+$ ions are stored and released. In Figure 4.40, the layered crystal structure of the $LiFePO_4$ electrode is shown. During the charging process, the $Li^+$ ions migrate through the electrolyte to the anode, where they are intercalated between the graphite layers. Upon discharge, the process is reversed, that is, intercalated lithium atoms in the graphite electrode are oxidized, deintercalated and move as $Li^+$ ions from the graphite electrode to the LFP electrode. The electrons flow through an external circuit to the LFP electrode.

The redox reactions for discharge can be expressed as follows. Upon discharge, at the graphite electrode (anode), we get:

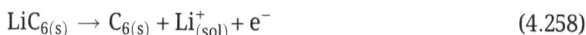

$$LiC_{6(s)} \rightarrow C_{6(s)} + Li^+_{(sol)} + e^- \tag{4.258}$$

At the LFP electrode (cathode), $Li^+$ ions are intercalated into the crystallographic LFP structure, and the transition metal (iron) undergoes a reduction, that is, trivalent Fe(III) is reduced to Fe(II). Thus, we obtain:

$$FePO_{4(s)} + Li^+_{(sol)} + e^- \rightarrow LiFePO_{4(s)} \tag{4.259}$$

Upon charging, the directions of the reactions at the anode and cathode are reversed. The overall redox reaction formula can be written as:

$$LiC_6 + FePO_4 \leftrightarrow C_6 + LiFePO_4 \tag{4.260}$$

In the case of the fully charged battery (left side of the reaction formula), $LiC_6$ is the lithium-intercalated graphite (anode material), and $FePO_4$ is the completely delithiated cathode. In the case of the fully discharged battery (right side of the reaction formula), $C_6$ is the delithiated graphite anode after all lithium is released, and $LiFePO_4$ is

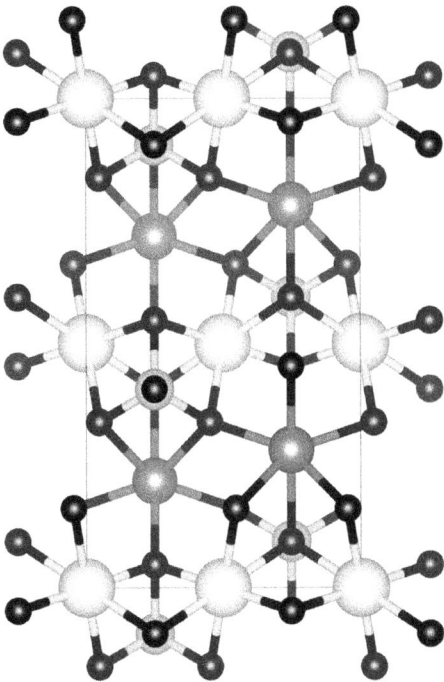

**Figure 4.40:** Crystal structure of LiFePO$_4$ (lithium: large white balls, phosphorus: small light gray balls located behind small black oxygen atoms, iron: medium-sized gray balls, oxygen: small black balls).

the fully lithiated cathode. This redox mechanism is highly reversible, which contributes to the excellent cycle life of LiFePO$_4$ batteries.

Finally, we want to mention a few words about the electrolytes. The electrolytes that are used in LFP batteries are typically non-aqueous liquid electrolytes, similar to those used in other LIBs. These electrolytes facilitate the migration of Li$^+$ ions between the anode and cathode during the charging and discharging processes, respectively. One can discriminate liquid or solid electrolytes.

The liquid electrolytes in LFP batteries are typically a mixture of a lithium salt and an organic solvent. The most common lithium salt used in LFP batteries is lithium hexafluorophosphate (LiPF$_6$). It provides high ionic conductivity and stability. Moreover, LiPF$_6$ functions well with a variety of organic solvents.

Alternatively, lithium salts like LiFSI (lithium bis(fluorosulfonyl)imide, LiF$_2$NO$_4$S$_2$), LiTFSI (lithium bis(trifluoromethanesulfonyl)imide, LiC$_2$F$_6$NO$_4$S$_2$), or LiDFOB (lithium difluoro(oxalato)borate, LiC$_2$O$_4$BF$_2$) can be used. However, these three lithium salts are less commonly used, or rather, employed only for special applications. LiFSI exhibits an improved thermal and electrochemical stability. LiTFSI provides an enhanced thermal stability and ionic conductivity, but it is less compatible with aluminum current collectors. LiDFOB can improve the SEI and cycling stability.

Various organic solvents with the ability to dissolve lithium salts can be used for the liquid electrolytes in LFP batteries. Common solvents are selected for their wide electrochemical stability windows. Various organic solvents are commonly used, each selected with regard to an optimization of the LFP battery system: (a) Ethylene carbonate (EC), that is, a cyclic carbonate with a high dielectric constant, which helps dissolve lithium salts effectively and forms a stable SEI layer on the anode surface; (b) dimethyl carbonate (DMC), a linear carbonate with low viscosity that enhances ion mobility and is mixed with EC to balance properties; (c) ethyl methyl carbonate (EMC), also a linear carbonate similar to DMC and often used in blends to optimize viscosity and conductivity; (d) diethyl carbonate (DEC), again a linear carbonate, used in mixtures for the improvement of low-temperature performance.

Electrolyte additives are often included to enhance performance, stability and safety. Some examples can be mentioned. For instance, lithium difluorophosphate ($LiPO_2F_2$) improves thermal stability and reduces gas generation at high temperatures. Another example is triphenyl phosphate ($C_{18}H_{15}O_4P$) that is a flame retardant added to the electrolyte to improve battery safety by reducing the risk of combustion. Vinylene carbonate ($C_3H_2O_3$) forms a protective SEI layer on the anode that is a passivating layer formed at the interface of the electrode and the electrolyte. VC improves the cycling stability and lifetime of the battery. During cycling it is stable, and permits fast lithium transport. Moreover, it is an electronic insulator.

LFP batteries are considered to be much safer than standard Li-ion batteries due to their unique chemical properties and thermal stability. However, while LFP batteries have minimized many safety concerns, no battery is entirely risk-free. Therefore, in the following paragraph, a comparison of both battery types is given – especially with regard to safety issues.

### 4.4.8 Comparison of lithium-ion and lithium iron phosphate batteries

Lithium iron phosphate ($LiFePO_4$) batteries (or LFP batteries) provide several significant advantages over standard LIBs. For a brief overview, several key points can be mentioned. In particular, LFP batteries are considered to be much safer than standard Li-ion batteries due to their chemical properties. In general, LFP batteries are thermally and chemically more stable than Li-ion batteries. However, although LFP batteries have minimized many safety concerns, no battery is entirely risk-free. Therefore, we will now briefly compare in a little more detail some safety issues and performance features that are associated with both battery types.

Standard LIBs can degrade quickly or fail at extreme temperatures (below 0 °C or above 45 °C), while LFP batteries have a wider operating temperature range, making them safer and more reliable in extreme conditions. Concerning thermal stability and the risk of fire or even explosion, LFP batteries are less prone to overheating and, thus, thermal runaway. Thermal runaway can be triggered by overheating. Especially, high temperatures can accelerate degradation and increase the risk of thermal runaway. Thermal runaway can result in fires or explosions, especially under improper handling or exposure to physical damage. LFP batteries catch fire to a significantly lesser extent than Li-ion batteries. This makes them safer, especially for applications where reliability is critical. The thermal runaway temperature of LFP batteries is above 250 °C, which is higher than that of LIBs. LFP batteries exhibit good thermal stability and, therefore, are highly resistant to overheating. Moreover, the cathodes in standard LIBs (e.g., Li-NMC cathodes) release oxygen when overheated, which can fuel fires. The LiFePO$_4$ cathode does not release oxygen under high temperatures; this significantly reduces the risk of fire or explosion. LFP batteries are considered to be one of the safest lithium battery chemistries and, thus, are suitable for high-temperature environments. Since LFP batteries have a wider operating temperature range, making them safer and more reliable in extreme conditions, they are particularly useful for outdoor and industrial applications where temperature fluctuations are common.

In standard operation, LFP batteries perform somewhat better in extreme temperatures compared to traditional Li-ion batteries. This holds true for both hot and cold temperatures. They typically can operate in the temperature range between about –20 and +60 °C, where the optimal performance range lies between 0 and 45 °C. This makes LFPs quite suitable for outdoor or industrial applications. In the case of Li-ion batteries, the operating temperature ranges from 0 to 45 °C for charging operations and –20 to 60 °C for discharging operations.

With regard to overcharging and short-circuiting, one can state that standard LIBs also exhibit lesser performance than LFP batteries. Overcharging or internal short circuits can cause thermal runaway. In Li-ion batteries, overcharging can lead to excessive heat generation, causing battery degradation or thermal runaway, while LFP batteries are less prone to overcharging. Even if they are overcharged, LFP batteries are less likely to experience dangerous chemical reactions. Internal short circuits that might originate from dendrite growth or manufacturing defects can cause fires or explosive reactions. This holds, in particular, for Li-ion batteries. But this is not the case if short-circuiting occurs in LFP batteries, due to their inherent stability and robust electrolytes.

Mechanical damage, such as punctures or deformation, can lead to internal short circuits and thermal runaway. The highly reactive nature of the materials in LIBs makes them vulnerable in accidents (e.g., car crashes). LFP batteries are more tolerant to mechanical damage. They are less likely to fail catastrophically when punctured or exposed to external stress. This makes them a preferred choice for applications requiring durability, such as electric vehicles and industrial energy storage.

Concerning battery operation, in general, LFP batteries last longer, with a lifespan of up to 2,000–3,000 charge cycles (or more) compared to around 500–1,000 cycles for standard Li-ion batteries. This makes LFP batteries more cost-effective over time. However, one has to annotate that the mentioned charge cycles of LIBs are typical values that can be optimized. For instance, the number of charge cycles can be enhanced if LIBs are charged to smaller maximal voltages. Moreover, LFP batteries have high charge and discharge efficiencies, often exceeding 90–95%, and they experience less energy loss, which is advantageous for applications like the storage of solar energy or in electric vehicles. Concerning the output of battery power, LFP batteries can deliver high currents for demanding applications without significant degradation. This makes them also suitable for power-intensive applications.

Finally, LFP batteries exhibit a quite eco-friendly composition. In particular, LiFePO$_4$, and thus, the LFP batteries are more environmentally friendly than LIBs since they don't contain cobalt or nickel, which are associated with environmental and ethical concerns in mining.

Altogether, although LFP batteries exhibit slightly higher upfront costs compared to standard LIBs, the longer lifespan and minimal maintenance requirements make LFP batteries more economical in the long run. Thus, they show a better cost-effectiveness over time. Despite these advantages, LFP batteries have some disadvantages compared to standard Li-ion batteries. In particular, they typically have lower energy densities than LIBs (e.g., LIBs with LiCoO$_2$ or Li-NMC electrodes). This means that they can store less energy per unit weight or volume. Therefore, LFP batteries are less suitable for applications where space or weight is critical, such as, for instance, compact portable electronics. Sacrificing some energy density for increased safety, LFP batteries are often used in electric vehicles, grid storage and industrial applications.

### 4.4.9 Sodium-ion batteries

Sodium-ion batteries (SIBs) are an emerging type of rechargeable battery technology that use sodium ions (Na$^+$) as charge carriers, instead of lithium ions (Li$^+$) used in standard lithium-ion (Li-ion) or LFP batteries. SIBs are of significant interest due to the abundant availability and low cost of sodium compared to lithium, making them a potential alternative to LIBs, especially for large-scale energy storage applications.

Speaking about SIBs, we can distinguish two principal types of SIBs, that is, aqueous and non-aqueous SIBs. Aqueous SIBs use water-based electrolytes like, for instance, sodium sulfate (Na$_2$SO$_4$) diluted in water. Non-aqueous SIBs use organic solvents (e.g., carbonate-based solvents) with dissolved sodium salts, such as sodium hexafluorophosphate (NaPF$_6$). With regard to an electrochemical stability window, a principal difference appears for these two types of SIBs. Aqueous SIBs are limited by water decomposition and, therefore, exhibit rather low cell voltages (~1.23 V), and thus, provide lower energy densities. Non-aqueous SIBs, on the other hand, allow for higher cell voltages and energy densities. In particular, the window for the cell voltage of non-aqueous SIBs is wider and lies between about 2.5 V and 4.0 V. Aqueous SIBs are safer than non-aqueous SIBs since they are not flammable. Moreover, they are environmentally friendlier. Non-aqueous SIBs are flammable and more toxic due to the organic solvents. On the other hand, aqueous SIBs may suffer from electrode corrosion and hydrogen and/or oxygen release. Non-aqueous SIBs exhibit a better long-term stability, but require a strict moisture control. As usual in life, nothing is for free. Anyway, for most of the technical applications non-aqueous SIBs are the more interesting ones.

Similar to LIBs, the SIB is not a specific type of battery but rather a quite large family of batteries with a wide variation of specific components (anodes, and especially cathodes), and thus, a large variation of corresponding chemical and electrochemical reactions, too. Na-ion batteries exhibit the following key features. In the anode the sodium ions are stored during the charging process, $Na^+$ ions are released from the anode during discharge. The cathode releases $Na^+$ ions during the charging process and accepts them during discharge. The electrolyte facilitates the migration of the $Na^+$ ions during charging and discharging cycles. A porous membrane (the separator) is used to physically separate the anode and the cathode while allowing ions to pass through. Of course, the electrolyte and the separator are insulators. During the charging and discharging cycles, the electrons have to migrate through the external circuit that connects the anode and the cathode, usually via any kind of load.

The principal setup of SIBs is similar to that of LIBs. The working principle of SIBs is also quite similar to that of LIBs, with the key difference being the use of sodium ions instead of lithium ions. When the battery is charged, $Na^+$ ions move from the cathode to the anode through the electrolyte. In parallel, electrons flow through the external circuit to the anode to balance the charge. During discharge, $Na^+$ ions move from the anode to the cathode through the electrolyte. Electrons flow back to the cathode through the external circuit, generating an electric current that can power a device.

Common anode materials for SIBs include hard carbon (C) and titanium-based oxides (e.g., $N_2Ti_6O_7$). Also, alloying materials like tin (Sn) or antimony (Sb) can be employed. Hard carbon is especially favored because it has a high reversible capacity for sodium storage. For the cathodes of SIBs, typically layered transition metal oxides are used – such as sodium cobalt oxide ($NaCoO_2$), sodium manganese oxide ($NaMnO_2$), sodium ferrite ($NaFeO_2$) and also sodium iron phosphate ($NaFePO_4$). Also, Prussian blue analogs can be used for cathode materials in SIBs. Of course, these materials are chosen based on their ability to intercalate and deintercalate $Na^+$ ions during discharge and charge cycles. As an example, the layered crystal structure of $NaCoO_2$ is shown in Figure 4.41.

Hard carbon is a type of non-graphitizable carbon that is widely used as an anode material in SIBs. Unlike graphite, hard carbon has a disordered and amorphous-like structure with micro- and nanopores and randomly oriented graphene layers. The randomly oriented graphene-like domains exhibit low graphitization even at very high temperatures (>2,000 °C). They are derived from carbonizing organic precursors (e.g., sucrose, pitch, wood, and polymers) at high temperatures, that is, typically in the range between 1,000 and 1,500 °C. The morphology of hard carbon contains both disordered regions that provide ion intercalation and nanopores or voids for ion absorption and storage. Sodium ions are larger than lithium ions. Therefore, the small interlayer spacing of about 0.34 nm is a little bit too tight $Na^+$ insertion. The

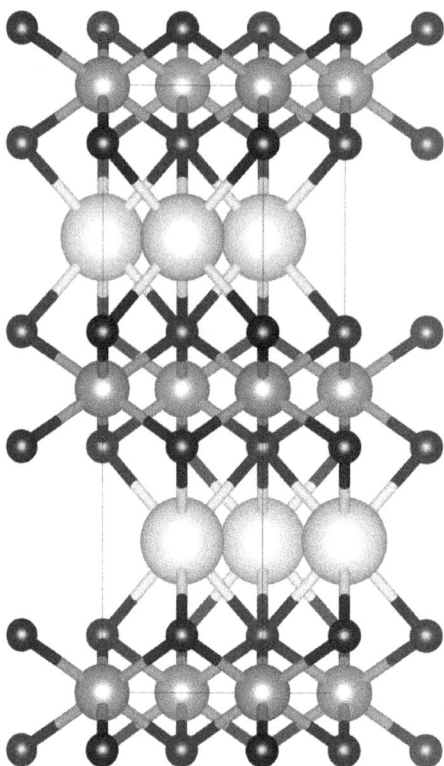

**Figure 4.41:** Crystal structure of NaCoO$_2$ (sodium: large light gray balls, cobalt: medium-sized gray balls, oxygen: small black balls).

larger interlayer spacing of hard carbon, about 0.37–0.40 nm, accommodates Na$^+$ ions much better than graphite; this comes from a combination of intercalation, adsorption and pore-filling mechanisms. Hence, hard carbon anodes exhibit a high reversible specific capacity, typically in the range between about 300–350 mAh/g. Moreover, hard carbon provides good cycling stability over hundreds to thousands of charging/discharging cycles. Due to the disordered and amorphous-like structure with micro- and nanopores and randomly oriented graphene layers, hard carbon can withstand volume changes during cycling better than crystalline materials and, therefore, it is rather tolerant to structural strain. And it is made from cheap and sustainable precursors, where a scalable mass production for commercial SIBs is possible. However, there are also some challenges to deal with. For instance, the optimization of pore structures and surface areas is critical to balance the capacity and efficiency of the SIBs; and also, irreversible reactions with the electrolyte can occur.

In SIBs, the electrolyte plays a crucial role in ensuring efficient migration of Na$^+$ ions between the anodes and cathodes. The compatibility with both electrodes is also very important. Most common are liquid electrolytes, where typically a sodium salt is dissolved in a non-aqueous organic solvent. Typically employed sodium salts are, for instance, sodium hexafluorophosphate (NaPF$_6$), sodium perchlorate (NaClO$_4$) and sodium bis[(trifluoromethyl)sulfonyl] imide with the chemical formula $C_2F_6NNaO_4S_2$ (abbreviated NaTFSI). $NaPF_6$ is most widely used. Organic carbonates are similar to those used in LIBs, for example, EC ($C_3H_4O_3$) and propylene carbonate ($C_4H_6O_3$). The latter one exhibits a good $Na^+$ solubility. Diethyl carbonate ($C_5H_{10}O_3$) and dimethyl carbonate ($C_3H_6O_3$) can be employed to adjust the ion conductivity. Aqueous electrolytes that are limited by the narrow electrochemical window of water can be useful for low-cost and safe – non-flammable – applications. Typically, sodium sulfate ($Na_2SO_4$), sodium chloride (NaCl) or sodium nitrate ($NaNO_3$) solutions are employed.

The electrochemical reactions in SIBs are, in principle, similar to the ones in standard LIBs or LFP batteries. During the charging and discharging cycles, sodium ions move between the cathode and anode, undergoing oxidation at one electrode and reduction at the other. During discharge, the stored chemical energy is converted into electrical energy. Sodium ions migrate from the anode to the cathode through the electrolyte, and electrons flow through the external circuit, powering the connected device. During charging, the reaction processes are reversed, and electrical energy is converted into chemical energy.

Upon discharge, at the anode, an oxidation reaction occurs, where typically hard carbon (C) releases $Na^+$ ions into the electrolyte, and the electrons ($e^-$) are released into the external electric circuit, creating an electric current that can power an electrical device. The oxidation reaction at the anode upon discharge is given by:

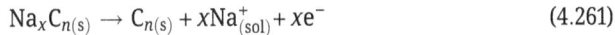

$$Na_xC_{n(s)} \rightarrow C_{n(s)} + xNa^+_{(sol)} + xe^- \tag{4.261}$$

Upon discharge, at the cathode, a reduction reaction occurs:

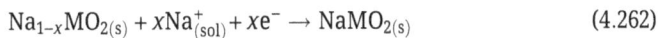

$$Na_{1-x}MO_{2(s)} + xNa^+_{(sol)} + xe^- \rightarrow NaMO_{2(s)} \tag{4.262}$$

Here, $MO_2$ represents the transition metal oxide (such as $NaCoO_2$, $NaMnO_2$, or $NaFeO_2$). Sodium ions are inserted into the cathode structure. Electrons flow from the external circuit to reduce the transition metal in $MO_2$ from a higher oxidation state to a lower oxidation state. The overall cell reaction during discharge is given by the following redox reaction:

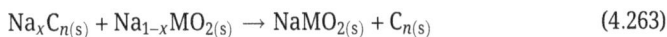

$$Na_xC_{n(s)} + Na_{1-x}MO_{2(s)} \rightarrow NaMO_{2(s)} + C_{n(s)} \tag{4.263}$$

Again, the sodium ions migrate from the anode (carbon) to the cathode (transition metal oxide). Electrons flow through the external circuit, generating electricity.

Upon charging, the redox reactions are reversed. Thus, in total, we get the bidirectional redox reaction:

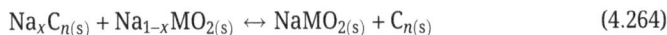

$$Na_xC_{n(s)} + Na_{1-x}MO_{2(s)} \leftrightarrow NaMO_{2(s)} + C_{n(s)} \tag{4.264}$$

If the arrow points to the right, the discharge reaction is described; and if the arrow points to the left, the charging reaction is described.

These redox reactions are crucial to the working of SIBs. The cathode material ($NaMO_2$) consists of a transition metal oxide, where "M" (a metal such as Co, Mn, or Fe) undergoes oxidation and reduction, changing oxidation states. During discharge, the transition metal undergoes a reduction, allowing sodium ions to be incorporated into the structure. During charging, the metal undergoes oxidation, releasing sodium ions into the electrolyte. The anode material (typically hard carbon) serves as a host for sodium ions. During discharge, the anode releases $Na^+$ ions into the electrolyte while undergoing oxidation. During charging, the anode undergoes reduction, taking in $Na^+$ ions and storing them for the next discharge cycle. $Na^+$ ions move through the electrolyte from the anode to the cathode during discharge and from the cathode to the anode during charging. Electrons flow through the external circuit from the anode to the cathode, and during discharge, they generate power.

A quite large variety of cathode materials were developed and investigated; and still, research on such materials is going on. Here, we present some representative examples. Different cathode materials in SIBs have slightly different redox reactions. First, we look at the sodium cobalt oxide ($NaCoO_2$) cathode. Upon the discharge of the battery, a reduction reaction occurs at the $NaCoO_2$ cathode:

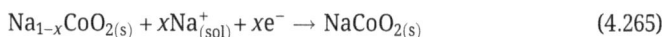

$$Na_{1-x}CoO_{2(s)} + xNa^+_{(sol)} + xe^- \rightarrow NaCoO_{2(s)} \qquad (4.265)$$

Upon charging, the reduction reaction (4.265) is reversed, and now a forced oxidation reaction occurs at the $NaCoO_2$ cathode:

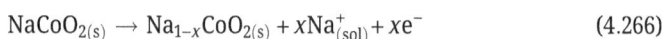

$$NaCoO_{2(s)} \rightarrow Na_{1-x}CoO_{2(s)} + xNa^+_{(sol)} + xe^- \qquad (4.266)$$

The reduction and oxidation reactions in the $NaCoO_2$ cathode involve cobalt oxidation and reduction, where the oxidation state of cobalt changes between tetravalent Co(IV) and trivalent Co(III). Upon discharge, cobalt in the $NaCoO_2$ cathode is reduced from the oxidation state Co(IV) to Co(III); upon charging, cobalt is oxidized, and the oxidation state changes from Co(III) to Co(IV). Using hard carbon (C) for the anode, in total, we get the bidirectional redox reaction:

$$Na_xC_{n(s)} + Na_{1-x}CO_{2(s)} \leftrightarrow NaCO_{2(s)} + C_{n(s)} \qquad (4.267)$$

where the discharge reaction points to the right, and the charging reaction points to the left.

Now, we look at a sodium iron phosphate ($NaFePO_4$) cathode. Upon the discharge of the battery, a reduction reaction occurs at the $NaFePO_4$ cathode:

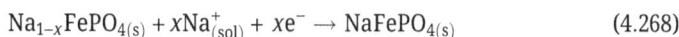

$$Na_{1-x}FePO_{4(s)} + xNa^+_{(sol)} + xe^- \rightarrow NaFePO_{4(s)} \qquad (4.268)$$

Upon charging, the reduction reaction (4.268) is reversed, and now a forced oxidation reaction occurs at the $NaFePO_4$ cathode:

$$NaFePO_{4(s)} \rightarrow Na_{1-x}FePO_{4(s)} + xNa^+_{(sol)} + xe^- \qquad (4.269)$$

The reduction and oxidation reactions in the $NaFePO_4$ cathode involve iron oxidation and reduction, where the oxidation state of iron changes between the divalent Fe(II) and trivalent Fe(III) oxidation states. Upon discharge, iron in the $NaFePO_4$ cathode is reduced from the trivalent oxidation state Fe(III) to the divalent Fe(II) state; upon charging, a forced oxidation of iron changes the divalent oxidation state Fe(II) to the trivalent state Fe(III). The total redox reaction for a SIB with a sodium iron phosphate ($NaFePO_4$) cathode and hard carbon (C) for the anode is given by

$$Na_xC_{n(s)} + Na_{1-x}FePO_{4(s)} \leftrightarrow NaFePO_{4(s)} + C_{n(s)} \qquad (4.270)$$

Another interesting example is cathodes made from Prussian blue analogs, particularly for the cathodes of aqueous SIBs. Prussian blue is a complex inorganic compound with the general chemical formula $Fe_4[Fe(CN)_6]_3 \cdot xH_2O$. Chemically, it is a mixed-valence compound composed of trivalent iron(III) and divalent iron(II) ions that are bridged by cyanide ligands. It has a face-centered cubic structure, where $Fe^{3+}$ and $Fe^{2+}$ ions are coordinated via cyanide groups ($-C \equiv N$). The Fe(III) centers are typically octahedrally coordinated by nitrogen atoms from the cyanide ligands, while Fe(II) ions are coordinated by carbon atoms. The $H_2O$ molecules in the chemical formula of Prussian blue represent crystal (or lattice) water, which is water incorporated into the solid structure but not directly bonded to the iron or cyanide groups as part of the main coordination framework. The value of $x$ in the general chemical formula of Prussian blue – $Fe_4[Fe(CN)_6]_3 \cdot xH_2O$ – is variable and usually ranges from 14 to 16 for insoluble Prussian blue. The water molecules occupy interstitial sites (voids) in the 3D lattice. They help stabilize the structure through hydrogen bonding and affect porosity and ionic conductivity.

Prussian blue analogs are a broad family of coordination compounds that share the same general crystal structure as Prussian blue but with different metal ions substituted into the framework. For example, the Prussian blue analog $NaFe[Fe(CN)_6]$ that is interesting for our considerations, can be used as a positive electrode for aqueous SIBs. In this case, the oxidation states of the Fe atoms and the framework structure matter. In common Prussian blue analogs, one Fe atom is located in the framework $(CN)_6$, where the iron atom is in the trivalent Fe(III) state, and the other one is in the divalent Fe(II) state and octahedrally coordinated with water or $Na^+$. Prussian blue analogs can typically accommodate up to two $Na^+$ ions per formula unit, though structural defects or water content may limit this.

Upon discharge, at the cathode, the following reduction reaction occurs:

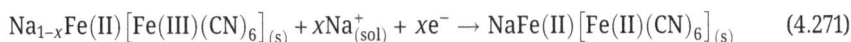

$$Na_{1-x}Fe(II)\left[Fe(III)(CN)_6\right]_{(s)} + xNa^+_{(sol)} + xe^- \rightarrow NaFe(II)\left[Fe(II)(CN)_6\right]_{(s)} \qquad (4.271)$$

In this reaction formula, the oxidation states of iron are indicated. For Prussian blue analogs, the redox-active species is typically the low-spin Fe(III) center in the $Fe(CN)_6$ framework. During discharge, $Na^+$ ions are inserted into the structure, and electrons reduce Fe(III) to Fe(II). In the fully discharged state, both iron centers are in the divalent Fe(II) state, as can be seen on the right side of reaction (4.272). For the charging process, we obtain the reversed reaction:

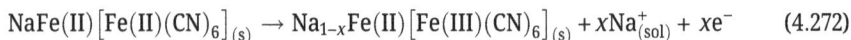

$$NaFe(II)\left[Fe(II)(CN)_6\right]_{(s)} \rightarrow Na_{1-x}Fe(II)\left[Fe(III)(CN)_6\right]_{(s)} + xNa^+_{(sol)} + xe^- \quad (4.272)$$

Using hard carbon again for the anode, the total redox reaction is given by

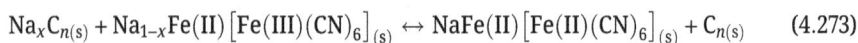

$$Na_xC_{n(s)} + Na_{1-x}Fe(II)\left[Fe(III)(CN)_6\right]_{(s)} \leftrightarrow NaFe(II)\left[Fe(II)(CN)_6\right]_{(s)} + C_{n(s)} \quad (4.273)$$

The advantages of SIBs can be summarized as follows. First of all, abundant and cheap materials are employed for their construction. Sodium is much more abundant and cheaper than lithium, making SIBs potentially more cost-effective. Sodium is commonly available from salt deposits and seawater, which is a significant advantage in terms of material sourcing. Due to the availability of sodium and the relative abundance of other raw materials, SIBs are considered to be more sustainable than LIBs in the long run. In addition, SIBs are potentially more environmentally friendly, as they do not rely on materials that are as difficult to mine or refine, such as cobalt or nickel, which are commonly found in LIBs. Finally, because of the lower cost of sodium, SIBs are seen as a promising option for large-scale grid energy storage applications, where the energy density requirements can be slightly lower compared to portable electronics.

Challenges of SIBs can be summarized as follows One of the main disadvantages of SIBs is their lower energy density compared to LIBs. Sodium ions are larger and heavier than lithium ions, which makes them less efficient at storing energy in the same amount of space. Due to the larger size of sodium ions and the need for different electrode structures, SIBs tend to be bulkier and heavier than LIBs which limit their application in portable devices. The cycle life (the number of charge and discharge cycles the battery can undergo before its capacity drops significantly) of SIBs is generally shorter than that of LIBs, especially when using certain cathode materials. However, improvements in materials and design are addressing this issue. SIBs may require different electrolytes than LIBs due to the size and chemical properties of sodium ions. Developing suitable, efficient and stable electrolytes for sodium-ion systems remains a key area of research.

SIBs are gaining attention as a more sustainable alternative to LIBs, but they still face several challenges, including unwanted side reactions that can degrade performance, safety and lifespan. Here are the most significant problematic side reactions that can occur in SIBs.

Electrolyte instability can occur at high or low potentials. This can cause gas evolution; for example, $CO_2$ or hydrocarbon gases might be released. Also, a loss of active sodium can happen. In particular, due to electrolyte instability, the formation of an unstable SEI can occur on the anode. SEI layers in SIBs tend to be less stable and more resistive than those in LIBs due to different ion solvation and interfacial chemistry.

The composition of electrolytes in non-aqueous SIBs typically uses solvents like EC, PC, DEC and others. Typically used salts are sodium hexafluorophosphate ($NaPF_6$), sodium perchlorate ($NaClO_4$) or sodium bis[(trifluoromethyl)sulfonyl] imide with the chemical formula $C_2F_6NNaO_4S_2$ (abbreviated NaTFSI). During initial cycling, solvents and salt anions decompose to form the SEI layer on the anode. For instance, looking at sodium hexafluorophosphate salt ($NaPF_6$) in EC with the chemical formula $(CH_2O)_2CO$ a reduction of EC occurs and sodium carbonate ($Na_2CO_3$) and ethylene gas ($C_2H_4$) are released:

$$EC + Na^+ + e^- \rightarrow Na_2CO_3 + C_2H_4 \tag{4.274}$$

or better:

$$(CH_2O)_2CO + Na^+ + e^- \rightarrow Na_2CO_3 + C_2H_4 \tag{4.275}$$

In the presence of trace moisture ($H_2O$), hexafluorophosphate salt reduces according to:

$$NaPF_6 + H_2O \rightarrow NaF + POF_3 + 2HF \tag{4.276}$$

In this reaction, sodium fluoride (NaF), phosphoryl fluoride ($POF_3$) and hydrogen fluoride (HF) are formed. Sodium fluoride, and in particular, hydrogen fluoride, is poisonous and rather problematic. HF can cause degradation of both the cathode and anode.

Thus, a formation of inorganic compounds occurs, for example, NaF and $Na_2CO_3$, as we have seen in reactions (4.276) and (4.277). Organic compounds can also be formed, for example, sodium ethylene dicarbonate with the chemical formula $(CH_2OCO_2)_2Na_2$ (abbreviated $Na_2EDC$), and others that form a passivating SEI. But the SEI in SIBs can be more unstable and resistive compared to the SEI in LIBs.

What could also be problematic is that a transition metal dissolution might occur at the cathode side of the SIB. This is prominent in layered oxides, for example, $NaMnO_2$. For example, the trivalent manganese Mn(III) can exhibit disproportionation according to:

$$2Mn(III) \rightarrow Mn(II) + Mn(IV) \tag{4.277}$$

or rather:

$$2Mn^{3+} \rightarrow Mn^{2+} + Mn^{4+} \tag{4.278}$$

The $Mn^{2+}$ ion is soluble in the electrolyte and can migrate to the anode. There, it can capture an electron and is reduced to metallic manganese, that causes plating of metallic manganese on the anode and catalyzes further decomposition of the SEI and the electrode.

$$Mn^{2+} + 2e^- \rightarrow Mn_{(metal)} \tag{4.279}$$

Moreover, hydrogen fluoride (HF) that is formed from $NaPF_6$ and $H_2O$ according to reaction (4.277) can support these processes by reacting with $MnO_2$:

$$MnO_2 + 4HF \rightarrow MnF_2 + 2H_2O + F_2 \tag{4.280}$$

Metallic sodium plating at the anode can also occur if the $Na^+$ ions are not intercalated fast enough in the anode.

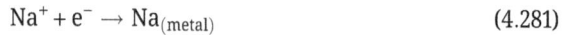

$$Na^+ + e^- \rightarrow Na_{(metal)} \tag{4.281}$$

This can happen at low temperatures or high currents. The metallic sodium can form dendrites, risking internal short-circuits.

While SIBs are in general considered to be safer than LIBs, they nevertheless can undergo thermal runaway processes under certain conditions. As we have already discussed in Section 4.4.6 for LIBs, thermal runaway is a self-accelerating exothermic process where heat generation outpaces heat dissipation, thus leading to thermal decomposition, gas release, fire, or even explosion. In general, SIBs tend to have higher onset temperatures and less energetic reactions, partly because sodium metal and its compounds are less reactive than the lithium analogs. Therefore, the gas generation is moderate and not as high as in LIBs. Especially, the flammability is lower in SIBs. In LIBs, the flammability is very high due to the electrolytes used and the reactivity of $Li^+$ ions; therefore, they exhibit a high explosive potential. The explosive potential in SIBs is significantly lower.

The reaction chain of a thermal runaway event in SIBs can be described as follows.

Initially, heat accumulations occur, for instance, due to overcharging, short circuits in the battery cell or mechanical abuse and destruction.

Exothermic breakdown of the SEI at the anode (e.g., hard carbon) begins in the temperature range between about 90 and 120 °C. For instance, organic sodium ethylene dicarbonate with the chemical formula $((CH_2OCO_2)_2Na_2)$ is the first substance that decomposes as temperatures increase. The exothermic reaction with heat release is given by:

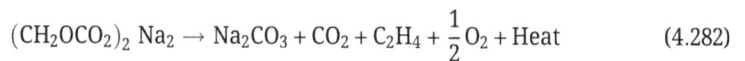

$$(CH_2OCO_2)_2\ Na_2 \rightarrow Na_2CO_3 + CO_2 + C_2H_4 + \frac{1}{2}O_2 + Heat \tag{4.282}$$

If sodium plating is present at the anode (hard carbon), sodium can oxidize violently with trace moisture, although it is less reactive than lithium. Through this exothermic reaction, sodium hydroxide (NaOH), hydrogen gas ($H_2$) and heat are released:

$$2Na + 2H_2O \rightarrow 2NaOH + H_{2(g)} + Heat \tag{4.283}$$

At temperatures between about 150 and 200 °C electrolyte decomposition occurs, where gas and more heat are released. Electrolytes (e.g., sodium hexafluoride) in SIBs decompose similarly like the ones (e.g., lithium hexafluoride $LiPF_6$) in LIBs. In case of sodium hexafluoride ($NaPF_6$), we obtain the decomposition reaction:

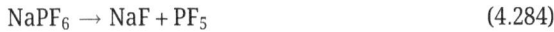

$$NaPF_6 \rightarrow NaF + PF_5 \tag{4.284}$$

Sodium fluoride (NaF) is a solid, and phosphorus pentafluoride ($PF_5$) is a gas. The latter is a strong Lewis acid that reacts with trace moisture ($H_2O$), resulting in the formation of two gases:

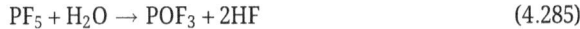

$$PF_5 + H_2O \rightarrow POF_3 + 2HF \tag{4.285}$$

Hydrogen fluoride (HF) is the more problematic one. Beside that it is very poisonous, it attacks the separator and electrode binders, and thus, it accelerates the degradation of the SIB. Phosphoryl fluoride ($POF_3$) – commonly also called phosphorus oxyfluoride – is a toxic gas that also reacts with water and acts as the precursor of fluorophosphoric acids by hydrolysis. The reaction sequence starts with the formation of difluorophosphoric acid ($HPO_2F_2$) and hydrogen fluoride:

$$POF_3 + H_2O \rightarrow HPO_2F_2 + HF \tag{4.286}$$

followed by the formation of monofluorophosphoric acid ($H_2PO_3F$), and, again, hydrogen fluoride:

$$HPO_2F_2 + H_2O \rightarrow H_2PO_3F + HF \tag{4.287}$$

Moreover, monofluorophosphoric acid reacts with water and forms phosphoric acid and hydrogen fluoride:

$$H_2PO_3F + H_2O \rightarrow H_3PO_4 + HF \tag{4.288}$$

Thus, fluorophosphoric acids are formed by hydrolysis accompanied by the formation of a significant amount of hydrogen fluoride. All these gases are corrosive and very toxic. These side reactions (4.286)–(4.289) can also be found in LIBs (see reactions (4.243)–(4.246)).

Oxygen gas ($O_2$) is released from the cathode (e.g., from layered $NaMO_2$ cathodes, where M stands for Ni, Mn, Fe) at temperatures between 200 and 250 °C. Oxygen release can occur according to the following formula, where M stands for the transition metal, or rather, a mixture of transition metals (Ni, Mn, and Fe):

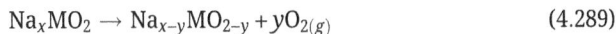

$$Na_xMO_2 \rightarrow Na_{x-y}MO_{2-y} + yO_{2(g)} \tag{4.289}$$

Besides the release of oxygen gas, this reaction can increase the risk of thermal runaway, for instance, if oxygen gas reacts with hydrogen gas. Exothermic reactions between the released oxygen and electrolytes can also occur, enhancing the thermal runaway process. As an example, we look at the reaction of EC with oxygen gas ($O_2$) that results in the formation of carbon dioxide ($CO_2$) and flammable gases ($CH_4$ and CO):

$$C_3H_4O_3 \rightarrow CH_4 + CO_2 + CO \tag{4.290}$$

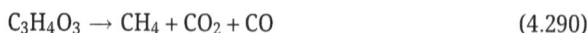

This decomposition reaction is the same as in LIBs (Section 4.4.6, reaction (4.238)). By the way, oxygen can also be released in the case of high-voltage charging of SIBs with cathodes made from layered transition metal oxides.

Finally, if the electrolyte and oxygen gas can strongly react, the battery cell can catch fire and even an explosion is possible.

One can summarize that SIBs may be safer than LIBs, but they are not immune to thermal runaway processes. The advantages of SIBs are, for instance, that sodium does not form dendrites as pronounced as lithium. Dendrites can create shorts and enhance the internal cell temperatures. Also, the SEI in SIBs is less flammable than the SEI in LIBs. The cathodes in SIBs release less oxygen than the Li-NMC cathodes ($LiNi_xMn_yCo_zO_2$, where $x + y + z = 1$); and finally, hard carbon does not store sodium as densely as graphite stores the highly reactive lithium. Disadvantages of SIBs are that they also employ flammable organic carbonate electrolytes and that the SEI still decomposes exothermally. Moreover, high-voltage cathodes (>4.0 V), like $NaNi_{0.5}Mn_{0.5}O_2$, can trigger thermal runaway due to a combination of chemical instability, oxygen release, and heat generation. One can also mention that the thermal conductivity of hard carbon is smaller than the one of graphite. Therefore, in SIBs, the local heating might be somewhat higher than in LIBs.

The lower cost of SIBs makes them ideal candidates for large-scale energy storage systems. These systems can help stabilize electrical grids by storing excess energy generated from renewable sources (e.g., solar energy, wind energy) and discharging it when demand is high. While SIBs currently face limitations in energy density, researchers are working to make them suitable for EVs, where both weight and energy storage are significant. However, SIBs probably will not completely replace lithium-ion technology in the near future. Thinking about consumer electronics, SIBs are less likely to be used in portable electronics (like smartphones or laptops) due to their lower energy density. However, as the technology improves, they might be used in less power-demanding devices or in specialized applications. Due to their cost-effectiveness, SIBs are considered a good option for backup power systems for residential or commercial applications.

While SIBs are still in the development phase and face several challenges, their advantages – such as low cost, abundant materials and sustainability – make them a promising alternative to LIBs. Research into improving their energy density, cycle life and overall performance continues, and SIBs could play a major role in the future of energy storage, particularly for large-scale applications. The potential for SIBs in grid storage and other large-scale uses could complement or, in some cases, replace LIBs in certain applications. However, for many high-performance applications, it is likely that LIBs will remain dominant for the foreseeable future.

# 5 Raw materials for lithium-ion batteries

In the last part of the book, we want to discuss a few aspects, which are not directly related to the physics and chemistry of batteries. However, those aspects have a very strong relevance concerning the steady growth of battery fabrication. In particular, in the following paragraphs, we want to emphasize on the example of lithium-ion batteries, and the importance of material reserves for future battery demands. Lithium-ion batteries are used for widespread applications. Due to their high capacity and high energy and power densities (see the Ragone diagram in Figure 4.1), they are the battery of choice for a lot of applications – especially concerning mobile communication and electric cars as well as potentially large-scale energy storage applications in frame of renewable energy production. Since these are significantly growing markets, it can be expected that an increasing demand for lithium-ion batteries and the demand for the corresponding raw materials will last for many years or even some decades too.

On the background of a rapidly growing demand of lithium-ion batteries, we briefly discuss the reserves and resources of the relevant raw materials. In particular, using the example of lithium-ion batteries, we briefly describe the exploitation of the most relevant raw materials – lithium, cobalt, nickel, manganese and graphite – and their chemical conversion to pure materials for the production of batteries.

## 5.1 Reserves and resources of raw materials

If one wants to discuss the reserves and/or resources of materials, one has to start with a precise explanation of some terms and definitions. Hence, we start our considerations about the exploitation of raw materials with some definitions and a summary about the correct wording; a timeline complements the definitions. Figure 5.1 gives an overview about the correct definitions and wording when discussing the consumption of raw materials or minerals in the past and in the future. In particular, the following definitions have to be taken into account.

The total potential of a certain raw material comprises the estimated complete amount of the material that can be obtained from the natural deposits on earth. Thus, the total potential covers the amount of a certain raw material that was used already in the past and all the remaining amount of this material on earth too.

The cumulative production represents the amount of raw material that was extracted from the ground since the beginning of the mining until today, that is, the cumulative production covers the past consumption.

The future consumption is described by two terms, that is, the reserves and the resources. Both terms are often confused and used in a misleading way. The precise definitions are the following ones: reserves are the assured raw materials that are still located in the mineral deposits and that can be economically extracted from the

https://doi.org/10.1515/9783111618531-005

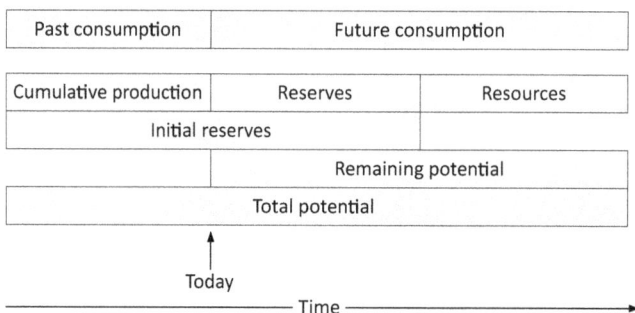

| Past consumption | Future consumption | | |
|---|---|---|---|

| Cumulative production | Reserves | Resources |
|---|---|---|

Initial reserves

Remaining potential

Total potential

↑
Today

———————— Time ————————→

**Figure 5.1:** Descriptions of past and future consumptions of raw materials. The mark "today" on the timescale separates the past from the future.

ground with today's technology. On the other hand, resources are raw materials that are assured in the ground but cannot be extracted with the today's technology, especially not under an economical point view. Moreover, resources are also raw materials that are not yet detected, but that might be expected in certain regions of the world due to geological estimates and expectations.

As shown in Figure 5.1, initial reserves comprise the cumulative production in the past and the more or less easy to get reserves. Moreover, the remaining potential comprises the reserves and the resources, that is, the complete amount of certain raw materials or minerals that are not extracted yet from the ground – that is, the easy to get ones as well as the hardly to extract raw materials.

According to the definition of the U.S. Geological Survey, data about reserves are dynamic. The reserves of a raw material may be reduced if it is mined or the extraction is not profitable anymore. One the other hand, the reserves can also increase, if additional deposits are developed. These new additional deposits could be either already known before, or they were recently discovered. The reserves can also be increased, if an improved technology or changed economic operating figures (e.g., increasing market prices) make the extraction of those raw materials feasible, whose extraction was not feasible before.

Thus, the supply of raw materials (minerals and ore) in the future originates first from the reserves, that is, from identified resources, whose extraction is technologically and economically feasible. Beyond that, future supplies can also come from resources in deposits that are not yet discovered and that will probably be discovered in the future (as can be estimated on the basis of our geological knowledge). Moreover, future supply of raw materials can also originate from recycled materials, that is, from materials that today are actually in use, or from minerals in waste disposal sites. Estimations of undiscovered deposits of minerals and ore significantly affect the strategies with regard to the future supply of raw materials.

On the base of the abovementioned definitions and annotations, we now start our considerations about the raw materials for lithium-ion batteries.

## 5.2 Lithium for lithium-ion batteries

The element lithium (Li) has the atomic number 3; it is an alkali metal with silver-white color in group 1 and period 2 of the periodic table of elements. According to the position of lithium in the periodic table of elements, it is the lightest metal and it is the lightest solid element too. Lithium is widely used in various industrial production processes, for example, as flux additives for the iron, steel and aluminum production, for heat-resistant glass and ceramics, lithium grease lubricants, and especially, for primary lithium batteries and secondary lithium-ion batteries. Lithium is also important for the treatment of mental diseases (bipolar disorder, major depression, etc.).

Lithium is highly reactive (like all alkali elements), and therefore, it does not appear as a pure element in nature. In the earth's crust, the concentration of lithium ranges between 20 and 70 ppm by weight. It usually occurs in ionic compounds – for instance, in pegmatitic minerals. Granitic pegmatites contain the largest amount of lithium-containing minerals, for example, spodumene or petalite. Such minerals are an important source of lithium. Moreover, the mass of lithium in the oceans can be estimated to a value of about 230 billion tons, where the mean lithium concentration in ocean water varies between about 0.14 and 0.25 ppm. Lithium concentrations can be much higher near hydrothermal vents, where concentration values up to 7 ppm are observed. In addition, lithium can be also found in salt lakes and brine pools.

Thus, lithium can be extracted from various silicate rocks and from salt lakes. Both lithium sources can be developed for the extraction of lithium for battery fabrication. Historically, lithium exploitation started with the extraction from hard rock. Later on, lithium or lithium compounds were predominantly extracted from water, that is, mineral springs, brine pools or salt lakes. Brine, that is, a highly concentrated solution of salt in water, recently became the most important source for lithium since the mining of lithium-containing rocks is in general more complex and more expensive.

Table 5.1 shows the current most important lithium mining countries. The order in the table is determined by the amount of reserves. Of these lithium mining countries, Chile has the largest actual reserves (9.3 million tons) followed by Australia (7.0 million tons) and Argentina (4.0 million tons).

An overview about the lithium resources is given in Table 5.2 (lithium in the oceans is not taken into account). By far the largest amount of lithium resources can be found in the so-called lithium triangle in South America, that is, in Bolivia, Argentina and Chile, that is, in salt lakes. The Salt Lake Salar de Uyuni in Bolivia is the largest lithium deposit in the world. The exploitation of this reservoir just starts, and it can be expected that Bolivia will be the biggest vendor for lithium carbonate in the future. Lithium carbonate ($Li_2CO_3$) is the basic material for most lithium-containing products.

Concerning the usage of lithium for battery technology, especially for lithium-ion batteries, the following two principle process chains have to be considered, depending

**Table 5.1:** Reserves and mine productions of lithium mining countries.

| Country | Lithium reserves (tons) | Lithium production (tons) | |
|---|---|---|---|
| | | 2023 | 2024 |
| Chile | 9,300,000 | 41,400 | 49,000 |
| Australia | 7,000,000 | 91,700 | 88,000 |
| Argentina | 4,000,000 | 8,630 | 18,000 |
| China | 3,000,000 | 35,700 | 41,000 |
| USA | 1,800,000 | – | – |
| Canada | 1,200,000 | 3,240 | 4,300 |
| Zimbabwe | 480,000 | 14,900 | 22,000 |
| Brazil | 390,000 | 5,260 | 10,000 |
| Portugal | 60,000 | 380 | 380 |
| Others (rounded) | 2,800,000 | – | – |
| **Total (rounded)** | **30,000,000** | 201,000 | 233,000 |

Note: Data from the U.S. Geological Survey, "Mineral Commodity Summaries" (January 2025) that are based on the information from governmental and industry sources.

**Table 5.2:** Resources of lithium.

| Country | Lithium resources (tons) | Country | Lithium resources (tons) |
|---|---|---|---|
| Bolivia | 23,000,000 | Serbia | 1,200,000 |
| Argentina | 23,000,000 | Peru | 1,000,000 |
| Chile | 11,000,000 | Russia | 1,000,000 |
| Australia | 8,900,000 | Zimbabwe | 860,000 |
| China | 6,800,000 | Spain | 320,000 |
| Canada | 5,700,000 | Portugal | 270,000 |
| Germany | 4,000,000 | Namibia | 230,000 |
| Congo | 3,000,000 | Ghana | 200,000 |
| Mexico | 1,700,000 | Austria | 60,000 |
| Brazil | 1,300,000 | Finland | 55,000 |
| Czechia | 1,300,000 | Kazakhstan | 45,000 |
| Mali | 1,200,000 | | |
| **Total (rounded)** | | **95,000,000** | |

Note: Data from the U.S. Geological Survey, "Mineral Commodity Summaries" (January 2025) that are based on the information from governmental and industry sources.

on the initial raw material (i.e., either lithium in brine or lithium in minerals). In principle, the supply of lithium for the production of lithium-ion batteries requires only a few process steps, as shown in Figure 5.2.

However, battery-grade lithium requires a high purity, that is, >99.5%. Therefore, the parent materials – lithium carbonate or lithium hydroxide – have to exhibit high quality, too. Hence, all process steps have to be carefully controlled. The production

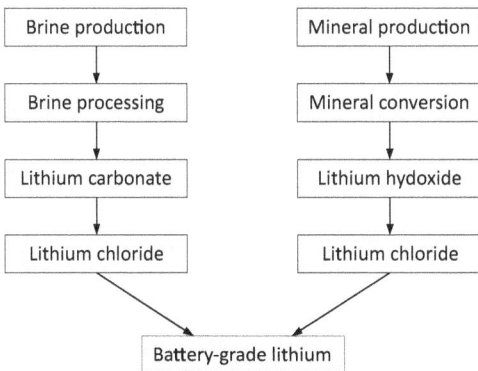

**Figure 5.2:** Principle process routes for battery-grade lithium (purity >99.5%) for lithium-ion batteries. The left process route shows the process steps, if the raw material is from brine. The right process route shows the corresponding process steps for minerals that contain lithium.

of battery-grade lithium is more complex, when lithium is extracted from mineral and rocks. Lithium production from brines requires less process steps, and especially, it consumes less energy. Therefore, from an economic point of view, lithium production from brines is more favorable.

### 5.2.1 Lithium extraction from brine

First, we look at the production of lithium that initially came from brine in salt lakes, that is, we concentrate on the left process route in Figure 5.2. Applying the lime soda evaporation process, lithium can be obtained from the brine in form of lithium carbonate.

The brine contains lithium chloride (LiCl). By water evaporation under sunlight, the concentration of lithium chloride can be significantly enhanced. For this purpose, the brine from saltwater or brine from the salt lake in pumped into shallow pools. However, the brine contains not only lithium chloride but also a mixture of a quite large variety of salts, for instance, different sulfates and chlorides of sodium, potassium, magnesium, calcium and boron. Those salts are also recovered by evaporation under sunlight in ponds that are filled with the brine.

After about 12–18 months the concentrations of the various salts in the brine are high, and further processing can be started. Concerning the production of high-purity lithium, for instance, magnesium dichloride ($MgCl_2$) is disturbing and has to be removed from the ponds. This can be done in the following way. Lime (CaO) is added that – together with water in the brine – forms calcium hydroxide ($Ca(OH)_2$):

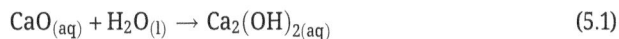

$$CaO_{(aq)} + H_2O_{(l)} \rightarrow Ca_2(OH)_{2(aq)} \tag{5.1}$$

Then, the calcium hydroxide reacts with magnesium dichloride, and magnesium hydroxide and calcium dichloride are formed:

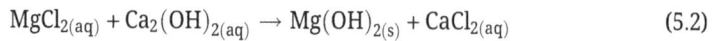

$$MgCl_{2(aq)} + Ca_2(OH)_{2(aq)} \rightarrow Mg(OH)_{2(s)} + CaCl_{2(aq)} \tag{5.2}$$

Calcium dichloride ($CaCl_2$) spontaneously reacts with the water in the brine and forms a calcium chloride hexahydrate ($CaCl_2(H_2O)_6$) complex:

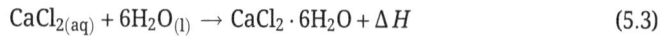

$$CaCl_{2(aq)} + 6H_2O_{(l)} \rightarrow CaCl_2 \cdot 6H_2O + \Delta H \tag{5.3}$$

This exothermal reaction releases energy; the change of the enthalpy is negative, that is, $\Delta H < 0$. Calcium chloride hexahydrate crystals dissolve in their own water of crystallization at 30 °C. Magnesium hydroxide is not soluble in water and can be filtered out. Since magnesium is also an important material of interest, $Mg(OH)_2$ can be refined and further processed with hydrochloric acid and water.

Hydrochloric acid has the chemical formula $H_2O$:$HCl$. It is an aqueous solution of gaseous hydrogen chloride ($HCl$). Hydrochloric acid is the salt of hydronium and chloride ($H_3O^+$ and $Cl^-$, respectively) that appears if gaseous $HCl$ is solved in water, that is,

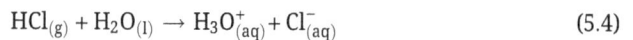

$$HCl_{(g)} + H_2O_{(l)} \rightarrow H_3O^+_{(aq)} + Cl^-_{(aq)} \tag{5.4}$$

Magnesium hydroxide and hydrochloric acid react according to the following formula:

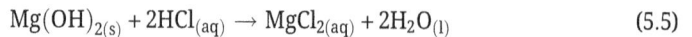

$$Mg(OH)_{2(s)} + 2HCl_{(aq)} \rightarrow MgCl_{2(aq)} + 2H_2O_{(l)} \tag{5.5}$$

or rather:

$$Mg(OH)_{2(s)} + 2H_3O^+_{(aq)} + 2Cl^-_{(aq)} \rightarrow Mg^{2+}_{(aq)} + 2Cl^-_{(aq)} + 4H_2O_{(l)} \tag{5.6}$$

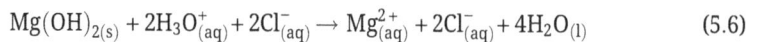

By further water evaporation in a vacuum evaporator, the concentration of magnesium and chloride ions ($Mg^{2+}$, $Cl^-$) in the solution are strongly increased, and finally, magnesium dichloride ($MgCl_2$) crystallizes and can be filtered out from the solution. Thus, purified $MgCl_2$ was extracted from the brine, and it can be further processed, for instance, by a fused-salt electrolysis process at 800 °C that finally delivers magnesium (Mg) and chlorine gas ($Cl_2$).

The extraction and purification of LiCl from the brine is done in a rather similar way. After water evaporation under sunlight for about 12–18 months, the LiCl concentration exceeds a critical value, that is, [LiCl] > 0.5%; and then sodium carbonate ($Na_2CO_3$) is added to the concentrated solution. A carbonate is an inorganic salt of carbonic acid ($H_2CO_3$); that is, sodium carbonate is the sodium salt of carbonic acid. If sodium carbonate is added to the brine, it reacts with the lithium chloride according to the reaction:

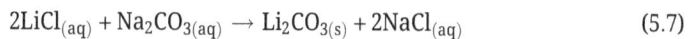

$$2LiCl_{(aq)} + Na_2CO_{3(aq)} \rightarrow Li_2CO_{3(s)} + 2NaCl_{(aq)} \tag{5.7}$$

or rather more in detail:

$$2Li^+_{(aq)} + 2Cl^-_{(aq)} + 2Na^+_{(aq)} + CO^-_{3(aq)} \rightarrow Li_2CO_{3(s)} + 2Na^+_{(aq)} + 2Cl^-_{(aq)} \qquad (5.8)$$

Since lithium chloride and sodium carbonate are dissociated in water, two $Li^+$ ions can combine with a negatively charged carbonate anion ($CO_3^{2-}$) and form lithium carbonate ($Li_2CO_3$). In this reaction, according to reactions (5.7) or (5.8), respectively, the hardly soluble lithium carbonate crystallizes and falls out; that is, $Li_2CO_3$ can be filtered out from the solution in the ponds. $Li_2CO_3$ is the basic product for the fabrication of the positive electrodes of lithium-ion batteries, for example, for the lithium cobalt dioxide ($LiCoO_2$) cathodes.

To approach the final goal, that is, the production of battery-grade metallic lithium, the lithium carbonate has to be treated with hydrochloric acid. Hence, processing continues with a reaction of $Li_2CO_3$ with hydrochloric acid – see eq. (5.3) – according to the following reaction:

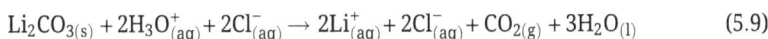

$$Li_2CO_{3(s)} + 2H_3O^+_{(aq)} + 2Cl^-_{(aq)} \rightarrow 2Li^+_{(aq)} + 2Cl^-_{(aq)} + CO_{2(g)} + 3H_2O_{(l)} \qquad (5.9)$$

In this reaction, carbon dioxide gas ($CO_2$) is formed and disappears to the ambient. Moreover, lithium chloride that is dissolved in water is formed again. The concentration of LiCl in the solution is already quite high. By further water evaporation in a vacuum evaporator, the lithium chloride concentration increases; and finally, LiCl crystallizes and can be filtered out from the solution. Now LiCl exhibits a purity that is appropriate for the production of battery-grade lithium.

Finally, by applying a fused-salt electrolysis process with molten LiCl at a temperature of about 450–500 °C, battery-grade metallic lithium can be obtained. Fused-salt electrolysis means that electrolysis is applied with a molten salt, that is, with molten LiCl. For this purpose, lithium chloride is mixed with potassium chloride (KCl); the LiCl:KCl ratio of the mixture is 55% to 45% resulting in a molten eutectic electrolyte. Potassium chloride lowers the fusion temperature and improves the ion conductivity of the molten salt mixture, that is, of the lithium ions, during electrolysis.

During the fused-salt electrolysis, at the positively charged electrode, a forced oxidation of chloride of the $C^-$ ions occurs, and chlorine gas ($Cl_2$) is set free:

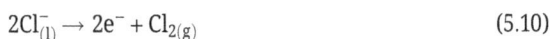

$$2Cl^-_{(l)} \rightarrow 2e^- + Cl_{2(g)} \qquad (5.10)$$

At the negatively charged electrode, a forced reduction of the positive $Li^+$ ions occurs, and metallic lithium is formed, which can be used for further production of lithium-ion batteries:

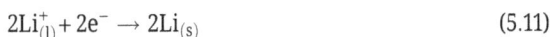

$$2Li^+_{(l)} + 2e^- \rightarrow 2Li_{(s)} \qquad (5.11)$$

Molten lithium rises to the surface. It is collected and has to be protected from oxidation, for instance, by wrapping it in paraffin wax.

## 5.2.2 Lithium extraction from minerals and rocks

In the following part, we discuss the right process route for the production of battery-grade lithium shown in Figure 5.2, that is, lithium extraction from minerals and rocks. Lithium can be found in many silicate minerals that consist of silicate groups and form rocks. They are the most important class of minerals since 90% of the earth's crust consists of silicate minerals and rocks. Silicate minerals are ionic compounds. Their anions are predominantly established by silicon and oxygen atoms, for example, $[SiO_4]^{4-}$, $[Si_2O_7]^{6-}$, $[Si_nO_{4n}]^{2n-}$, $[Si_{4n}O_{11n}]^{6n-}$ and $[Al_xSi_yO_{(2x+2y)}]^{x-}$. Lithium appears in many silicate rocks; however, usually the lithium concentration is rather low. Moreover, the deposits are rather small. Hence, most of the silicate minerals currently play not a big role for the lithium production.

Concerning lithium production, pegmatites with lithium-containing minerals are interesting. Pegmatite is an igneous or magmatic rock with interlocking crystals. Important lithium-containing pegmatitic minerals are shown in Table 5.3.

Spodumene and petalite are the mostly used mineral for lithium extraction; lepidolite and hectorite are also used. Although all these minerals have a significantly higher lithium content than saltwater from salt lakes, the lithium production from mined minerals is much more expensive than the extraction from brine. The reason is that the extraction of lithium from minerals like spodumene or petalite is more complex and requires much more energy.

Lithium extraction from minerals will be exemplary discussed for spodumene $LiAlSi_2O_6$ since spodumene is most commonly used for the production of lithium. After mining, the ore is crushed, grinded and beneficiated. The separation of the lithium-containing minerals from the mixed ore can be achieved, for instance, by physical separations by taking advantage of the physical, magnetic or electrical properties of the various minerals, that is, dry and wet screening, magnetic, electrostatic or gravity/density separation. Floatation is applied to achieve high-grade spodumene concentrates.

Two pyrometallurgical processes have to be considered for the lithium extraction from spodumene, that is, the calcination and roasting. Calcination means that a thermal decomposition of materials in furnaces or reactors occurs. On the other hand, roasting means that thermal solid–gas reactions occur like oxidation, reduction, chlorination or sulfation reactions.

Two phases of spodumene are known: the low-temperature phase (α-spodumene) crystallizes in the monoclinic system, while the high-temperature phase (β-spodumene) crystallizes in the tetragonal system. At temperatures above 900 °C, the α-spodumene phase converts to β-spodumene. Spodumene naturally occurs in the monoclinic alpha form (α-spodumene). However, the tetragonal beta form (β-spodumene) is essential for the extraction of lithium. Therefore, α-spodumene has to be converted to β-spodumene in order to be susceptible for the lithium extraction processes. This conversion is called decrepitation. Thus, after mining and segregating pretreatments, a calcination of spodumene is applied at high temperatures, for example, above 1,000 °C, to convert the min-

**Table 5.3:** Lithium-containing minerals (the negatively charged anions are written in square brackets).

| Mineral | Formula | Lithium concentration | |
|---------|---------|-----------------------|-----------|
| | | **Weight (%)** | **Atom (%)** |
| Spodumene | $LiAl[Si_2O_6]$ | 3.73 | 10.00 |
| Petalite | $LiAl[Si_4O_{10}]$ | 2.27 | 6.25 |
| Lepidolite | $K(Li,Al)_3[(Al,Si)_4O_{10}](F,OH)_2$ | 4.07 | 10.98 |
| Amblygonite | $(Li,Na)Al[PO_4](F,OH)$ | 3.44 | 9.09 |
| Hectorite | $Na_{0.3}(Mg,Li)_3[Si_4O_{10}](F,OH)_2$ | 1.38 | 3.79 |
| Cryolithionite | $Li_3Na_3[AlF_6]_2$ | 5.60 | 15.00 |
| Triphylite | $LiFe^{2+}[PO_4]$ | 4.40 | 14.29 |
| Zinnwaldite | $KLiFe^{2+}[(AlSi)_4O_{10}](F,OH)_2$ | 1.59 | 4.88 |

Note: Data from "Handbook of Mineralogy," www.handbookofmineralogy.org (2020), and "Mineralienatlas," www.mineralienatlas.de (2020).

eral into the beta phase. Due to the phase transformation, the crystal structure of spodumene expands by about 30%. β-Spodumene has a lower density ($2.4$ g/cm$^3$) than α-spodumene ($3.1$ g/cm$^3$). Moreover, it is friable and exhibits numerous surface defects and internal cracks. Therefore, the reactivity of β-spodumene upon further processing is better than the one of α-spodumene.

If calcination is applied at a temperature of about 1,050–1,100 °C, and if calcium carbonate is used as an additive material, several reactions occur and have to be considered. First, calcium carbonate decomposes to calcium oxide and carbon dioxide:

$$CaCO_{3(s)} \rightarrow CaO_{(s)} + CO_{2(g)} \tag{5.12}$$

This reaction occurs already at temperatures above 800 °C. Carbon dioxide is released to the ambient. At the same time, α-spodumene converts to β-spodumene. Therefore, almost simultaneously to the decomposition of calcium carbonate and to the conversion of the α-spodumene phase to β-spodumene, the next step arises, and β-spodumene reacts with calcium oxide (CaO), according to the following reaction:

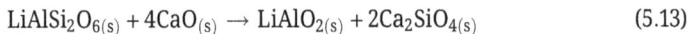

$$LiAlSi_2O_{6(s)} + 4CaO_{(s)} \rightarrow LiAlO_{2(s)} + 2Ca_2SiO_{4(s)} \tag{5.13}$$

This reaction formula and the further processes should become clearer if we multiply both sides of reaction (5.13) with 2 and use the following notation:

$$Li_2O \cdot Al_2O_3 \cdot 4SiO_{2(s)} + 8CaO_{(s)} \rightarrow Li_2O \cdot Al_2O_{3(s)} + 4(2CaO \cdot SiO_2)_{(s)} \tag{5.14}$$

Thus, two units of β-spodumene ($2LiAlSi_2O_6$ is now written as $LiO_2 \cdot Al_2O_3 \cdot 4SiO_2$) react with eight units of calcium oxide (8CaO) and form two units of lithium aluminate ($2LiAlO_2$ or rather $Li_2O \cdot Al_2O_3$) and four units of calcium silicate ($4Ca_2SiO_4$ or rather $2CaO \cdot SiO_2$).

After calcination, the material is cooled down to a temperature below 100 °C. It is crushed again and leached with water at about 90 °C for several hours. During the leaching process, the remaining calcium oxide in the material reacts with water and forms calcium hydroxide (Ca(OH)$_2$):

$$CaO_{(s)} + H_2O_{(l)} \rightarrow Ca(OH)_{2(s)} \tag{5.15}$$

Thereafter, the calcium hydroxide reacts with the lithium aluminate and forms lithium hydroxide (LiOH) and calcium aluminate (CaAl$_2$O$_4$ or rather CaO·Al$_2$O$_3$):

$$Li_2O \cdot Al_2O_{3(s)} + Ca(OH)_{2(s)} \rightarrow 2LiOH_{(s)} + CaO \cdot Al_2O_{3(s)} \tag{5.16}$$

Lithium hydroxide can be dissolved in water and washed out. Afterward the water can be evaporated again in a vacuum evaporator, and the LiOH concentration in the solution strongly increases; and finally, LiOH crystallizes and can be filtered out.

Lithium hydroxide can be further processed with hydrochloric acid. Lithium hydroxide reacts with hydrochloric acid according to the following formula:

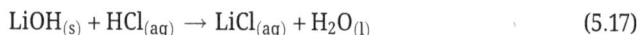

$$LiOH_{(s)} + HCl_{(aq)} \rightarrow LiCl_{(aq)} + H_2O_{(l)} \tag{5.17}$$

or rather:

$$LiOH_{(s)} + H_3O^+_{(aq)} + Cl^-_{(aq)} \rightarrow Li^+_{(aq)} + Cl^-_{(aq)} + 2H_2O_{(l)} \tag{5.18}$$

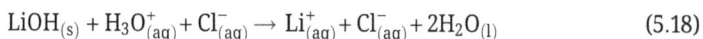

Now we can continue in a similar way as before in case of the lithium production from brine. By water evaporation in a vacuum evaporator, the lithium chloride concentration can be strongly increased; and finally, LiCl crystallizes and can be filtered out. Again, LiCl has an appropriate purity for the production of battery-grade lithium. Finally, by applying a fused-salt electrolysis process with molten LiCl at a temperature of about 450–500 °C, battery-grade metallic lithium can be obtained, as was explained earlier. Fused-salt electrolysis provides chlorine gas (Cl$_2$) at the positive electrode according to eq. (5.10); and metallic molten lithium can be obtained at the negative electrode according to eq. (5.11).

We also want to mention an alternative process route for the lithium extraction from spodumene. After roasting at temperatures above 1,000 °C for several hours to convert the α-spodumene phase to β-spodumene, the material is cooled down. Due to the conversion from α-spodumene to β-spodumene, the material is brittle with a lot of internal cracks and surface defects that make it susceptible to an efficient reaction with hot sulfuric acid. At temperatures below 100 °C, β-spodumene is mixed with sulfuric acid (H$_2$SO$_4$) and heated again to about 200 °C. An exothermic reaction occurs that delivers lithium sulfate (Li$_2$SO$_4$):

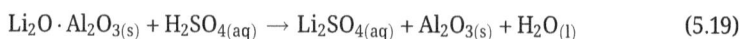

$$Li_2O \cdot Al_2O_{3(s)} + H_2SO_{4(aq)} \rightarrow Li_2SO_{4(aq)} + Al_2O_{3(s)} + H_2O_{(l)} \tag{5.19}$$

After this reaction, the material is cooled down again. At temperatures just below 100 °C, water is added since Li$_2$SO$_4$ is highly soluble in water and can be washed out.

An adjacent treatment with sodium hydroxide (NaOH) that results in a formation of lithium hydroxide (LiOH) and a subsequent carbonation of LiOH with carbon dioxide ($CO_2$) results in a formation of lithium carbonate ($Li_2CO_3$):

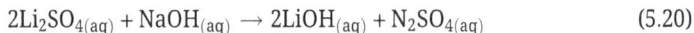

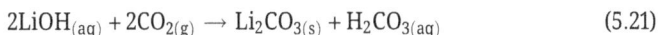

$$2Li_2SO_{4(aq)} + NaOH_{(aq)} \rightarrow 2LiOH_{(aq)} + N_2SO_{4(aq)} \tag{5.20}$$

$$2LiOH_{(aq)} + 2CO_{2(g)} \rightarrow Li_2CO_{3(s)} + H_2CO_{3(aq)} \tag{5.21}$$

The further processing of lithium carbonate occurs – as we mentioned in Section 5.2.1 – with a hydrochloric acid treatment according to the following reaction (see also reactions (5.4) and (5.9)):

$$Li_2CO_{3(s)} + 2HCl_{(aq)} \rightarrow 2LiCl_{(aq)} + CO_{2(g)} + H_2O_{(l)} \tag{5.22}$$

In this reaction, carbon dioxide gas ($CO_2$) is released and disappears to the ambient. Moreover, lithium chloride that is dissolved in water can be crystallized by water evaporation in a vacuum evaporator and filtered out from the solution. By applying a final fused-salt electrolysis process with molten LiCl at a temperature of about 450–500 °C, battery-grade metallic lithium can be obtained.

## 5.3 Electrode materials for lithium-ion batteries

In the following paragraphs, we briefly discuss the most important electrode materials for lithium-ion batteries, that is, in case of the positive electrode, the elements cobalt, nickel and manganese, and for the negative electrode, graphite. Of course, cobalt, nickel, manganese and graphite are not only used for batteries.

In advance, according to the U.S. geological survey ("Mineral Commodity Summaries," 2020), we can summarize that concerning the production of lithium-ion batteries, the most critical raw material is cobalt with reserves of less than 7 million tons. Speaking about resources, the situation is somewhat better, that is, one can expect about 25 million tons of terrestrial cobalt resources. Moreover, additional 120 million tons – or even more – of cobalt resources can be expected on the ocean floors in manganese nodules. Anyway, cobalt can be partly replaced by nickel and/or manganese. In case of nickel, the reserves are about 90 million tons. Moreover, the identified land-based resources contain at least 130 million tons of nickel, if we take into account nickel deposits with an average 1% nickel concentration. Manganese reserves are also rather small. However, the resources are much larger – especially, if one takes into account the manganese nodules on the ocean floor. The amount of manganese in the manganese nodules comprises an estimated value of 500 billion tons.

Thus, since cobalt is a rather seldom and expensive material, there is a strong tendency to reduce the fraction of cobalt in the electrodes of lithium-ion batteries. Lithium nickel manganese cobalt dioxides have become more important for the positive electrode in lithium-ion batteries, where instead of $LiCoO_2$ electrodes mixed oxides with the general composition $LiNi_xMn_yCo_zO_2$ are used. Concerning the crystallographic structure, $LiNi_xMn_yCo_zO_2$ electrodes are closely related to $LiCoO_2$ electrodes since $x + y + z = 1$.

Below we briefly summarize the situation of cobalt, nickel and manganese reserves and production, where the latter two materials partly replace cobalt in the positive electrodes of lithium-ion batteries. Moreover, we also briefly describe the situation for graphite, that is, the most common raw material for the negative electrode.

### 5.3.1 Cobalt reserves and exploitation

The element cobalt (Co) has the atomic number 27; it is a transition metal with silver-gray color in group 9 and period 4 (i.e., in the d-block) of the periodic table of elements. It is ferromagnetic and exhibits similar properties like iron. The radioactive isotope $^{60}Co$ is an emitter of γ-rays, that is, it emits highly energetic radiation that is applied for radiotherapy or for the sterilization of medical supplies and waste. Gamma-ray exposure can also be used for the sterilization of foods (cold pasteurization). Moreover, cobalt is used for the production of high-performance alloys (superalloys), for example, for jet engines or gas turbines. In fact, the application of cobalt for superalloys is the prevailing usage of cobalt. Cobalt-based alloys are corrosion and wear resistant. Therefore, they are also appropriate for durable orthopedic implants. With this regard, cobalt shows the same useful properties like titanium. Cobalt salt pigments are also well known for coloring, for example, for the production of blue-colored glass in the past. Some cobalt compounds are important catalysts for various industrial processes. Finally, cobalt is also an essential component in vitamin $B_{12}$ that contains a cobalt ion ($Co^+$) in the center of a so-called corrin ring.

Cobalt is a widespread element in the earth's crust; its concentration in the earth's crust is about 0.0029%. However, cobalt does not occur in elemental form. It is contained in many minerals, but mostly, in those minerals the cobalt concentration is rather low. Cobalt is very often associated with nickel; and it can be frequently found together with copper, silver, iron or uranium. Cobalt can also be found in iron meteorites together with nickel. Usually cobalt is extracted as a byproduct of nickel or copper mining from a variety of different ores in several countries. Cobalt alone is only extracted from arsenide ores in Morocco and Canada. More than 50% of cobalt production in the world comes from the mining of nickel ores.

Table 5.4 shows the most important cobalt mining countries. The order in the table is determined by the amount of reserves. The largest reserves are located in the Democratic Republic of the Congo (6.0 million tons) and in Australia (1.7 million tons). Congo is also the world's largest producer of cobalt, and it supplies about 70% of the world cobalt production. In general, cobalt is mined as a byproduct of copper and nickel. China is the world's leading producer of refined cobalt (more than 75%), and it is also the biggest consumer. Actually, most of the Chinese consumption of cobalt is used by the battery industry, that is, especially for lithium-ion batteries, where $LiCoO_2$ is widely used for the cathodes of lithium-ion batteries. It is to add that cobalt is also used for nickel–metal hydride batteries since a cobalt additive improves the oxidation of nickel.

**Table 5.4:** Reserves and mine productions of cobalt mining countries.

| Country | Cobalt reserves (tons) | Cobalt production (tons) | |
|---|---|---|---|
| | | 2023 | 2024 |
| Congo | 6,000,000 | 175,000 | 220,000 |
| Australia | 1,700,000 | 5,220 | 3,600 |
| Indonesia | 640,000 | 19,000 | 28,000 |
| Cuba | 500,000 | 3,300 | 3,500 |
| Philippines | 260,000 | 3,800 | 3,800 |
| Russia | 250,000 | 8,700 | 8,700 |
| Canada | 220,000 | 4,220 | 4,500 |
| Madagascar | 100,000 | 4,000 | 2,600 |
| Turkey | 91,000 | 2,500 | 2,700 |
| USA | 70,000 | 500 | 300 |
| Papua New Guinea | 62,000 | 3,070 | 2,800 |
| New Caledonia | – | 2,570 | 1,500 |
| Others (rounded) | 800,000 | 6,080 | 6,200 |
| **Total (rounded)** | **11,000,000** | 238,000 | 284,000 |

Note: Data from the U.S. Geological Survey, "Mineral Commodity Summaries" (MCS, January 2025) that are based on the information from governmental and industry sources.

Many commercial mobile electronic products (cellphones, smartphones, laptops, tablets, etc.) use lithium-ion batteries with lithium–cobalt dioxide electrodes; and it is very likely that lithium-ion batteries that use cobalt-based cathodes remain the predominantly used battery type for this purpose. Moreover, it can be expected that the increasing market for lithium-ion batteries powering electronic cars will drive a very strong additional demand for cobalt too.

As mentioned earlier, cobalt is very often associated with nickel; cobalt is also frequently found together with copper, silver, iron or uranium. Important minerals containing cobalt are, for instance, cobaltite (CoAsS), glaucodot ((Co,Fe)AsS) or skutterudite ($CoAs_3$).

Usually, cobalt is extracted from nickel or copper ores. Details of the extraction can differ since they are depending on the specific composition of the ore. At first, roasting converts iron sulfides (FeS, $FeS_2$) that are present in the ore to iron oxides ($Fe_2O_3$). Adding silica (quartz, $SiO_2$), the iron oxide can be slagged to iron silicates (e.g., $Fe_2SiO_4$ and $FeSiO_3$). The remaining matte contains cobalt, nickel, copper and still iron sulfides or iron arsenate ($FeAsO_4$). The sulfates and arsenates can be leached out with water, and the related metal oxides remain. This mixture of metal oxides is treated with sulfuric acid ($H_2SO_4$) and/or hydrochloric acid (HCl); and nickel, iron and cobalt are dissolved in an aqueous solution. Only copper is not dissolved and can be separated. By a further treatment with chlorinated lime that is added to the solution, cobalt can be selectively removed from the solution.

In particular, cobalt reacts with hypochlorite and transforms to cobalt hydroxide ($Co(OH)_2$) that falls out. Hypochlorite is an ion with the chemical formula $ClO^-$; it can combine with various cations to form hypochlorites that are the salts of hypochlorous acid. Hypochlorous acid (HClO) is usually formed if chlorine gas ($Cl_2$) reacts with water:

$$Cl_{2(g)} + H_2O_{(l)} \rightarrow HCl_{(aq)} + HClO_{(aq)} \tag{5.23}$$

or rather:

$$Cl_{2(g)} + H_2O_{(l)} \rightarrow 2H^+_{(aq)} + Cl^-_{(aq)} + ClO^-_{(aq)} \tag{5.24}$$

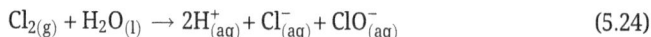

In the actual process for cobalt extraction, the chlorine for the formation of hypochlorous acid originates from the chlorinated lime that was added to the aqueous solution providing hypochlorite. Thus, divalent cobalt(II) ions in the solution can react with hypochlorite ($ClO^-$):

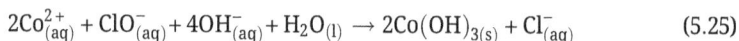

$$2Co^{2+}_{(aq)} + ClO^-_{(aq)} + 4OH^-_{(aq)} + H_2O_{(l)} \rightarrow 2Co(OH)_{3(s)} + Cl^-_{(aq)} \tag{5.25}$$

In this reaction, cobalt hydroxide ($Co(OH)_3$) falls out. During a subsequent heat treatment, $Co(OH)_3$ is converted to divalent/trivalent cobalt(II,III) oxide ($Co_3O_4$). Applying a final heat treatment with the execution of a reduction reaction with coke, cobalt(II,III) oxide is reduced to metallic cobalt; and carbon dioxide is released to the ambient:

$$Co_3O_{4(s)} + 2C_{(s)} \rightarrow 3Co_{(s)} + 2CO_{2(g)} \tag{5.26}$$

## 5.3.2 Nickel reserves and exploitation

The element nickel (Ni) has the atomic number 28; it is a silver-colored transition metal with a golden tinge positioned in the group 10 and period 4 (i.e., in the d-block) of the periodic table of elements. It is ferromagnetic and similar to iron and cobalt. Nickel is used for many industrial applications and products, that is, for the production of stainless steel and special alloys (permalloy, invar, elinvar), magnets, coins and batteries (NiCd, NiMH, lithium-ion batteries with mixed oxide electrodes). Nickel is an excellent alloy metal for many applications. In particular, it is used in nickel steels and nickel cast irons, where it is responsible for the increase of the tensile strength, the toughness and the elastic limit of the alloys. Moreover, nickel alloys are often used as catalysts.

The concentration of nickel in the earth's crust is about 0.008%; however, elemental nickel is rare. Nickel can be found in sulfide minerals, for instance, in millerite (NiS) and pentlandite ((Fe,Ni)$_9$S$_8$), or in the arsenide mineral like nickeline (NiAs) and others. In iron meteorites, nickel can be also found. Nickel is often associated with cobalt as mentioned earlier.

Table 5.5 shows the most important nickel mining countries. The order in the table is determined by the amount of reserves. The largest nickel reserves are located in Indonesia, Australia and Brazil; Indonesia is actually also the country with the biggest mining activities followed by the Philippines.

**Table 5.5:** Reserves and mine productions of nickel mining countries.

| Country | Nickel reserves (tons) | Nickel production (tons) | |
|---|---|---|---|
| | | 2023 | 2024 |
| Indonesia | 55,000,000 | 2,030,000 | 2,200,000 |
| Australia | 24,000,000 | 149,000 | 110,000 |
| Brazil | 16,000,000 | 82,700 | 77,000 |
| Russia | 8,300,000 | 210,000 | 210,000 |
| New Caledonia | 7,100,000 | 231,000 | 110,000 |
| Philippines | 4,800,000 | 413,000 | 330,000 |
| China | 4,400,000 | 117,000 | 120,000 |
| Canada | 2,200,000 | 159,000 | 190,000 |
| USA | 310,000 | 16,400 | 8,000 |
| Others (rounded) | >9,100,000 | 340,000 | 300,000 |
| **Total (rounded)** | **>131,000,000** | 3,750,000 | 3,700,000 |

Note: Data from the U.S. Geological Survey, "Mineral Commodity Summaries" (MCS, January 2025) that are based on the information from governmental and industry sources.

Nickel can be extracted from nickel–copper ores. Nickel extraction processes are rather complicated, and the details of the extraction differ and depend on the specific composition of the ore. In general, floatation processes can enhance the nickel concentration in ores from sulfite deposits. Subsequent roasting and smelting is applied. In a first roasting step, iron sulfides (FeS, FeS$_2$) that are present in the ore are converted to iron oxides (Fe$_2$O$_3$). Adding silica (quartz, SiO$_2$), the iron oxide can be slagged to iron silicates (e.g., Fe$_2$SiO$_4$, FeSiO$_3$); hence, the sulfide and iron concentration in the material is reduced. The remaining matte is slowly cooled down and various nickel and copper phases are formed, that is, nickel sulfide (Ni$_3$S$_2$), copper(I) sulfide (Cu$_2$S) and metallic nickel/copper alloy. Those phases can be mechanically separated. Subsequent roasting transforms nickel sulfide to nickel oxide (NiO).

Further purification and extraction of metallic nickel can be obtained, for instance, by a three-step carbonyl or Mond process (named after the chemist Ludwig Mond). In the first step, nickel oxide reacts with syngas that consists primarily of hydrogen ($H_2$) and carbon monoxide (CO); the particular reactor occurs between nickel oxide and hydrogen according to the following formula:

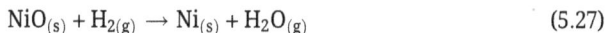

$$NiO_{(s)} + H_{2(g)} \rightarrow Ni_{(s)} + H_2O_{(g)} \tag{5.27}$$

Reaction (5.27) occurs at 200 °C. The hydrogen that is enclosed in the syngas reduces nickel oxide to metallic nickel and water vapor. Still, the nickel is impure; and iron and/ or cobalt are also formed. In a second step, the impure nickel reacts at about 50 or 60 °C with the carbon monoxide in the syngas and forms gaseous nickel tetracarbonyl:

$$Ni_{(s)} + 4CO_{(g)} \rightarrow Ni(CO)_{4(g)} \tag{5.28}$$

The solid impurities are left behind. In a final third step, the nickel tetracarbonyl gas and the syngas are heated to 230 °C. At this temperature, nickel tetracarbonyl decomposes to (pure) metallic nickel and carbon monoxide:

$$Ni(CO)_{4(g)} \rightarrow Ni_{(s)} + 4CO_{(g)} \tag{5.29}$$

Favorable for this process is the fact that carbon monoxide readily reacts with nickel to form nickel carbonyl; and in addition, this reaction is reversible. Moreover, in the described extraction process no other element forms a carbonyl under the applied conditions at rather low temperatures.

### 5.3.3 Manganese reserves and exploitation

The element manganese (Mn) has the atomic number 25; it is a silver metallic-colored transition metal in group 7 and period 4 (i.e., in the d-block) of the periodic table of elements. The properties of manganese resemble those of iron, but manganese exhibits paramagnetic behavior; it is not a ferromagnetic metal like iron. Manganese is employed mostly in steel industry; it is a major component of low-cost stainless steel. Manganese is also important for the production of corrosion-resistant aluminum alloys for aerospace applications. And, of course, manganese is an important raw material for batteries, as we have frequently discussed in this book.

The concentration of manganese in the earth's crust is about 0.095%. Behind iron and titanium, it is the third most common transition metal. Manganese does not occur in elemental form. In general, land-based manganese resources are large; however, they are irregularly distributed. Therefore, the reserves are not so big. Important manganese ore is pyrolusite ($MnO_2$), braunite ($MnMn_6O_8SiO_4$) and others. Most important for mining is the mineral pyrolusite.

Table 5.6 shows the most important manganese mining countries. The order in the table is determined by the amount of reserves. The manganese reserves shown in Table 5.6 do not consider manganese nodules on the ocean floor. The largest reserves of manganese are located in South Africa and Australia; most of the manganese is actually mined in South Africa followed by Gabon and Australia.

**Table 5.6:** Reserves and mine productions of manganese mining countries.

| Country | Manganese reserves (tons) | Manganese production (tons) | |
| --- | --- | --- | --- |
| | | 2023 | 2024 |
| South Africa | 560,000 | 7,300 | 7,400 |
| Australia | 500,000 | 2,860 | 2,800 |
| China | 280,000 | 767 | 770 |
| Brazil | 270,000 | 580 | 590 |
| Gabon | 61,000 | 4,490 | 4,600 |
| India | 34,000 | 744 | 800 |
| Ghana | 13,000 | 818 | 820 |
| Malaysia | n.a. | 410 | 410 |
| Cote d'Ivoire | n.a. | 357 | 360 |
| Others (rounded) | Small | 1,230 | 1,300 |
| **Total (rounded)** | **1,700,000** | 19,600 | 20,000 |

Note: Data from the U.S. Geological Survey, "Mineral Commodity Summaries" (MCS, January 2020) that are based on the information from governmental and industry sources. Data that was not available (n.a.) was set to zero.

Manganese ore that is supposed to be used for battery production should have a minimal manganese concentration of 44%; the concentrations of copper, nickel and cobalt should be low. A reduction process with carbon cannot extract pure manganese with good efficiency since stable carbides are formed (e.g., $Mn_7C_3$). Hence, metallic manganese is predominantly produced by hydrometallurgy, where manganese ore is oxidized, leached and finally electrolyzed. By electrolysis, divalent manganese(II) sulfate solutions are electrolyzed according to the following reaction:

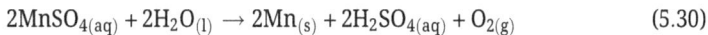

$$2MnSO_{4(aq)} + 2H_2O_{(l)} \rightarrow 2Mn_{(s)} + 2H_2SO_{4(aq)} + O_{2(g)} \tag{5.30}$$

At the cathode, $Mn^{2+}$ ions are reduced to metallic manganese; at the anode, oxygen gas is formed.

### 5.3.4 Graphite reserves and exploitation

The element carbon (C) has the atomic number 6 and is located in the group 14 (carbon group) and period 2 of the periodic table of elements. Carbon has several allotropes, that is, it can exist in the same physical state (e.g., the solid state) in several different forms. The allotropes of carbon are diamond and graphite, the macromolecules fullerenes, carbon nanotubes and cyclo[18]carbon, and graphene with a two-dimensional hexagonal lattice. Graphite is the allotrope that exhibits the most stable form of carbon. The physical properties of the carbon allotropes can be very different. Diamond is the hardest natural material with the highest thermal conductivity, too. Moreover, it is an electrical insulator (wide bandgap material). On the other hand, graphite is a soft material that exhibits a very good electrical conductivity.

Graphite is mined in many countries around the world; and it is used for various industrial applications – besides its usage for pencils. The largest graphite reserves are located in China and Brazil (Table 5.7). By far, the biggest producer of graphite is China. For further processing and usage, natural graphite usually needs beneficiation by floatation and leaching with hydrofluoric or hydrochloric acid. However, concerning the application of graphite for electrodes, synthetic graphite is usually employed.

**Table 5.7:** Reserves and mine productions of graphite mining countries.

| Country | Graphite reserves (tons) | Graphite production (tons) | |
|---|---|---|---|
| | | 2023 | 2024 |
| China | 81,000,000 | 1,210,000 | 1,270,000 |
| Brazil | 74,000,000 | 66,300 | 68,000 |
| Madagascar | 27,000,000 | 63,000 | 89,000 |
| Mozambique | 25,000,000 | 98,000 | 75,000 |
| Tanzania | 18,000,000 | 13,200 | 25,000 |
| Russia | 14,000,000 | 15,000 | 20,000 |
| Vietnam | 9,700,000 | 2,500 | 2,000 |
| India | 8,600,000 | 25,600 | 27,800 |
| Turkey | 6,900,000 | 2,800 | 3,100 |
| Canada | 5,900,000 | 5,470 | 20,000 |
| Mexico | 3,100,000 | 1,300 | 900 |
| North Korea | 2,000,000 | 8,100 | 8,100 |
| South Korea | 1,800,000 | 9,620 | 9,600 |
| Sri Lanka | 1,500,000 | 3,000 | 3,300 |
| Norway | 600,000 | 6,480 | 7,000 |
| Ukraine | n.a. | 1,670 | 1,200 |
| Others (rounded) | >10,000 | 700 | 700 |
| **Total (rounded)** | **290,000,000** | **1,530,000** | **1,630,000** |

Note: Data from the U.S. Geological Survey, "Mineral Commodity Summaries" (MCS, January 2025) that are based on the information from governmental and industry sources. Data that was not available (n.a.) was set to zero.

The Acheson process that is named after the American chemist Edward Goodrich Acheson was initially developed for the production of silicon carbide (SiC). During the silicon carbide formation process, synthetic graphite was found as an unintentional side reaction product. The Acheson process could be optimized for the production of high-purity synthetic graphite from solid amorphous carbons. The main reaction step in this process is the elimination of silica ($SiO_2$).

In particular, a mixture of silica or quartz sand (i.e., silicon dioxide ($SiO_2$)) and powdered coke is heated up to about 2,500 °C for more than 1 day. Various reactions occur in parallel. Carbon and silicon dioxide form silicon monoxide and carbon monoxide gas:

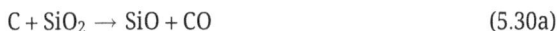

$$C + SiO_2 \rightarrow SiO + CO \tag{5.30a}$$

Silicon dioxide melts at 1,713 °C and starts boiling above 2,200 °C (under normal pressure). Silicon monoxide melts at about 1,700 °C and starts boiling at 1,880 °C. Silicon dioxide and carbon monoxide form silicon monoxide and carbon dioxide gas:

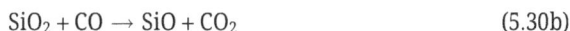

$$SiO_2 + CO \rightarrow SiO + CO_2 \tag{5.30b}$$

Carbon and carbon dioxide form carbon monoxide:

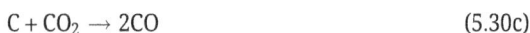

$$C + CO_2 \rightarrow 2CO \tag{5.30c}$$

Moreover, silicon carbide is formed too:

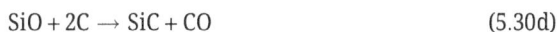

$$SiO + 2C \rightarrow SiC + CO \tag{5.30d}$$

Above 2,300 °C, silicon carbide decomposes and graphite remains:

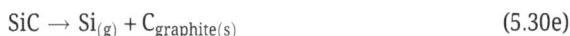

$$SiC \rightarrow Si_{(g)} + C_{graphite(s)} \tag{5.30e}$$

Finally, high-purity graphite obtained can be used for electrodes in lithium-ion batteries.

# Further reading

A selection of books, articles and review papers for further reading, and deeper studies are listed below (the author's choice).

## Chapter 1: Introduction

[1]  J. Hansen, "Storms of My Grandchildren: The Truth About the Coming Climate Change Catastrophe and Our Last Chance to Save Humanity", Bloomsbury, 2009.
[2]  D. Wallace-Wells, "The Uninhabitable Earth: Life after Warming", Tim Duggan Books, 2019.
[3]  K. Mertens, "Photovoltaics: Fundamentals, Technology and Practice", John Wiley and Sons Ltd., 2019.
[4]  B. Sørensen, "Renewable Energy", Elsevier Academic Press, 3rd Edition, 2004.
[5]  R. Ehrlich, "Renewable Energy – A First Course", CRC Press, 2013.
[6]  R. A. Huggins, "Energy Storage", Springer, 2010.
[7]  K. Krishner, K. Schönleber, "Physics of Energy Conversion", Walter De Gruyter, 2015.
[8]  G. Pistoia, B. Liaw (Editors), "Behaviour of Lithium-Ion Batteries in Electric Vehicles", Springer, 2018.
[9]  T. R. Hawkins, O. M. Gausen, A. H. Strømman, "Environmental impacts of hybrid and electric vehicles – a review", The International Journal of Life Cycle Assessment, vol. 17, pp. 997–1014, 2012.
[10] E. J. Cairns, P. Albertus, "Batteries for electric and hybrid-electric vehicles", Annual Review of Chemical and Biomolecular Engineering, vol. 1, pp. 299–320, 2010.
[11] B. Sarlioglu, C. T. Morris, D. Han, S. Li, "Driving toward accessibility: A review of technological improvements for electric machines, power electronics, and batteries for electric and hybrid vehicles", IEEE Industry Applications Magazine, vol. 23, issue 1, pp. 14–25, 2017.
[12] T. A. Faunce, J. Prest, D. Su, S. J. Hearne, F. Iacopi, "On-grid batteries for large-scale energy storage: Challenges and opportunities for policy and technology", MRS Energy & Sustainability: A Review Journal, vol. 5, doi: 10.1557/mre.2018.11, 2018.
[13] H. Chen, T. N. Cong, W. Yang, C. Tan, Y. Li, Y. Ding, "Progress in electrical energy storage systems: a critical review", Progress in Natural Science, vol. 19, pp. 291–312, 2009.

## Chapter 2: Thermodynamics

[14] F. Reif, "Fundamentals of Statistical and Thermal Physics", Waveland Press, 2009.
[15] D. C. Giancoli, "Physics for Scientists and Engineers with Modern Physics", Pearson Education, 4th Edition, 2019.
[16] R. Holyst, A. Poniewierski, "Thermodynamics for Chemists, Physicists and Engineers", Springer, 2012.
[17] G. Astarita, "Thermodynamics – An Advanced Textbook for Chemical Engineers", Springer, 1989.
[18] F. Reif, "Statistische Physik und Theorie der Wärme", Walter de Gruyter, 3rd Edition, 1987.
[19] L. Bergmann, C. Schaefer, "Lehrbuch der Experimentalphysik Bd. 1, Mechanik, Akkustik, Wärmelehre", Walter de Gruyter, 9th Edition, 1974.

https://doi.org/10.1515/9783111618531-006

## Chapter 3: Basics of electrochemistry

[20] P. W. Atkins, "Physical Chemistry", Oxford University Press, 6th Edition, 1998.

[21] T. Engel, P. Reid, "Physical Chemistry", Pearson Education, 2006.

[22] R. A. Huggins, "Energy Storage", Springer, 2010.

[23] N. Lawrence, J. Wadhawan, R. Compton, "Foundations of Physical Chemistry", Oxford University Press, 1999.

[24] T. Minami, M. Tatsumisago, M. Wakihara, C. Iwakura, S. Kohjiya, I. Tanaka (Editors), "Solid State Ionics for Batteries", Springer, 2005.

[25] J. Rumble (Editor), "CRC Handbook of Chemistry and Physics", CRC Press, 100th Edition, 2019.

[26] G. Wedler, H.-J. Freund, "Lehrbuch der Physikalischen Chemie", Wiley-VCH, 6th Edition, 2012.

[27] E. Zirngiebl, "Einführung in die angewandte Elektrochemie", Salle + Sauerländer Verlag, 1993.

[28] C. H. Hamann, W. Vielstich, "Elektrochemie", Wiley-VCH, 4th Edition, 2005.

[29] C. Czeslik, H. Seemann, R. Winter, "Basiswissen Physikalischen Chemie", Vieweg + Teubner, 4th Edition, 2010.

[30] J. Hoinkis, "Chemie für Ingenieure", Wiley-VCH, 14th Edition, 2016.

[31] D. V. Ragone, "Review of battery systems for electrically powered vehicles", SAE Technical Paper, doi: 10.4271/680453, 1968.

[32] S. C. Lee, W. Y. Jung, "Analogical understanding of the Ragone plot and a new categorization of energy devices", Energy Procedia, vol. 88, pp. 526–530, 2016.

## Chapter 4: Batteries

[33] J. F. Daniell, "An Introduction to the Study of Chemical Philosophy: Being a Preparatory View of the Forces Which Concur to the Production of Chemical Phenomena", J. W. Parker, 2nd Edition, 1843.

[34] R. A. Huggins, "Advanced Batteries – Materials Science Aspects", Springer, 2009.

[35] C. Julien, A. Mauger, A. Vijh, K. Zaghib, "Lithium Batteries", Springer, 2016.

[36] G. A. Nazri, G. Pistoia (Editors), "Lithium Batteries – Science and Technology", Springer, 2009.

[37] V. Pop, H. J. Bergveld, D. Danilov, P. P. L. Regtien, P. H. L. Notten, "Battery Management Systems", Springer, 2008.

[38] L. C. Trueb, P. Rüetschi, "Batterien und Akkumulatoren", Springer, 1998.

[39] T.-K. Ying, X.-P. Gao, W.-K. Hu, F. Wu, D. Noréus, "Studies on rechargeable NiMH batteries", International Journal of Hydrogen Energy, vol. 31, issue 4, pp. 525–530, 2006.

[40] R. Korthauer (Editor), "Lithium-Ion Batteries – Basics and Applications", Springer, 2019.

[41] T. R. Jow, K. Xu, O. Borodin, M. Ue (Editors), "Electrolytes for Lithium and Lithium-Ion Batteries", Springer, 2014.

[42] D. Schröder, "Analysis of Reaction and Transport Processes in Zinc Air Batteries", Springer Vieweg, 2016.

[43] C. Mikolajczak, M. Kahn, K. White, R. T. Long, "Lithium-Ion Batteries Hazard and Use Assessment", Springer, 2011.

[44] W. Baumann, A. Muth, "Batterien – Daten und Fakten zum Umweltschutz", Springer, 1997.

[45] A. Al-Haj Hussein and I. Batarseh, "A review of charging algorithms for nickel and lithium battery chargers", IEEE Transactions on Vehicular Technology, vol. 60, issue 3, pp. 830–838, 2011.

[46] C.-M. Park, J.-H. Kim, H. Kim, H.-J. Sohn, "Li-alloy based anode materials for Li secondary batteries", Chemical Society Reviews, vol. 39, issue 8, pp. 3115–3141, 2010.

[47] B. Writer, "Lithium-Ion Batteries – A Machine-Generated Summery of Current Research", Springer, 2019.

[48] G. J. May, A. Davidson, B. Monahov, "Lead batteries for utility energy storage: A review", Journal of Energy Storage, vol. 15, pp. 145–157, 2018.

[49] P. T. Moseley, D. A. J. Rand, K. Peters, "Enhancing the performance of lead-acid batteries with carbon – In pursuit of an understanding", Journal of Power Sources, vol. 295, pp. 268–274, 2015.

[50] L. Wang, H. Zhang, W. Zhang, G. Cao, H. Zhao, Y. Yang, "Enhancing the cycle performance of lead-acid battery anodes by lead-doped porous carbon composite and graphite additives", Materials Letters, vol. 206, pp. 113–116, 2017.

[51] R. M. Dell, "Batteries – fifty years of materials development", Solid State Ionics, vol. 134, pp. 139–158, 2000.

[52] A. Kraytsberg, Y. Ein-Eli, "Review on Li-air batteries – opportunities, limitations and perspective", Journal of Power Sources, vol. 196, pp. 886–893, 2011.

[53] S. P. S. Badwal, S. S. Giddey, C. Munnings, A. I. Bhatt, A. F. Hollenkamp, "Emerging electrochemical energy conversion and storage technologies", Frontiers in Chemistry, vol. 2, doi: 10.3389/fchem.2014.00079, 2014.

[54] O. Haas, E. J. Cairns, "Electrochemical energy storage", Annual Reports on Progress in Chemistry, Sect. C, vol. 95, pp. 163–197, 1999.

[55] H. D. Yoo, E. Markevich, G. Salitra, D. Sharon, D. Aurbach, "On the challenge of developing advanced technologies for electrochemical energy storage and conversion", Materials Today, vol. 17, issue 3, pp. 110–121, 2014.

[56] N. Nitta, F. Wu, J. T. Lee, G. Yushin, "Li-ion battery materials: present and future", Materials Today, vol. 18, issue 5, pp. 252–264, 2015.

[57] M. Winter, R. J. Brodd, "What are batteries, fuel cells, and supercapacitors", Chemical reviews, vol. 104, issue 10, pp. 4245–4269, 2004.

[58] Y. Liu, Q. Sun, W. Li, K. R. Adair, J. Li, X. Sun, "A comprehensive review on recent aluminum-air batteries", Green Energy & Environment, vol. 2, pp. 246–277, 2017.

[59] W. Martin, Y. Tian, J. Xiao, "Understanding diffusion and electrochemical reduction of $Li^+$ ions in liquid lithium metal Batteries", Journal of The Electrochemical Society, vol. 168, pp. 060513, 2021.

[60] J.-C. Zhang, Z.-D. Liu, C.-H. Zeng, J.-W. Luo, Y.-D. Deng, X.-Y. Cui, Y.-N. Chen, "High-voltage $LiCoO_2$ cathodes for high-energy-density lithium-ion battery", Rare Metals, vol. 41, pp. 3946–3956, 2022.

[61] Y. Lyu, X. Wu, K. Wang, Z. Feng, T. Cheng, Y. Liu, M. Wang, R. Chen, L. Xu, J. Zhou, Y. Lu, B. Guo, "An overview on the advances of $LiCoO_2$ cathodes for lithium-ion batteries", Advanced Energy Materials, vol. 11, issue 2, doi: 10.1002/aenm.202000982, 2021.

[62] N. H. Kwon, D. Mouck-Makanda, K. M. Fromm, "A review: Carbon additives in LiMnPO4- and $LiCoO_2$-based cathode composites for lithium ion batteries", Batteries, vol. 4, doi: 10.3390/batteries4040050, 2018.

[63] J. Wang, Q. Zhang, J. Sheng, Z. Liang, J. Ma, Y. Chen, G. Zhou, H.-M. Cheng, "Direct and green repairing of degraded $LiCoO_2$ for reuse in lithium-ion batteries", National Science Review, vol. 9, issue 8, doi: 10.1093/nsr/nwac097, 2022.

[64] Y. Shi, G. Chen, Z. Chen, "Effective regeneration of $LiCoO_2$ from spent lithium-ion batteries: direct approach towards high-performance active particles", Green Chemistry, vol. 20, issue 4, pp. 851–862, 2018.

[65] M. Bianchini, M. Roca-Ayats, P. Hartmann, T. Brezesinski, J. Janek, "There an back again – the journey on $LiNiO_2$ as a cathode active material", Angewandte Chemie International Edition, vol. 58, issue 31, pp. 10434–10458, 2019.

[66] R. Koutavarapu, M. Cho, J. Shim, M. C. Rao, "Structural and electrochemical properties of $LiNiO_2$ cathodes prepared by solid state reaction method", Ionics, vol. 26, pp. 5991–6002, 2020.

[67] P. Kalyani, N. Kalaiselvi, "Various aspects of $LiNiO_2$ chemistry: A review", Science and Technology of Advanced Materials, vol. 6, issue 6, pp. 689–703, 2005.

[68]  M. Kim, L. Zou, S.-B. Son, I. D. Bloom, C. Wang, G. Chen, "Improving LiNiO$_2$ cathode performance through particle design and optimization", Journal of Materials Chemistry A, vol. 10, pp. 12890–12899, 2022.

[69]  B. Dong, A. D. Poletayev, J. P. Cottom, J. Castells-Gil, B. F. Spencer, C. Li, P. Zhu, Y. Chen, J.-M. Price, L. L. Driscoll, P. K. Allan, E. Kendrick, M. S. Islam, P. R. Slater, "Effects of sulfate modification of stoichiometric and lithium-rich LiNiO$_2$ cathode materials", Journal of Materials Chemistry A, vol. 12, pp. 11390–11402, 2024.

[70]  M. Mishra, K. P. C. Yao, "The emergence of a robust lithium gallium oxide surface layer on gallium-doped LiNiO2 cathodes enables extended cycling stability", Materials Advances, vol. 5, pp. 7016–7027, 2024.

[71]  A.-H. Marincas, F. Goga, S.-A. Dorneau, P. Ilea, "Review on synthesis methods to obtain LiMn$_2$O$_4$-based cathode materials for Li-ion batteries", Journal of Solid State Electrochemistry, vol. 24, pp. 473–497, 2020.

[72]  H. Xia, Z. Luo, J. Xie, "Nanostructured LiMn$_2$O$_4$ and their composites as high-performance cathodes for lithium-ion batteries", Progress in Natural Science: Materials International, vol. 22, issue 6, pp. 572–584, 2012.

[73]  T.-F. Yi, Y.-R. Zhu, X.-D. Zhu, J. Shu, C.-B. Yue, A.-N. Zhou, "A review of recent developments in the surface modification of LiMn$_2$O$_4$ as cathode material of power lithium-ion battery", Ionics, vol. 15, pp. 779–784, 2009. Erratum to: "A review of recent developments in the surface modification of LiMn$_2$O$_4$ as cathode material of power lithium-ion battery", Ionics, vol. 15, p. 785, 2009.

[74]  O. Nyamaa, H.-M. Jeong, G.-H. Kang, J.-S. Kim, K.-M. Goo, I.-G. Baek, J.-H. Yang, T.-H. Namb, J.-P. Noh, "Enhanced LiMn$_2$O$_4$ cathode performance in lithium-ion batteries through synergistic cation and anion substitution", Materials Advances, vol. 5, pp. 2872–2887, 2024.

[75]  K. Ragavendran, H. Xia, P. Mandal, A. K. Arof, "Jahn–Teller effect in LiMn$_2$O$_4$: influence on charge ordering, magnetoresistance and battery performance", Physical Chemistry Chemical Physics, vol. 19, pp. 2073–2077, 2017.

[76]  C. Y. Ouyang, S. Q. Shi, M. S. Lei, "Jahn–Teller distortion and electronic structure of LiMn$_2$O$_4$", Journal of Alloys and Compounds, vol. 474, issues 1–2, pp. 370–374, 2009.

[77]  W.-W. Liu, D. Wang, Z. Wang, J. Deng, W.-M. Lau, Y. Zhang, "Influence of magnetic ordering and Jahn–Teller distortion on the lithiation process of LiMn$_2$O$_4$", Physical Chemistry Chemical Physics, vol. 19, pp. 6481–6486, 2017.

[78]  S. Oswald, H. A. Gasteiger, "The structural stability limit of layered lithium transition metal oxides due to oxygen release at high state of charge and its dependence on the nickel content", Journal of The Electrochemical Society, vol. 170, p. 030506, 2023.

[79]  R. Younesi, G. M. Veith, P. Johansson, K. Edström, T. Vegge, "Lithium salts for advanced lithium batteries: Li-metal, Li-O$_2$, and Li-S", Energy & Environmental Science, vol. 8, pp. 1905–1922, 2015.

[80]  H. Banerjee, C. P. Grey, A. J. Morris, "Stability and redox mechanisms of Ni-rich NMC cathodes: insights from first-principles many-body calculations", Chemistry of Materials, vol. 36, pp. 6575–6587, 2024.

[81]  G. Houchins, V. Viswanathan, "Towards ultra low cobalt cathodes: A high fidelity computational phase search of layered Li-Ni-Mn-Co oxides", Journal of The Electrochemical Society, vol. 167, p. 070506, 2020.

[82]  X. Sun, X. Luo, Z. Zhang, F. Meng, J. Yang, "Life cycle assessment of lithium nickel cobalt manganese oxide (NCM) batteries for electric passenger vehicles", Journal of Cleaner Production, vol. 273, p. 123006, 2020.

[83]  H. Adenusi, G. A. Chass, S. Passerini, K. V. Tian, G. Chen, "Lithium batteries and the solid electrolyte interphase (SEI) – progress and outlook", Advanced Energy Materials, vol. 23, doi: 10.1002/aenm.202203307, 2023.

[84] S.-J. Yang, N. Yao, X.-Q. Xu, F.-N. Jiang, X. Chen, H. Liu, H. Yuan, J.-Q. Huang, X.-B. Cheng, "Formation mechanism of the solid electrolyte interphase in different ester electrolytes", Journal of Materials Chemistry A, vol. 9, pp. 19664–19668, 2021.

[85] I. A. Shkrob, Y. Zhu, T. W. Marin, D. Abraham, "Reduction of carbonate electrolytes and the formation of solid-electrolyte interface (SEI) in lithium-ion batteries. 2. Radiolytically induced polymerization of ethylene carbonate", Journal of Physical Chemistry C, vol. 117, issue 38, pp. 19270–19279, 2013.

[86] F. Fasulo, A. B. Muñoz-García, A. Massaro, O. Crescenzi, C. Huang, M. Pavone, "Vinylene carbonate reactivity at lithium metal surface: first-principles insights into the early steps of SEI formation", Journal of Materials Chemistry A, vol. 11, pp. 5660–5669, 2023.

[87] Y. Yang, J. Xiong, S. Lai, R. Zhou, M. Zao, H. Geng, Y. Zhang, Y. Fang, C. Li, J. Zhao, "Vinyl ethylene carbonate as an effective SEI-forming additive in carbonate-based electrolyte for lithium-metal anodes", ACS Applied Materials & Interfaces, vol. 11, issue 6, pp. 6118–6125, 2019.

[88] K. Xu, "Electrolytes and interphases in Li-Ion batteries and beyond", Chemical Reviews, vol. 114, issue 23, pp. 11503–11618, 2014.

[89] X. Feng, M. Ouyang, X. Liu, L. Lu, Y. Xia, X. He, "Thermal runaway mechanism of lithium ion battery for electric vehicles: A review", Energy Storage Materials, vol. 10, pp. 246–267, 2018.

[90] Y. Dai, A. Panahi, "Thermal runaway process in lithium-ion batteries: A review", Next Energy, vol. 6, p. 100186, 2025.

[91] Z. An, L. Jia, Y. Ding, C. Dang, X. Li, "A review on lithium-ion power battery thermal management technologies and thermal safety", Journal of Thermal Science, vol. 6, pp. 391–412, 2017.

[92] Y. Yang, R. Wang, Z. Shen, Q. Yu, R. Xiong, W. Shen, "Towards a safer lithium-ion batteries: A critical review on cause, characteristics, warning and disposal strategy for thermal runaway", Advances in Applied Energy, vol. 11, p. 100146, 2023.

[93] X. Feng, D. Ren, X. He, M. Ouyang, "Mitigating thermal runaway of lithium-ion batteries", Joule, vol. 4, pp. 743–770, 2020.

[94] Y. Tesfamhret, H. Liu, Z. Chai, E. Berg, R. Younesi, "On the manganese dissolution process from $LiMn_2O_4$ cathode materials", ChemElectroChem, vol. 8, pp. 1516–1523, 2021.

[95] W. Zeng, F. Xia, J. Wang, J. Yang, H. Peng, W. Shu1, Q. Li, H. Wang, G. Wang, S. Mu, J. Wu, "Entropy-increased $LiMn_2O_4$-based positive electrodes for fast-charging lithium metal batteries", Nature Communications, vol. 15, doi: 10.1038/s41467-024-51168-1, 2024.

[96] Y.-Q. Sun, X.-T. Luo, Y.-S. Zhu, X.-J. Liao, C.-J. Li, "$Li_3PO_4$ electrolyte of high conductivity for all-solid-state lithium battery prepared by plasma spray", Journal of the European Ceramic Society, doi: 10.1016/j.jeurceramsoc.2022.04.010, 2022.

[97] S. Shahid, M. Agelin-Chaab, "A review of thermal runaway prevention and mitigation strategies for lithium-ion batteries", Energy Conversion and Management: X, doi: 10.1016/j.ecmx.2022, p. 100310, 2022.

[98] A. Manthiram, "A reflection on lithium-ion battery cathode chemistry", Nature Communications, doi: 10.1038/s41467-020-15355-0, 2020.

[99] A. Manthiram, J. B. Goodenough, "Lithium insertion into $Fe_2(MO_4)_3$ frameworks: Comparison of $M$ = W with $M$ = Mo", Journal of Solid State Chemistry, vol. 71, pp. 349–360, 1987.

[100] A. Manthiram, J. B. Goodenough, "Lithium insertion into $Fe_2(SO_4)_3$ frameworks", Journal of Power Sources, vol. 26, pp. 403–408, 1989.

[101] Y.-G. Guo, J.-S. Hu, L.-J. Wan, "Nanostructured materials for electrochemical energy conversion and storage devices", Advanced Materials, vol. 20, pp. 2878–2887, 2008.

[102] K. Inoue, S. Fujieda, K. Shinoda, S. Suzuki, Y. Waseda, "Chemical state of iron of $LiFePO_4$ during charge-discharge cycles studied by in-situ x-ray absorption spectroscopy", Materials Transactions, vol. 51, issue 12, pp. 2220–2224, 2010.

[103] C. Masquelier, L. Croguennec, "Polyanionic (phosphates, silicates, sulfates) frameworks as electrode materials for rechargeable Li (or Na) batteries", Chemical Reviews, vol. 113, issue 8, pp. 6552–6591, 2013.

[104] N. Yabuuchi, H. Yoshida, S. Komaba, "Crystal structures and electrode performance of alpha-NaFeO$_2$ for rechargeable sodium batteries", Electrochemistry, vol. 80, issue 10, pp. 716–719, 2012.

[105] B. L. Ellis, L. F. Nazar, "Sodium and sodium-ion energy storage and materials science", Current Opinion in Solid State and Materials Science, vol. 16, pp. 168–177, 2012.

[106] K. M. Abraham, "How comparable are sodium-ion batteries to lithium-ion counterparts", ASC Energy letters, vol. 5, pp. 3544–3547, 2020.

[107] J.-Y. Hwang, S.-T. Myung, Y.-K. Sun, "Sodium-ion batteries: present and future", Chemical Society Reviews, vol. 46, pp. 3529–3614, 2017.

[108] A. R. Nurohmah, S. S. Nisa, K. N. R. Stulasti, C. S. Yudha, W. G. Suci, K. Aliwarga, H. Widiyandari, A. Purwanto, "Sodium-ion battery from sea salt: a review", Materials for Renewable and Sustainable Energy, vol. 11, pp. 71–89, 2022.

[109] N. Yabuuchi, K. Kubota, M. Dahbi, S. Komaba, "Research development on sodium-ion batteries", Chemical Reviews, vol. 114, issue 23, pp. 11636–11682, 2014.

[110] J. Meng, G. Jia, H. Yang, M. Wang, "Recent advances for SEI of hard carbon anode in sodium-ion batteries: A min review", Frontiers in Chemistry, vol. 10, doi: 10.3389/fchem.2022.986541, 2022.

## Chapter 5: Raw materials for lithium-ion batteries

[111] N. N. Greenwood, A. Earnshaw, "Chemistry of the Elements", Pergamon Press, 1984.

[112] A. Hollemann, E. Wiberg, "Lehrbuch der Anorganischen Chemie", Walter de Gruyter, 102nd Edition, 1987.

[113] U.S. Geological Survey, "Mineral Commodity Summaries 2025", U.S. Geological Survey, 2025.

[114] J. R. Davis, "Nickel, Cobalt, and Their Alloys", ASM International, 2000.

[115] S. Al-Thyabat, T. Nakamura, E. Shibata, A. Iizuka, "Adaptation of minerals processing operations for lithium-ion (LiBs) and nickel metal hydride (NiMH) batteries recycling: critical review", Minerals Engineering, vol. 45, pp. 4–17, 2017.

[116] M. F. Ashby, "Materials and the Environment: Eco-Informed Material Choice", Butterworth-Heinemann, 2009.

[117] D. Segal, "Materials for the 21st Century", Oxford University Press, 2017.

# Index

https://doi.org/10.1515/9783111618531-007

www.ingramcontent.com/pod-product-compliance
Lightning Source LLC
Chambersburg PA
CBHW082109220326
41598CB00066BA/5939